Clinical Cases in Psychiatry: Integrating Translational Neuroscience Approaches

Alfredo Carlo Altamura • Paolo Brambilla
Editors

Clinical Cases in Psychiatry: Integrating Translational Neuroscience Approaches

Springer

Editors
Alfredo Carlo Altamura
Department of Pathophysiology and Transplantation
University of Milan
Milan, Italy

Department of Neuroscience and Mental Health, IRCCS Fondazione Ca' Granda Ospedale Maggiore Policlinico
Milan, Italy

Paolo Brambilla
Department of Pathophysiology and Transplantation
University of Milan
Milan, Italy

IRCCS "E. Medea" Scientific Institute
Bosisio Parini (Lc), Italy

ISBN 978-3-030-06253-8 ISBN 978-3-319-91557-9 (eBook)
https://doi.org/10.1007/978-3-319-91557-9

This Springer imprint is published by the registered company Springer Nature Switzerland AG
The registered company address is: Gewerbestrasse 11, 6330 Cham, Switzerland

Foreword

In recent years, the term translational medicine has been used in common parlance to describe the application of both clinical and basic research to solving clinical problems with the ultimate goal of developing truly personalized medicine. Most assume that the translation is largely from the laboratory to the patient hence the term "bench to bedside" but laboratory research on disease processes and the development of treatments requires clinical observations and often biologic samples from patients. Indeed, there are numerous examples where solutions for clinical problems have gone from the patient to the laboratory and back to the patient within a research program. Thus, translational research really needs to be bedside to laboratory and back to the patient.

A fallout of the growth in technology available for application to research of clinical problems—e.g., genetics, brain imaging—has been that a great deal of findings seem to hang in the air with no clear clinical application that can be easily gleaned from the findings. While these data may ultimately find an application in the medical hemisphere, one wonders at times whether we are generating data much faster than we can integrate them. All of this leaves much of translational neurosciences unintegrated and often confusing for the practitioner. Indeed, how best to incorporate the neurosciences into clinical training and practice has remained a problem for the field. Until now! This text is a brilliant example of how we can build neuroscience into clinical perspectives in psychiatry.

Clinical Cases in Psychiatry: Integrating Translational Neuroscience Approaches edited by Altamura and Brambilla really gives you the clinical to laboratory to clinical perspective by incorporating both case descriptions and research findings to the teaching of the latest in the diagnosis and treatment of a number of common clinical problems. The text has 16 chapters that cover a wide range of psychopathology. Each chapter reviews the scope of the problem and then presents a case that highlights the issues. It then segues to how neuroscience can help our understanding of the pathogenesis of the disorder and its treatment. Examples of the clinical areas covered include many interesting areas—e.g., early-onset psychosis, psychosis in the elderly, psychotic bipolar disorder. Much of the neuroscience presented revolves around available brain imaging technologies. The last three chapters are oriented toward specific treatments approaches (e.g., r-TMS, long acting antipsychotics), but even in these chapters, there are clearly presented and interesting case reports that really highlight how best to apply these treatments.

Many texts try to present information that can be used by clinicians for patients. My own books in psychopharmacology present such material. However, what is so unique here is that clinical cases are a core feature of each chapter that helps to organize the rest of the material. The neuroscience discussed is cutting edge but germane to the clinical situation. This allows the reader, practitioner, and trainee to immediately integrate neuroscience data into clinical approaches. The book is truly unique and is a must for anyone interested in clinical psychiatry. Kudos to the authors for each of their chapters and to the editors, Altamura and Brambilla, for conceiving and executing this superb book.

Kenneth T. Norris, Jr Professor of Psychiatry and Behavioral Sciences
Stanford University School of Medicine
Stanford, CA, USA

Alan F. Schatzberg

Contents

Early-Onset Psychosis with Adolescence Onset

1

Gabriele Panza and Silvia Paletta

Abstract

There is often diagnostic uncertainty in the first episode of adolescent-onset psychosis. The differential diagnosis may include primary psychosis, bipolar disorder, and psychotic disorder due to a general medical condition or a substance-induced psychotic disorder.

The recent escalating substance use in adolescence and its association with psychotic symptoms in regular users has fuelled concerns. A large number of studies have reported a link between adolescent substance use and the development of stable psychosis in early adulthood. A further complication is the high rates of concomitant substance use by subjects with a psychotic illness which, especially in young users with an early-phase psychotic disorder, can make diagnosis difficult.

We have presented a clinical case of an adolescent male who was admitted to our inpatient psychiatric unit with a psychotic onset. This patient also reported significant substance use.

This case highlights three main issues in primary care: recognizing early signs of emerging psychosis; complexities associated with comorbid substance use; and patient management both in the acute and post-acute phases.

While it is often difficult to distinguish substance-induced psychosis from primary psychosis, especially early in the course of treatment, this differential diagnosis has important implications for treatment planning. This clinical case emphasizes the importance of assessment and treatment planning in the early stages of psychotic disorder, because at onset the symptom picture is often

G. Panza (✉) · S. Paletta
Department of Neurosciences and Mental Health, Fondazione IRCCS Ca' Granda Ospedale Maggiore Policlinico, University of Milan, Milan, Italy
e-mail: gabriele.panza@policlinico.mi.it

A. C. Altamura, P. Brambilla (eds.), *Clinical Cases in Psychiatry: Integrating Translational Neuroscience Approaches*,
https://doi.org/10.1007/978-3-319-91557-9_1

unclear and a proper match of diagnosis with treatment may be critically important for outcome.

Longitudinal follow-up of patients initially presenting with psychosis and substance use comorbidity is warranted by the occurrence of heavy substance misuse overlying presentation of psychotic symptoms, adding greater complexity to the diagnostic process, and the greater instability of substance-induced psychosis diagnoses.

Keywords
Early-onset psychosis · Adolescence · Substance-induced psychosis

1.1 Introduction

Over the past decade, psychiatric illnesses and, in particular, psychotic disorders in children and adolescents have been increasingly acknowledged. It is unclear whether psychotic disorders in younger individuals are actually becoming more prevalent or whether educational efforts directed at enhanced screening and treatment have improved case detection. In light of this, efficacious and safe interventions should be employed to maximize beneficial effects and to minimize medication-related complications [1].

As recently defined, psychosis denotes the presence of either delusions (false implausible beliefs) or hallucinations (false perceptions involving any sensory modality) [2]. Broader definitions of psychosis include manifestations of thought disorders, behavioral disorganization, or catatonia. Family members, particularly parents, are often concerned that children who manifest delusions or hallucinations might have schizophrenia. Fortunately, most forms of psychosis in children and adolescents are not a result of schizophrenia [1].

Population-based surveys show that the prevalence of psychotic symptoms may be far greater than had been previously considered, with a meta-analysis suggesting a prevalence rate of 5–8% in the general population (which is nearly 10 times higher than the prevalence of diagnosed psychotic disorders) [3]. Prevalence rates of such symptoms may be even higher among young people. Large, population-based studies surveying psychotic symptoms among adolescents have found rates of 9–14% in interview-based studies and rates greater than 25% in some studies using self-report questionnaires [4–7].

A longitudinal, questionnaire-based study of suicide prevention in high school students found an unexpectedly robust association between experiencing auditory hallucinations and subsequent suicide attempts at follow-up (34% of students who previously endorsed psychopathology and hallucinations endorsed suicide attempts at follow-up) [8].

Prominent psychotic symptoms (i.e., hallucinations and/or delusions) can also be caused by the effects of a psychoactive substance, such as cocaine or amphetamines. A substance may induce psychotic symptoms during intoxication (while the individual is under the influence of the drug) or during withdrawal (after an individual stops using the drug).

Over the last few years, with the increasing diffusion of substance abuse, there has been an increase in cases of substance-induced psychosis in adolescents [9]. Not all of these cases evolve into primary psychosis; therefore it is important to monitor these patients over time to see if it is the onset of a more pernicious psychotic disease.

Cannabis is one of the most widely used illicit drugs among adolescents, and most users first experiment with it during adolescence. Adolescence is a critical phase for brain development, characterized by neuronal maturation and rearrangement processes, such as myelination, synaptic pruning, and dendritic plasticity. The endocannabinoid system plays an important role in fundamental brain developmental processes such as neuronal cell proliferation, migration, and differentiation. Therefore changes in endocannabinoid activity during this specific developmental phase, induced by the psychoactive component of marijuana, Δ9-tetrahydrocannabinol, might lead to subtle but lasting neurobiological changes that can affect brain functions and behavior [10].

Prospective studies estimate that cannabis use is associated with a twofold increase in later schizophrenia outcomes, and early, adolescent-onset cannabis use is associated with a higher risk [11], possibly because individuals who begin to use cannabis when the brain is still developing are most vulnerable to its deleterious effects [12, 13]. Nonetheless, the vast majority of young people who use cannabis do not develop psychosis, suggesting the hypothesis that, if cannabis is indeed causal, some individuals may be genetically vulnerable to its effects. The presence of such a gene-environment interaction is indicated by the finding that the association between cannabis and psychosis outcomes is most marked in subjects with an established vulnerability to psychosis [14, 15]. Some studies have demonstrated that a functional polymorphism of the catechol-*O*-methyltransferase (COMT) gene moderates the association between cannabis use and the risk of developing psychosis [16].

Recently, five large-scale longitudinal studies and a systematic review have shown that cannabis use in adolescence is associated with a two- to threefold increase in the relative risk of later developing schizophrenia [11]. Furthermore short-term psychotic reactions, particularly in naive users, have also been reported. Thomas [17] describes that one in seven people reported psychotic-like symptoms. Such reactions are usually acute, transient and self-limited, however very unpleasant ("hearing voices, becoming convinced that someone is trying to harm you or that you are persecuted") [18]. But cannabinoids are considered able to trigger long-lasting psychotic decompensations in predisposed individuals, which may in part account for the epidemiological association described between cannabis consumption and psychotic disorders [19, 20].

1.2 Clinical Evaluation

Written consent for the performance of care should be signed by the parents/legal guardians of the underage patient at the beginning of hospitalization, explaining to them in detail the diagnostic and therapeutic procedures that may be performed. In addition, there is the presence of a dedicated educator who follows the patient during hospitalization.

A comprehensive clinical evaluation is necessary to determine what may be the causes of the psychotic symptoms and to rule out possible etiologies (Box 1.1). A wide-range workout including psychiatric, neurologic, and medical assessment, as well as blood, urine, brain, cardiac, and imaging studies, should be carried out.

A very accurate anamnestic interview should be held on admission, with the help of relatives or caregivers. The interview should include previous psychiatric or neurological history, ongoing medical conditions, and pharmacological history.

Clinical acuity of the presentation and intervention impact should guide the evaluation, according to the history and initial examination.

Plasma levels of daily medications prescribed to patients should be assessed, for the measurable drugs. Many psychiatric and nonpsychiatric treatments may be the cause of agitation, confusion, delusions, and hallucinations if the dosage is above a certain threshold.

A computed tomography (CT) scan should be performed in the case of altered/clouded sensorium and other neurological symptoms of acute onset, to exclude neurological/neurosurgical conditions, such as stroke and intracranial hemorrhage. After the exclusion of the most acute and dangerous causes, a full psychiatric assessment should be performed. Overall, it is of fundamental importance to conduct a full brain imaging assessment, consisting of magnetic resonance imaging (MRI) and positron emission tomography (PET), that may greatly help in the diagnostic processes [21].

Another major issue of concern is treatment, which might be guided by a multidisciplinary endeavor, including pharmacological and non-pharmacological interventions. The choice and dosage of pharmacological medications should be guided not only by efficacy but also by potential side effects and unwanted interactions with other medications. Drug starting doses are usually lower compared to those recommended for adults and should be titrated up or down slowly according to clinical response and side effect onset. Overall, treating younger patients presents more difficulty because of greater sensitivity to drugs and their side effects and variation in the pharmacokinetic parameters [22].

Box 1.1 Exams and Assessment

- Accurate interview with the patient and caregivers
- Physical examination with vital signs and electrocardiography
- Blood and urine exams, looking for infections, alteration in electrolytes, and uncontrolled metabolic conditions (complete blood count, thyroid function, liver and renal function, glucidic and lipidic profiles)
- Plasma levels of prescribed medications and drugs of abuse
- Brain imaging: CT scan in acute; MRI and PET to assess neurodegenerative disorders

1.3 Case Presentation

A 16-year-old boy was an inpatient of our psychiatric unit for a psychotic episode with delusions. No previous psychiatric history had been reported before the actual admission, and he had no family psychiatric history. He reported no history of any major medical illnesses and was not on any regular medication. He had one younger sister. He was born of a full-term normal delivery, and his mother said that his developmental milestones were normal. Patient reported no neglect or abuse as a child.

Regarding premorbid temperament, his mother described him as a cheerful and sociable child; the patient was frequenting peers, but never making significant friendships. Also, motor restlessness and difficulty in following the rules were reported.

From 7 to 12 years, he was followed by the services of pediatric neuropsychiatry to facilitate meetings with his father, who during this time period was imprisoned.

At the time of the admission, he was not taking any medications.

Nevertheless a history of substance abuse was reported by his parents. Furthermore he reported a recent history of stressful events (family conflicts) prior to the onset of symptoms. In that period the patient was attending high school with poor performance.

From the age of 14 years, he had started using cannabinoids (6–7 g/week) and taken amino acids and creatine during gym workouts.

At the age of 16 years, the patient began to express irritability and verbal aggression against his mother.

After a few weeks, his parents observed the onset of a symptomatology characterized by behavioral abnormalities (opposing and bizarre attitudes), initial insomnia, logorrhea, auditory hallucinations, and delusional, both mystical-megalomanic and persecutory, behaviors toward his parents and the karate teacher.

After a few days, the mother decided to call the ambulance which accompanied the patient to the emergency room of our hospital because of the worsening of symptoms. At the time of evaluation in the emergency room, he showed dysphoric mood, psychomotor acceleration, severe insomnia, and aggressiveness; persecutory and jealousy delusions were also referred by the patient. He showed idiosyncratic dissociation, persecutory thoughts, emotional dissonance, and behavioral abnormalities.

On entering our inpatient unit, the patient appeared disoriented, poorly cooperative, uninhibited, and agitated on the psychomotor level. The facial expression was faded and perplexed and the speech disorganized. It was likely that there were hearing disassociations and the content of thought was characterized by persecution and mystic thought. Illness criticism was absent.

During his stay in our inpatient unit, blood and urine exams were performed. No significant data emerged from the blood exams, except for an occasional finding of a D-dimer value above the normality threshold ($\leq$250 ng/ml D-dimer units). For this reason a thoracic computerized tomography (CT) was performed, and it led to the exclusion of embolic or phlogistic pathology. The urine drug screening was positive for cannabinoids (150 ng/ml).

We performed encephalic magnetic resonance imaging (MRI), electroencephalography (EEG), and fluorodeoxyglucose positron emission tomography (FDG-PET). The EEG resulted negative for anomalies, except for unspecific theta waves in the frontotemporal electrodes. The MRI did not result in pathological alterations of the overvoltage of supratentorial or subtentorial parenchymal signal, except for a minute hyperintensity signal (in FLAIR) precentral subcortical to the left of possible vascular genesis (Fig. 1.1). The PET highlighted a cortical glucidic

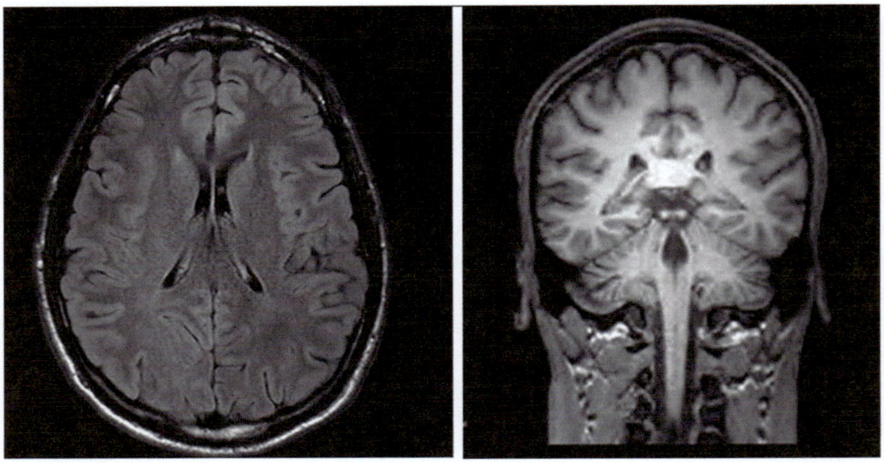

Fig. 1.1 The ventricular system is in axis, of normal size. Seasonal liquids are standard for bed base and convexity. No pathological alterations of the overvoltage of supratentorial or subtentorial parenchymal signal are allowed except for a *minute hyperintensity signal (in FLAIR) precentral subcortical to the left of possible vascular genesis.* There are no obvious gross abnormalities in the Willis polygon vessels

metabolism increased to the crawler, at the prefrontal level and at the nuclei of the base (Fig. 1.2).

During hospitalization the patient underwent a clinical assessment based on the administration of Brief Psychiatric Rating Scale (BPRS) [23], Positive and Negative Syndrome Scale (PANSS) [24], and Global Assessment of Functioning Scale (GAF) [25] (Fig. 1.3).

As a pharmacologic treatment, he was administered intramuscular haloperidol up to 10 mg/day, followed by an intramuscular therapy with olanzapine up to 20 mg/day and daily oral levomepromazine up to 300 mg and zuclopenthixol up to 70 mg. The patient presented a good response in terms of reduction of delusions and psychomotor acceleration, as presented in Fig. 1.3. After a few days of treatment, he became more aware of his condition and accepted both the stay in the acute ward and the treatments. However, a full compliance with pharmacological treatment was not completely achieved as the patient still didn't show a full insight into his disease. From the clinical presentation and the response to treatment, a diagnosis of

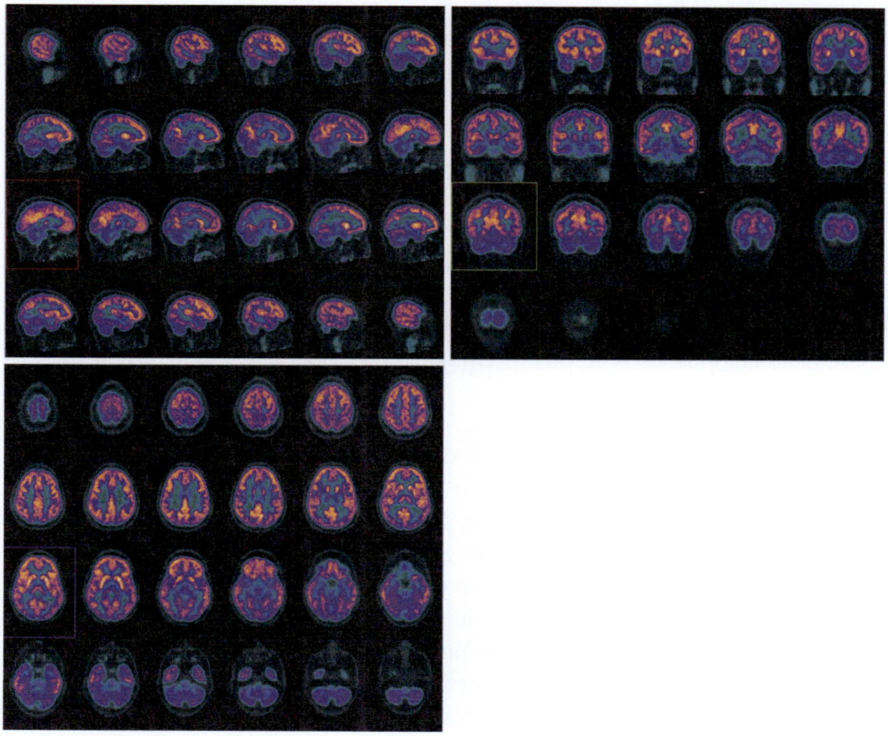

Fig. 1.2 The analysis of tomoscintigraphic PET images highlights *cortical glucidic metabolism modestly increased to the entire crawler bilaterally, and at the upper and midpoint bilateral rounds, and more orbital-frontally and slightly reduced to both cerebellar lobes.* No significant alterations or asymmetries of the distribution of the marked glucose analogue in the other supratentorial or subtentorial cortical encephalic structures are investigated. At the subcortical level, metabolism of the thalamus is bilaterally conserved and symmetric, whereas that at the nuclei of the base is increased

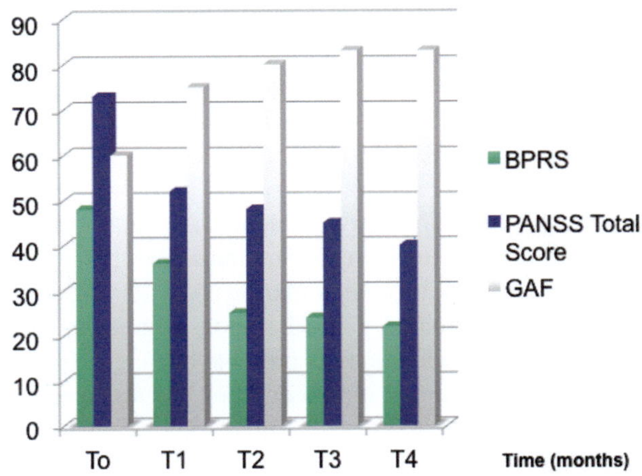

Fig. 1.3 Psychopathological Rating Scale (mean ± SD) time course

substance-induced psychosis was made, also on the basis of acute onset of delusions (mainly of a persecutory nature) within a period of <2 weeks. Cannabinoids were considered the likely precipitant as the onset of the symptoms followed its initiation. Although a functional origin of symptoms could not be completely excluded (i.e., first episode of a functional mental illness like schizophrenia), the apparent absence of any prodromal symptoms, previous psychiatric history, or family psychiatric history made this unlikely.

At discharge, the patient managed to recover fair global functioning, however lower than previously. He started to attend school again, undergoing psychiatric examinations for a few months. However, also complicated by his young age and by the presence of a poorly supportive and constituent family context, he didn't achieve a complete compliance with pharmacological treatment and a full insight into his disease, so his adherence to the visits became more and more irregular and he interrupted pharmacological therapy.

Due to his poor compliance, several successive admissions to our inpatient unit followed. Unlike the first hospitalization, toxicological examinations were always negative for drugs. Progressively mystical-megalomanic ideas, persecutory ideation (especially with regard to the father) and reference, suspicion, and worsening anguish appeared. There was a further worsening of the clinical picture with reduced sleep, logorrhea, worsening persecution, and anxiety. The repeated psychotic exacerbations and gradually shorter intercritical periods caused a progressive impairment of social and scholastic functioning.

As a consequence of poor compliance, a program to dismiss the patient with a depot therapy was effected. After discharge, he accepted monthly long-acting injection (LAI) with olanzapine pamoate (405 mg every 28 days). The patient showed a response to treatment after 1 month and reached a gradual remission after 4 months.

Every access for the LAI administration consisted of clinical evaluation, including safety, tolerability, and determination of olanzapine plasma levels.

Clinical efficacy was evaluated by BPRS [23] and PANSS [24]. Safety and tolerability measures included vital sign measurements, physical examination findings such as weight measurement, 12-lead electrocardiograms, and clinical laboratory tests as low-density lipoprotein (LDL), high-density lipoprotein (HDL), total cholesterol (CHO), triglycerides (Try), glucose, aspartate aminotransferase (AST), alanine aminotransferase (ALT), and gamma-glutamyl transpeptidase (GGT). Each one was tested every month.

The patient did not report any significant adverse effects, and changes in blood examinations did not appear during the follow-up period. Olanzapine plasma levels ranged from 41 ng/mL to 67 ng/mL.

During the various admissions, clinical observation of the psychopathological framework revealed that the initial diagnosis of substance-induced psychosis was not incorrect; meanwhile a primary psychosis was outlined that cannabis had contributed to showing.

1.4 Literature Review

The *Diagnostic and Statistical Manual of Mental Disorders*, 5th edition (DSM-5) [2], defines psychosis as having delusions or hallucinations, with the hallucinations occurring in the absence of insight into their pathological nature. A broader definition includes symptoms such as disorganized speech and grossly disorganized or catatonic behavior. Semper and McClellan [26] refer to the DSM definition of psychosis for children and also note that negative symptoms such as alogia, amotivation, and anhedonia can be present. Cognitive and mood symptoms may also be present. The DSM-5 applies the same diagnostic criteria for psychotic disorders in children and adolescents as for adults.

The *DSM-5* necessary criteria for diagnosis of schizophrenia and of substance-induced psychotic disorder are reported in Boxes 1.2 and 1.3, respectively [2].

Box 1.2 *DSM-5* Criteria for Diagnosis of Schizophrenia

(A) Two (or more) of the following symptoms, each present for at least a 1-month (or longer) period of time (or less if treated effectively).
At least one of these symptoms must be (1), (2), or (3):
1. Delirium
2. Hallucinations
3. Disorganized speech (e.g., frequent derailment or inconsistency)
4. Grossly disorganized or catatonic behavior
5. Negative symptoms (i.e., diminished emotional expression)

(B) For a significant part of the time since the onset of the disorder, the level of operation in one or more of the major areas, such as work, interpersonal relationships, or self-care, is markedly below the level reached before the

(continued)

Box 1.2 (continued)

onset (or when the onset is in childhood or adolescence, the inability to reach the expected interpersonal, school, or work level is manifested).

(C) Continuous signs of disturbance persist for at least 6 months. This period of 6 months must include at least 1 month of symptoms (or not if treated effectively) which meet Criterion A (i.e., active phase of symptoms) and may include periods of prodromal or residual symptoms. During these prodromal or residual periods, the signs of the disorder can only be highlighted by negative symptoms or by two or more symptoms listed in Criterion A present in attenuated form (e.g., extravagant convictions, unusual perceptual experiences).

(D) Schizoaffective disorder and depressive disorder or bipolar disorder with psychotic characteristics were excluded because (1) there were no greater or manic depressive episodes in concomitance with the active phase of the symptoms or, (2) if episodes of mood alteration occurred during the active phase of the symptoms, they have manifested themselves for a minority of the total duration of active and residual phases of the disease.

(E) Disturbance is not attributable to the physiological effects of a substance (e.g., substance abuse, medication) or to another medical condition.

(F) If there is a history of autism spectrum disorder or a communication disorder (childhood onset), the additional diagnosis of schizophrenia is only postponed if there are hallucinations for at least 1 month (or less if treated effectively) or predatory delusions, in addition to the other symptoms of schizophrenia.

Box 1.3 *DSM-5* Criteria for Diagnosis of a Substance-Induced Psychotic Disorder

(A) Presence of one or both of the following symptoms:
1. Delusions
2. Hallucinations

(B) There is evidence from the history, physical examination, or laboratory findings of both (1) and (2):
1. The symptoms in Criterion A developed during or soon after substance intoxication or withdrawal or after exposure to a medication.
2. The involved substance/medication is capable of producing the symptoms in Criterion A.

(C) The disturbance is not better explained by a psychotic disorder that is not substance/medication-induced. Such evidence of an independent psychotic disorder could include the following:

(continued)

Box 1.3 (continued)

The symptoms preceded the onset of the substance/medication use; the symptoms persist for a substantial period of time (e.g., about 1 month) after the cessation of acute withdrawal or severe intoxication; or there is other evidence of an independent non-substance/medication-induced psychotic disorder (e.g., a history of recurrent non-substance/medication-related episodes).

(D) The disturbance does not occur exclusively during the course of a delirium.

(E) The disturbance causes clinically significant distress or impairment in social, occupational, or other important areas of functioning.

There are relatively few epidemiological studies on adolescent-onset psychosis. It is generally held that the incidence of psychosis, and especially of schizophrenia, increases markedly during the teenage years, with a preponderance of male over female patients [27]. Risk factors for early-onset psychosis (before age 18) are related to the interaction of environmental and biological factors in vulnerable individuals. Weiden and Buckley suggest that urban residence, social adversity, and early substance use, specifically cannabis use, are environmental risk factors that may increase the risk of developing schizophrenia during adolescence and young adulthood [28]. A positive family history, birth complications, advanced parental age, childhood developmental abnormalities, and early infections are some of the biological factors that have been reported [29]. Dalman et al. reviewed exposure to infections in early life as a risk factor for the development of psychosis [30].

The risk of developing psychotic forms is greater in those who abuse it earlier on. The duration of abuse also affects psychotic risk (cumulative effect). The highest psychotic risk correlates with high-potency cannabis varieties, such as "skunk" [31].

There is often diagnostic uncertainty in the first episode of adolescent-onset psychosis. The differential diagnosis may include primary psychosis (schizophrenia), bipolar disorder, psychotic disorder due to a general medical condition, or a substance-induced psychotic disorder. Delusional disorder and brief psychotic disorder are less common [32].

Some medical conditions (such as temporal lobe epilepsy or Huntington's chorea) can produce psychotic symptoms, and, since individuals are likely to be taking medications for these conditions, it can be difficult to determine the cause of the psychotic symptoms. If the symptoms are determined to be due to the medical condition, then a diagnosis of a psychotic disorder due to a general medical condition is warranted. Table 1.1 shows psychiatric conditions associated with psychotic episodes in children and adolescents.

In adolescence it is of great importance to be able to distinguish primary psychosis from substance-induced psychosis. While there are no absolute means of determining substance use as a cause, a good patient history that includes careful assessment of onset and course of symptoms, along with that of substance use, is

Table 1.1 Psychiatric conditions associated with psychotic episodes in children and adolescents

Alcohol intoxication/withdrawal
Attention-deficit/hyperactivity disorder
Autism spectrum disorders
Bipolar disorder
Brief reactive psychosis
Catatonia
Delirium
Delusional disorders
Factitious disorders
Major depressive disorder
Malingering
Obsessive-compulsive disorder
Parasomnias
Personality disorders
Post-traumatic stress disorder
Schizoaffective disorder
Schizophrenia
Schizophreniform disorder
Severe stress

Based on Freudenreich and Goff [33]

imperative. Often, the patient's testimony is unreliable, necessitating the gathering of information from family, friends, coworkers, employment records, medical records, and the like.

Substance-induced psychotic disorder displays psychotic symptoms (hallucinations not recognized by the individual as substance-induced or delusions). In order to justify this diagnosis, the psychotic symptoms must occur within a month after substance intoxication or withdrawal, or as a result of medication that caused the symptoms. However, the diagnosis is not made if the symptoms occurred before the substance or medication was ingested or are more severe than could be reasonably caused by the amount of substance involved. If the disorder persists for more than a month after withdrawal of the substance, the diagnosis becomes increasingly questionable, and a diagnosis of primary psychosis, or the existence of a medical condition, becomes more plausible [34].

Psychotic symptoms induced by substance intoxication usually subside once the substance is eliminated. Symptoms persist depending on the half-life of the substances (i.e., how long it takes before the substance is no longer present in an individual's system). Symptoms, therefore, can persist for hours, days, or weeks after a substance is last used.

1.4.1 Neurotransmitters Involved in Early-Onset Primary Psychosis

The neurotransmitter most commonly implicated in the pathophysiology of psychosis is dopamine. Drugs that increase dopaminergic receptor activity (e.g., cocaine, amphetamines) may induce an acute psychotic episode, whereas drugs that block postsynaptic D2 receptors help to alleviate psychotic symptoms. Furthermore, persons with psychosis also have fewer D1 receptors in the prefrontal cortex [35]. Disturbances in a variety of other neurotransmitters such as glutamate, serotonin, and γ-aminobutyric acid (GABA) have been implicated in the pathophysiology of primary psychosis. Researchers are also interested in glutamate because phencyclidine (PCP or angel dust) is an NMDA/glutamine antagonist; NMDA receptor dysfunction can also cause psychotic symptoms [36]. Preliminary studies suggest that GABA may also play a role in the development of chronic psychotic syndromes. Overall, disrupted neurotransmission and cognitive functions are key components in the pathophysiology of psychosis; however, no single neurotransmitter is clearly responsible for the onset and progression of psychosis.

1.4.2 Neurodevelopmental and Neurobiological Abnormalities Associated with Early-Onset Primary Psychosis

It is widely agreed that primary psychosis is a neurodevelopmental disorder; no single neurodevelopment model, however, explains the pathophysiology of the illness. Several studies have implicated complications during pregnancy and delivery as risk factors [37]. The combination of genetic risk and evidence of acquired damage has suggested a neurodevelopmental theory with early central nervous system abnormalities that contribute to an increased vulnerability to schizophrenia later in life. Hypoxia-associated obstetrical complications also appear to increase the odds of developing earlier-onset primary psychosis [38].

As in adults, early-onset primary psychosis shows bilateral enlargement of the lateral ventricles on neuroimaging. However, unlike adults, the abnormalities in brain morphology evolve during adolescence. Rapoport et al. [39] reported significantly reduced frontal and temporal gray matter volumes in adolescents compared with those observed in healthy age-matched controls. Moreover, youths with early-onset primary psychosis appear to lose more cortical gray matter compared to children who suffer from transient psychosis [40]. Subsequent studies from this group have shown that the healthy siblings of afflicted patients also have reductions in cerebral volume and gray matter [41]. In a systematic review of 66 articles comparing brain volume in patients with a first psychotic episode with the volume seen in healthy controls, a meta-analysis demonstrated that whole-brain and hippocampal volumes are reduced and that ventricular volumes are increased in affected patients compared to healthy controls [42]. Future improvements in neuroimaging technology hold the potential to reveal more about neurobiological disturbances and their correlates in schizophrenia.

1.4.3 Substance-Induced Psychosis Versus Primary Psychosis

Diagnostic certainty, in early-phase psychotic disorder, is often difficult to achieve and is challenged further when psychosis co-occurs with the use of alcohol or drugs. Diagnostic change over time has been observed in longitudinal studies of primary psychotic disorders. A change in diagnosis from a substance-induced psychosis to a primary psychosis can reflect the evolution of an illness, the availability of new information about onset or course, or unreliable diagnostic assessments. Psychotomimetic drug use may precipitate a schizophrenia-like illness or may evolve into a chronic psychotic disorder over time.

The primary psychosis versus substance-induced psychosis distinction was remarkably stable over the 1-year follow-up period. Caton et al. [43] observed a change in diagnostic category from substance-induced psychosis at baseline to primary psychotic disorder at the 1-year follow-up in about 25% of those diagnosed with substance-induced psychosis at baseline.

Greater instability in substance-induced psychosis diagnoses compared with primary psychosis diagnoses has been observed [44].

A differentiation between substance-induced disorder and a psychiatric disorder may be aided by the following:

- Time of onset: if symptoms began prior to substance use, it is most likely a psychiatric disorder.
- Substance use patterns: if symptoms persist for three months or longer after substance is discontinued, a psychiatric disorder is probable.
- Consistency of symptoms: symptoms more exaggerated than one would expect with a particular substance type and dose most likely amount to a psychiatric disorder.
- Family history: a family history of mental illness may indicate a psychiatric disorder.
- Response to substance abuse treatment: Clients with both psychiatric and substance use disorders often have serious difficulty with traditional substance abuse treatment programs and relapse during or shortly after treatment cessation.
- Patient's stated reason for substance use: those with a primary psychiatric diagnosis and secondary substance use disorder will often indicate they "medicate symptoms," for example, drink to dispel auditory hallucinations, use stimulants to combat depression, and use depressants to reduce anxiety or soothe a manic phase. While such substance use most often exacerbates the psychotic condition, it does not necessarily mean it is a substance-induced psychotic disorder [34, 45].

While it is often difficult to distinguish substance-induced from primary psychoses, especially early in the course of treatment, this differential diagnosis has important implications for treatment planning. The issues of assessment and treatment planning are particularly important in the early stages of psychotic disorder, because this is a time when the symptom picture is often unclear and a proper match of diagnosis with treatment may be critically important for outcome [45].

Longitudinal follow-up of patients initially presenting with psychosis and substance use comorbidity is warranted by the occurrence of heavy substance misuse overlying presentation of psychotic symptoms, adding greater complexity to the diagnostic process, and the greater instability of substance-induced psychosis diagnoses.

Unfortunately, psychological tests are not always helpful in determining if a psychotic disorder is caused by substance use or is being exacerbated by it. However, evaluations, such as the MMPI-2 (Minnesota Multiphasic Personality Inventory-2) [46], MAC-R scale (MacAndrew Alcoholism Scale-Revised) [47], or the Wechsler Memory Scale-Revised [48], can be useful in making a differential diagnosis.

1.4.4 Treatment

The diagnostic distinction between a substance-induced and a primary psychotic disorder is critically important, because each disorder requires a different treatment.

For example, subjects with drug-induced psychosis may need different medications, no medications, or brief medication therapy, and they may be more susceptible to the adverse effects of antipsychotic medications. Although psychotomimetic drug use may precipitate a chronic schizophrenic illness, an accurate diagnostic assessment is particularly significant in the early stages of psychotic disorder, when the diagnostic picture is often clouded by the presence of substance use and differential therapeutics are appropriate [45].

Furthermore pharmacologic treatment for psychosis in youth remains an area of active research, with several large multisite trials completed within the past few years. Six-week, randomized, placebo-controlled studies of second-generation antipsychotics for early-onset substance-induced psychosis have been completed thus far. They showed that aripiprazole, olanzapine, paliperidone, quetiapine, and risperidone—but not ziprasidone—were superior to placebo, with regard to improvement in the Positive and Negative Syndrome Scale (PANSS) total score [49–52].

In terms of atypical antipsychotic dosing, typically, these medications are initiated at a low dose and gradually titrated upward to achieve efficacy. Risperidone, for example, is generally started at 0.25 mg twice a day and can be increased every 1 to 2 days in association with close observation. In general, olanzapine and quetiapine are more sedating; their dosing is usually started at 2.5 to 5 mg/day and 25 to 50 mg/day, respectively. As most antipsychotic medications have a relatively long half-life, they typically do not need to be administered more than twice daily. For patients maintained on antipsychotics, clinicians should monitor weight, vital signs, and relevant laboratory results (e.g., triglycerides, cholesterol, fasting glucose, and prolactin levels).

Newer atypical antipsychotics (e.g., asenapine, iloperidone, lurasidone) may eventually have a role in the treatment of patients with childhood-onset or early-onset psychosis and in patients deemed at risk of developing psychosis due to genetic and/or environmental factors. However, at present, there is little evidence to support the safety, tolerability, or efficacy of these agents in the pediatric population.

In sum, current pharmacologic research for early-onset and childhood-onset psychosis focuses on the identification of antipsychotic agents that will provide optimal efficacy without also causing undue adverse events. Many typical and atypical antipsychotics have been shown (based on open trials, randomized clinical trials, and head-to-head comparisons) to be efficacious in the treatment of early-onset and childhood-onset schizophrenia [53]; however, they are well known to cause side effects such as weight gain, extrapyramidal symptoms, and metabolic abnormalities.

Nonadherence to oral antipsychotic medications is one of the most significant clinical challenges in the treatment of schizophrenia. Nonadherence rates may be as high as 50% in the first year of treatment and 75% during the first 2 years of treatment. Despite evidence that continuous antipsychotic treatment is more effective than interrupted treatment, long-acting therapy use in the United States remains low. Barriers to increased long-acting therapy use include physicians' reluctance to administer injectable medications, confusing reimbursement procedures, and the unfounded belief that patients would reject an offer of this treatment modality [54].

In clinical situations wherein adherence with oral preparations has proven difficult, it is reasonable especially in older adolescents to consider using a long-acting medication. Risperidone long-acting therapy has been shown to be efficacious and well tolerated in the treatment of schizophrenia, including those with newly diagnosed schizophrenia [55, 56]. Short-term and long-term studies found that risperidone long-acting therapy was associated with low rates of discontinuation related to adverse events and few reports of pain from injections [57]. Evidence for newer long-acting injectables (e.g., olanzapine, paliperidone) remains limited [58]. The available evidence suggests that long-acting injectable antipsychotics can be used safely and effectively in early stages of the illness and that they may be associated with better outcomes than with oral medications. However, this is largely supported by evidence from naturalistic cohort studies and by a small number of controlled trials of risperidone long-acting injection. Given the paucity of data of long-acting therapy in the adolescent population, the use of these preparations in adolescents remains empiric. Table 1.2 provides the side effect risk profiles of common second-generation antipsychotic medications used in children and adolescents [59–62].

Clinical Points

- Hallucinatory experiences and delusions occur frequently in the pediatric population. While not a benign occurrence, psychotic symptoms do not necessarily foreshadow the future development of schizophrenia.

Table 1.2 Side effect risk profiles of common second-generation antipsychotic medications used in children and adolescents

Second-generation antipsychotics (trade name)	Anticholinergic	Diabetes	Hyperlipidemia	Hyperprolactinemia	Hypotension	Sedation	Tardive dyskinesia	Weight gain
Aripiprazole (Abilify)	0	+	+	0	+	+	+	+
Asenapine (Saphris)	0	+	+	+	0	+	+	+
Clozapine (Clozaril)	+++	+++	++/+++	0	+++	+++	0	+++
Iloperidone (Fanapt)	0	+	+	+	+	+	+	+
Lurasidone (Latuda)	0	+	+	+	0	+	+	+
Olanzapine (Zyprexa)	++	+++	++/+++	+/++	+	+/++	+	+++
Paliperidone (Invega)	0	+	+/++	++	+	+	+	++
Quetiapine (Seroquel)	0/+	++	+/++	0	++	++	0	++
Rjsperidone (Risperda])	0	+	+/++	++	++	+	+	++
Ziprasidone (Geodon)	0	+	+	0	+	+	+	+

Based on Celano et al. [59] and Correll et al. [60]

Symbols: 0 = low risk, + = mild risk, ++ = moderate risk, +++ = high risk

- Psychotic symptoms in children and adolescents can occur in the context of a bevy of psychiatric disorders (schizophrenia, depression, anxiety, bipolar disorder, attention-deficit/hyperactivity disorder, posttraumatic states, and autism spectrum disorders) or can be secondary to a wide variety of medical conditions and substance abuse. A comprehensive clinical evaluation is necessary to determine what may be the causes of the psychotic symptoms and to rule out possible etiologies.
- The onset of psychosis is usually preceded by a period of nonpsychotic symptoms known as prodromal symptoms. In recent years, there has been substantial research in early intervention efforts (e.g., with psychotherapy or antipsychotic medicines) focused on the early stages of primary psychosis and on young people with prodromal symptoms. Presently, outcome data are insufficient to draw definitive conclusions.
- The finding that the association between substance abuse and psychosis outcomes is most marked in subjects with an established vulnerability to psychosis suggests the presence of the aforesaid gene-environment interaction.
- The pharmacologic treatment of psychotic symptoms in a pediatric population is similar in many ways to the treatment of infection with antibiotics: the clinician needs to choose the appropriate medication at a sufficient dose and then await therapeutic results while monitoring for potential side effects.

Self-Assessment Questionnaire

1. **What are the main symptoms of early-onset psychosis?**
   (A) Euphoria, high mood, hyperactivity, decreased need to sleep
   (B) Feelings of guilt, suicidal ideation, depressed mood
   (C) Delusions, hallucinations, thought disorders, behavioral disorganization
   (D) Inattention, hyperactivity, impulsivity
2. **Which is the most frequent differential diagnosis that could be confusing in early presenting psychotic symptoms?**
   (A) ADHD
   (B) Bipolar Disorder
   (C) Delirium
   (D) Substance-induced psychosis
3. **Which neurotransmitters are most commonly implicated in psychosis?**
   (A) Glutamate
   (B) **Dopamine**
   (C) γ-aminobutyric acid (GABA)
   (D) Serotonin
4. **In order to justify the diagnosis of substance-induced psychosis, when must the psychotic symptoms occur?**
   (A) **Within a month after substance intoxication or withdrawal**
   (B) 1 year after substance intoxication or withdrawal
   (C) 6 months after substance intoxication or withdrawal

(D) 3 months after substance intoxication or withdrawal

5. **Which neurodevelopmental and neurobiological abnormalities are associated with early-onset primary psychosis compared to early-onset substance-induced psychosis?**
   (A) Reduced frontal and temporal gray matter volumes
   (B) Reduced ventricular volumes
   (C) Cerebellum volume modifications
   (D) Increased hippocampal volumes
6. **Which is the best therapeutic option for the stabilization of early-onset psychosis?**
   (A) Long-acting injectable antipsychotic
   (B) Valproic acid
   (C) Lithium
   (D) Antipsychotic

References

1. Stevens JR, Prince JB, Prager LM, ,Stern TA. Psychotic disorders in children and adolescents: a primer on contemporary evaluation and management. Prim Care Companion CNS Disord. 2014; 16(2): PCC.13f01514.
2. American Psychiatric Association. Diagnostic and statistical manual of mental disorders: DSM-5. Washington: American Psychiatric Association; 2013.
3. Van Os J, Linscott RJ, Myin-Germeys I, et al. A systematic review and meta-analysis of the psychosis continuum: evidence for a psychosis-proneness-persistence-impairment model of psychotic disorder. Psychol Med. 2009;39(2):179–95.
4. Bartels-Velthuis AA, Jenner JA, van de Willige G, et al. Prevalence and correlates of auditory vocal hallucinations in middle childhood. Br J Psychiatry. 2010;196(1):41–6.
5. Horwood J, Salvi G, Thomas K, et al. IQ and non-clinical psychotic symptoms in 12-year-olds: results from the ALSPAC birth cohort. Br J Psychiatry. 2008;193(3):185–91.
6. Kelleher I, Harley M, Murtagh A, et al. Are screening instruments valid for psychotic-like experiences? A validation study of screening questions for psychotic-like experiences using in-depth clinical interview. Schizophr Bull. 2011;37(2):362–9.
7. Laurens KR, Hodgins S, Maughan B, et al. Community screening for psychotic-like experiences and other putative antecedents of schizophrenia in children aged 9–12 years. Schizophr Res. 2007;90(1–3):130–46.
8. Kelleher I, Corcoran P, Keeley H, et al. Psychotic symptoms and population risk for suicide attempt: a prospective cohort study. JAMA Psychiat. 2013;70(9):940–8.
9. Schoeler T, Petros N, Di Forti M, Pingault JB, Klamerus E, Foglia E, Small A, Murray R, Bhattacharyya S. Association between continued Cannabis use and risk of relapse in first-episode psychosis: a quasi-experimental investigation within an observational study. JAMA Psychiatr. 2016;73(11):1173–9.
10. Malone DT, Hill MN, Rubino T. Adolescent cannabis use and psychosis: epidemiology and neurodevelopmental models. Br J Pharmacol. 2010;160(3):511–22.
11. Arseneault L, Cannon M, Witton J, Murray RM. Causal association between cannabis and psychosis: examination of the evidence. Br J Psychiatry. 2004;184:110–7.
12. Pistis M, Perra S, Pillolla G, Melia M, Muntoni AL, Gessa GL. Adolescent exposure to cannabinoids induces long-lasting changes in the response to drugs of abuse of rat midbrain dopamine neurons. Biol Psychiatry. 2004;56:86–94.

13. Pope HG, Gruber AJ, Hudson JI, Cohane G, Huestis MA, Yurgelun-Todd D. Early-onset cannabis use and cognitive deficits: what is the nature of the association? Drug Alcohol Depend. 2003;69:303–10.
14. Henquet C, Krabbendam L, Spauwen J, Kaplan C, Lieb R, Wittchen H-U, et al. (2004): Prospective cohort study of cannabis use, predisposition for psychosis, and psychotic symptoms in young people. BMJ. 330:11. doi:10.1136/bmj.38267.664086.63 [Available: BMJ. Published 1 Dec 2004].
15. Verdoux H, Gindre C, Sorbara F, Tournier M, Swendsen JD. Effects of cannabis and psychosis vulnerability in daily life: an experience sampling test study. Psychol Med. 2003;33:23–32.
16. Caspi A, Moffitt TE, Cannon M, McClay J, Murray R, Harrington HL, Taylor A, Arseneault L, Williams B, Braithwaite A, Poulton R, Craig IW. Moderation of the effect of adolescent-onset Cannabis use on adult psychosis by a functional polymorphism in the catechol-O-Methyltransferase gene: longitudinal evidence of a gene X environment interaction. Biol Psychiatry. 2005;57:1117–27.
17. Thomas H. A community survey of adverse effects of cannabis use. Drug Alcohol Depend. 1996;42:201–7.
18. Hall W. Cannabis and psychosis. Drug Alcohol Rev. 1998;17:433–44.
19. Johns A. Psychiatric effect of cannabis. Br J Psychiatry. 2001;178:116–22. https://doi.org/10.1192/bjp.178.2.116.
20. Van Os J, Bak M, Hanssen M, Bijl RV, de Graaf R, Verdoux H. Cannabis use and psychosis: a longitudinal population-based study. Am J Epidemiol. 2002;156:319–27.
21. Birur B, Kraguljac NV, Shelton RC, Lahti AC. Brain structure, function, and neurochemistry in schizophrenia and bipolar disorder—a systematic review of the magnetic resonance neuroimaging literature. NPJ Schizophr. 2017;3:15. https://doi.org/10.1038/s41537-017-0013-9. Published Online 3 Apr 2017
22. Dragogna F, Oldani L, Buoli M, Altamura AC. A case of severe psychosis induced by novel recreational drugs. Version 1. F1000Res. 2014;3:21. https://doi.org/10.12688/f1000research.3-21.v1. Published Online 22 Jan 2014
23. Overall LE, Gorham DR. The brief psychiatric rating scale. Psychol Rep. 1962;10:799–812.
24. Kay SR, Opler LA, Fiszbein A. Positive and negative syndrome scale (PANSS) rating manual. New York: Albert Einstein College of Medicine; 1986.
25. Endicott J, Spitzer RL, Fleiss J, et al. The global assessment scale: a procedure for measuring overall severity of psychiatric disturbance. Arch Gen Psychiatry. 1976;33:766–71.
26. Semper TF, McClellan JM. The psychotic child. Child Adolesc Psychiatr Clin North Am. 2003;12:679–91.
27. Clark AF. Proposed treatment for adolescent psychosis; schizophrenia and schizophrenia-like psychoses. Adv Psychiatr Treat. 2001;7:16–23.
28. Weiden PJ, Buckley PF. Understanding and treating 'first-episode' schizophrenia. Psychiatr Clin North Am. 2007;30(3):481–506.
29. Messias EL, Chen CY, Eaton WW. Epidemiology of schizophrenia: review of findings and myths. Psychiatr Clin North Am. 2007;30(3):323–38.
30. Dalman C, Alleback P, Gunnel D, et al. Infections in the CNS during childhood and the risk of subsequent psychotic illness: a cohort study of more than one million Swedish subjects. Am J Psychiatry. 2008;165(1):59–65.
31. Arendt M, Rosenberg R, Foldager L, Perto G, Munk-Jørgensen P. Cannabis-induced psychosis and subsequent schizophrenia-spectrum disorders: follow-up study of 535 incident cases. Br J Psychiatry. 2005;187:510–5.
32. Reimherr JP, McClellan JM. Diagnostic challenges in children and adolescents with psychotic disorders. J Clin Psychiatry. 2004;65(Suppl. 6):5–11.
33. Freudenreich O, Goff DC. Psychotic patients. In: Stern TA, Fricchione GH, editors. Massachusetts General Hospital handbook of general hospital psychiatry. Amsterdam: Elsevier Science; 2010.

34. Kaplan Harold I, Sadock BJ. Kaplan and Sadock's synopsis of psychiatry: behavioral sciences, clinical psychiatry. 8th ed. Baltimore: Williams and Wilkins; 2002.
35. Okubo Y, Suhara T, Suzuki K, et al. Decreased prefrontal dopamine D1 receptors in schizophrenia revealed by PET. Nature. 1997;385(6617):634–6.
36. Clinton SM, Haroutunian V, Davis KL, et al. Altered transcript expression of NMDA receptor-associated postsynaptic proteins in the thalamus of subjects with schizophrenia. Am J Psychiatry. 2003;160(6):1100–9.
37. Byrne M, Agerbo E, Bennedsen B, et al. Obstetric conditions and risk of first admission with schizophrenia: a Danish national register based study. Schizophr Res. 2007;97(1–3):51–9.
38. Fish B, Kendler KS. Abnormal infant neurodevelopment predicts schizophrenia spectrum disorders. J Child Adolesc Psychopharmacol. 2005;15(3):348–61.
39. Rapoport JL, Giedd JN, Blumenthal J, et al. Progressive cortical change during adolescence in childhood-onset schizophrenia: a longitudinal magnetic resonance imaging study. Arch Gen Psychiatry. 1999;56(7):649–54.
40. Rapoport JL, Gogtay N. Childhood onset schizophrenia: support for a progressive neurodevelopmental disorder. Int J Dev Neurosci. 2011;29(3):251–8.
41. Gogtay N, Sporn A, Clasen LS, et al. Structural brain MRI abnormalities in healthy siblings of patients with childhood-onset schizophrenia. Am J Psychiatry. 2003;160(3):569–71.
42. Steen RG, Mull C, McClure R, et al. Brain volume in first-episode schizophrenia: systematic review and meta-analysis of magnetic resonance imaging studies. Br J Psychiatry. 2006;188 (6):510–8.
43. Caton CLM, Hasin DS, Shrout PE, et al. Stability of early-phase primary psychotic disorders with concurrent substance use and substance-induced psychosis. Br J Psychiatry. 2007;190:105–11.
44. Whitty P, Clarke M, Mc Tigue O, et al. Diagnostic stability four years after a first episode of psychosis. Psychiatr Serv. 2005;56:1084–8.
45. Caton CLM, Drake RE, Hasin DS, et al. Differences between early phase primary psychotic disorders with concurrent substance use and psychotic disorders with concurrent substance use and substance-induced psychosis. Arch Gen Psychiatry. 2005;62:137–45.
46. Hathaway SR, McKinley JC. Manual for the Minnesota multiphasic personality inventory. Minneapolis: University of Minnesota Press; 1942.
47. MacAndrew C. The differentiation of male alcoholic outpatients from non-alcoholic psychiatric outpatients by means of the MMPI. Q J Stud Alcohol. 1965;26:238–46.
48. Wechsler D. Manual for the Wechsler memory scale-revised. San Antonio: The Psychological Corporation; 1987.
49. Findling RL, Robb A, Nyilas M, et al. A multiple-center, randomized, double-blind, placebo-controlled study of oral aripiprazole for treatment of adolescents with schizophrenia. Am J Psychiatry. 2008;165(11):1432–41.
50. Kryzhanovskaya L, Schulz SC, McDougle C, et al. Olanzapine versus placebo in adolescents with schizophrenia: a 6-week, randomized, double-blind, placebo-controlled trial. J Am Acad Child Adolesc Psychiatry. 2009;48(1):60–70.
51. Haas M, Unis AS, Armenteros J, et al. A 6-week, randomized, double-blind, placebo-controlled study of the efficacy and safety of risperidone in adolescents with schizophrenia. J Child Adolesc Psychopharmacol. 2009;19(6):611–21.
52. Singh J, Robb A, Vijapurkar U, et al. A randomized, double-blind study of paliperidone extended-release in treatment of acute schizophrenia in adolescents. Biol Psychiatry. 2011;70 (12):1179–87.
53. Lasser RA, Schooler NR, Kujawa M, et al. A new psychosocial tool for gaining patient understanding and acceptance of long-acting injectable antipsychotic therapy. Psychiatry (Edgmont). 2009;6(4):22–7.
54. Parellada E, Andrezina R, Milanova V, et al. Patients in the early phases of schizophrenia and schizoaffective disorders effectively treated with risperidone long-acting injectable. J Psychopharmacol. 2005;19(Suppl):5–14.

55. Kane JM, Eerdekens M, Lindenmayer JP, et al. Long-acting injectable risperidone: efficacy and safety of the first long-acting atypical antipsychotic. Am J Psychiatry. 2003;160(6):1125–32.
56. Simpson GM, Mahmoud RA, Lasser RA, et al. A 1-year double-blind study of 2 doses of long-acting risperidone in stable patients with schizophrenia or schizoaffective disorder. J Clin Psychiatry. 2006;67(8):1194–203.
57. Lindenmayer JP, Jarboe K, Bossie CA, et al. Minimal injection site pain and high patient satisfaction during treatment with long-acting risperidone. Int Clin Psychopharmacol. 2005;20(4):213–21.
58. Emsley R, Chiliza B, Asmal L, et al. Long-acting injectable antipsychotics in early psychosis: a literature review. Early Interv Psychiatry. 2013;7(3):247–54.
59. Celano CM, Beach SR, Huffman JC, et al. Side effects profiles of psychotropic medications from cardiovascular and other side effects of psychotropic medications. In: Stern TA, Herman JB, Gorrindo T, editors. The Massachusetts General Hospital psychiatry update and board preparation. 3rd ed. Boston: MGH Psychiatry Academy Publishing; 2012.
60. Correll CU, Penzner JB, Parikh UH, et al. Recognizing and monitoring adverse events of second-generation antipsychotics in children and adolescents. Child Adolesc Psychiatr Clin N Am. 2006;15(1):177–206.
61. Freudenreich O. Psychotic disorders: a practical guide. Philadelphia: Lippincott Williams and Wilkens; 2008.
62. Schulz SC, Findling RL, Wise A, Friedman L, Kenny J. Child and adolescent schizophrenia. Psychiatr Clin North Am. 1998;21(1):43–56.

Deficit Schizophrenia

2

R. A. Paoli and M. Grottaroli

Abstract

Schizophrenia patients show symptoms in the framework of negative, positive, affective, disorganized, and cognitive dimensions. In particular, a relationship between neurocognitive impairment, negative symptoms, and disorganization has been detected, featuring the so-called deficit schizophrenia. In this context, we here present a clinical case of severe schizophrenia characterized by neurocognitive dysfunction, altered functionality, negative symptoms, and imaging abnormalities.

Keywords

Negative symptoms · Deficit subtypes · Disorganizative dimension · Neurocognitive impairment · Functional neuroimaging · Poor outcome

2.1 Introduction

The concept of schizophrenia was introduced by Eugen Bleuler in 1908. He coined the term "schizophrenia", or more precisely "the group of schizophrenias," because in his opinion "the breaking up or splitting of psychic functioning is an excellent symptom of the whole group" [1]. Moreover, Bleuler identified avolition (i.e., lack of motivation or initiative) as central to schizophrenia. As is known, the term schizophrenia is still used, but throughout the seven DSM (*Diagnostic and Statistical Manual of Mental Disorders*) editions, the definition of schizophrenia evolved, sometimes to encompass the undeniable heterogeneity of schizophrenia symptoms.

R. A. Paoli (✉) · M. Grottaroli
Fondazione IRCCS Ca' Granda Ospedale Maggiore Policlinico, Department of Neurosciences and Mental Health, University of Milan, Milan, Italy
e-mail: riccardo.paoli@policlinico.mi.it; riccardo.paoli@guest.unimi.it

A. C. Altamura, P. Brambilla (eds.), *Clinical Cases in Psychiatry: Integrating Translational Neuroscience Approaches*,
https://doi.org/10.1007/978-3-319-91557-9_2

The previous DSM-IV indeed specified five schizophrenia subtypes (paranoid, disorganized, catatonic, undifferentiated, and residual type). This notwithstanding, traditional schizophrenia subtypes have been demonstrated as not being stable over time and not responsive to specific treatments. In the DSM-5, these subtypes have been removed because of their low reliability and poor validity but also because of the growing tendency to focus on a dimensional rather than a categorical perspective [2]. Nevertheless, patients suffering from schizophrenia show differences in terms of clinical symptoms, daily-life difficulties, illness course, cognitive impairment, treatment response, and biological and neuroimaging features. Sometimes the clinical picture of schizophrenia may be associated with a specific medical condition or substance abuse. For this reason, clinicians should also ask about recent head injury or trauma, seizures, cerebrovascular disease, and headaches. They should also exclude oncologic causes, thyrotoxicosis, encephalitis, and porphyria. Testing for human immunodeficiency virus infection and syphilis should also be considered [3].

Beyond these aspects of differential diagnosis, regarding the dismissal of categorical subtypes of schizophrenia with the DSM-5, the diagnostic dimensional approach is also partially implied by the conceptualization of the negative, positive, disorganized, and affective dimension [4]. These symptomatological clusters have been related in various degrees to cognitive impairment, unsatisfactory functioning, treatment response, and outcome of the disease, allowing the identification of worsening subtypes of schizophrenia [5] that are hypothesized as deficit subtypes of schizophrenia [6].

While positive dimension has been less related to neurocognitive impairment [7], more than a year's persistence of negative symptoms (affective flattening, alogia, avolition-apathy, asociality, attentional impairment), in spite of an adequate pharmacologic treatment, has been stated as the core of a supposed separate disease, with its specific risk factors and its specific biological features. In the past, Carpenter et al. [8] defined this syndrome as deficit schizophrenia, empathizing the persistence of decrease in emotional range, poverty of speech, loss of interest, and the loss of drive with, compared to schizophrenia, positive anamnesis for worsening premorbid functioning, poor insight, long-term disability, and bad prognosis [9]. To this purpose, Kirkpatrick et al. [10] strengthened the diagnostic importance of distinguishing primary enduring negative symptoms (clinical core of deficit schizophrenia) from secondary ones, which often occur in the context, or as a consequence, of concurrent positive, depressive, and/or extrapyramidal symptoms.

This negative enduring and unremitting schizophrenia was also mentioned as "Kraepelinian schizophrenia" [11]. In his manuscripts, Kraepelin indeed described how some schizophrenic patients exhibit "failure of mental activities," "weakening of volition," "loss of mastery over volition," and "loss of ability for independent action" [12]. Underlining that schizophrenia is a processual progressive disease that leads to neurocognitive defects, Kraepelin used the term "dementia praecox," in order to remark the putative neuropathological aspects of this syndrome. We now know that some forms of schizophrenia are associated with progressive structural brain abnormalities, affecting both gray and white matter [13, 14]. Other schizophrenic patients exhibited a progressive reduction in frontal lobe white matter volume, with a concomitant increase in frontal lobe cerebrospinal fluid volume. As partially predicted by Kraepelin, schizophrenic patients with poor outcome had a greater lateral ventricular enlargement over time, while enlargement in frontal lobe cerebrospinal fluid volume has been associated with greater negative symptoms severity [15].

Most studies have hinted that deficit schizophrenia is related to a more severe neurocognitive impairment [16] [17], which may also sustain a group of signs and symptoms, named as psychomotor slowing and emphasized as characterizing enduring negative schizophrenia [18]. During psychomotor tasks the behavior is governed by a number of neurocognitive processes (not only based on motor skill learning ideas and cognitive control of actions) which, if impaired, may cause psychomotor slowing, proved to be related to the patients' social, clinical, and functional outcome. Selecting writing tasks that were able to generate measurements for different subprocesses of psychomotor functioning (planning, initiation, and execution), Bervoets et al. [19] sought to explain the relationship between psychomotor slowing and neurocognition and predominant symptoms in deficit schizophrenia. Interestingly, they found that negative symptoms were found to be mainly associated with difficulties in the initiation of fine motor movements, whereas planning and execution deficit were independent of the symptomatology. Compared to non-deficit patients, deficit ones performed significantly worse on social and global cognition and language [20]. Yu et al. also compared clinical deficit schizophrenia to non-deficit types [21]: both schizophrenia subgroups had overall more severe cognitive impairments than the controls, while patients with deficit schizophrenia performed worse on every neuropsychological measure except the Stroop interference, during the homonymous test in which the patient must say the color of a word (on PC monitor) but not the name of the word. The authors speculated that reduced sustained attention might be the key impaired cognitive domain for negative subtype of deficit schizophrenia.

During the past decades, a relationship between neurocognitive impairment and disorganization was also detected. Considering schizophrenia and DSM-5 classification [22], international experts emphasized five key symptoms of schizophrenia: (1) delusions, (2) hallucinations, (3) disorganized speech, (4) disorganized or catatonic behavior, and (5) negative symptoms. As noted, the adjective "disorganized" recurs: a disorganized behavior means casual sequence of activities, formal thought disorder, loosened thought associations, and schizophasia. Schizophrenia has indeed been linked to the disorganized dimension since 1881 when Ewald Hecker, a German psychiatrist, described a specific syndrome, hebephrenia, characterized by an early-onset and severe disorganized symptoms, quickly progressing to functional and cognitive impairment. Several features of hebephrenia led to the diagnosis of the disorganized subtype (DSM-IV TR) [23], consisting in disorganized speech, disorganized behavior (distortion of idea production, distortion of language, and activities), and inappropriate affectivity. In their meta-analysis [24], Ventura et al. examined the strength of the relationship of neurocognition with schizophrenic symptom clusters across a range of neurocognitive domains (attention, reasoning, speed of processing, verbal and visual memory, working memory). The extent of the relationship between neurocognition and disorganization was significantly larger than the correlation between neurocognition and reality distortion. Disorganized patients also showed deficits in theory of mind [25] and in the ability of integration of contextual stimuli [26] with low-level visual integration processes and impaired visual closure task [27]. Disorganized patients have also been found to exhibit reduced spatial working memory performances [28] with, on Multiple Errands Test and Hotel Task (an ecological test) [29], a greater number of errors compared to the other diagnostic groups [5].

The clinical case we are going to describe aims to show how in serious schizophrenia specific symptomatological dimensions join in different ways with neurocognitive deficits, involving at different level executive functions, working memory, attention span, and sensory-motor coordination, and all together seemingly related to the increase over the years of functional neuroimaging alterations.

2.2 Case Presentation

Mr. B. is 24 years old. He graduated from middle school; he was never been employed, and he lives with his parents. He was brought to the emergency room by his parents, who were worried about the fact that he had begun talking to the television. They also revealed his poor functioning during the previous months: he didn't attend the local religious voluntary organization, he lost all his friends, and he spent his days at home.

During the interview at the emergency room, the patient is evasive: he speaks in monosyllables, he shows poor interest in his psychic conditions, and he also looks scarcely interested in his parents' worries. He looks untidy and bizarre: his long hair is dirty, and he is wearing an orange hat which he pulls over his eyes. He says he can't take it off; he says "saints" talk to him through that hat. He speaks with a low voice, and sometimes he stops talking, exclaiming "not now!". He watches a laptop monitor, and then he falls into silence and begins staring at it. After that, he starts touching the tip of his shoes by alternating his right hand on his left foot and vice versa. He keeps doing this for half an hour, refusing to provide any explanation. His mother says he has recently taken to making that sequence of movements. She made him come to the hospital that day because he'd gotten into a physical fight with his father just because he turned on the television. The patient burst into tears exclaiming that in that way the "saints" couldn't contact him.

The patient displayed a variety of signs: delusions, hallucinations, disorganization, as well as a recent increase in aggressive behavior. He also showed a significant decline in his functioning. Despite having a caregiver motivated to put him on treatment, Mr. B. exhibited no insight into his psychic conditions, being also poorly collaborative about undergoing therapies, stating he would take only orange pills. Furthermore, a substantial part of the interview was impaired by his lack of attention, as he appeared disturbed by acoustic hallucinations. So, considering all these aspects, the patient was hospitalized in the psychiatry unit.

Family anamnesis was positive for schizophrenia (uncle, maternal line), but negative for substance abuse. The patient is a smoker (30 cigarettes/day); negative findings for alcohol and substance use were collected. His personal history revealed eutocic delivery, with a birth weight of 3.7 kg and a normal psychomotor development. Family members described Mr. B. as quite a shy child. They reported that he hadn't difficulties in studying, and he had an intermediate academic performance at first. He didn't develop an outgoing character, but he had a few close friends among his schoolmates. Things worsened after graduation. Without a scheduled scholastic routine, the patient used to spend a lot of time at home, showing little interest in responding to calls or emails from his mates. After a month, he quit his job at the supermarket, because his mind wandered easily, not allowing him to focus on his activities. Worried about his social isolation, Mr. B.'s mother forced him to participate in her voluntary religious activities, although unsuccessfully. At home, he began exhibiting strange behaviors, such as wearing the hat (only the orange one), speaking about "saints," not eating certain foods because "he didn't like their colors," and turning around the TV and PC screens. He cared less about his appearance, and surprisingly, on one occasion, he shaved only half of his beard, without giving any reasonable explanation.

At the first psychic examination after admission to the ward, he appeared to be time and space oriented, though perplexed. He was restless, getting up uninterruptedly from the chair, running away, and then returning to the room. His speech was fragmentary, with partially loose associations and occasional neologisms. He was partly reticent to elaborate his contents during the interview. Nevertheless, he admitted he was hearing voices of "saints," who were accustomed to communicating with him through television and other monitors.

2.2.1 Physical Examination

The patient was alert and oriented. His vital parameters were collected: blood pressure 130/80 mmHg, respiration rate 22, and heart rate 90 bpm. Head was normocephalic, pupils were equal and reactive, and nostrils were pervious. Neck was without lymphadenopathy. The cardiac rate and rhythm were regular; at the auscultation, normal breathing sounds, with no crackles or wheezes. The abdomen was mildly adipose, with bowel sounds heard. Extremities were without cyanosis, clubbing, or edema; the skin was warm and dry. Blood count, renal, and hepatic function were on limits. TSH reflex was average. HBV, HCV, HIV, and *Treponema pallidum* antibodies were negative. At the EKG, sinus rhythm and QTC < 430 milliseconds.

2.2.2 Psychometric Evaluation

The patient underwent the Brief Psychiatric Rating Scale (BPRS), with a total score of 58 (Box 2.1). BPRS can be divided into five factors: hostility (HOST), anxiety depression (ANDP), anergy (ANER), thought disorder (THOT), and activity (ACTV). As can be seen, the patient got higher scores on items related to thought alterations and disorganized activities.

Box 2.1 As can Be Seen, THOT Is the Most Detected Factor

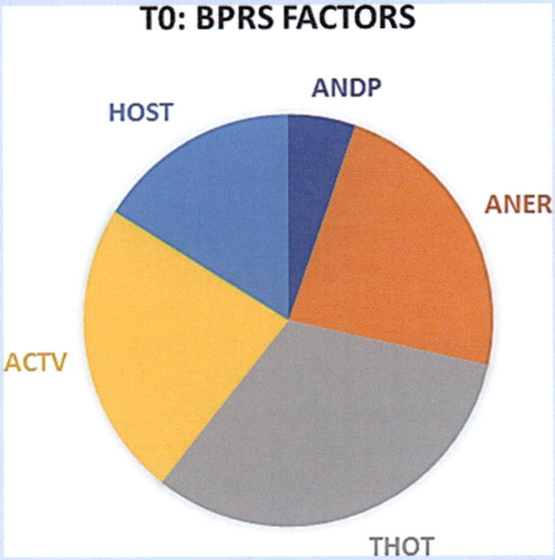

The Positive and Negative Syndrome Scale (PANSS) was also administered (Box 2.2), with a score of 95 and prevalence of disorganized dimension: Mr. B. lost his train of thought during conversations, beginning to talk to "saints." Tangentially jumping from one argument to another, apparently at random, he made loose associations of topics too, gave answers to unrelated questions, and made ritual movements like begging the "saints" to absolve him.

Box 2.2 PANSS Can Be Subdivided into Five Dimensions

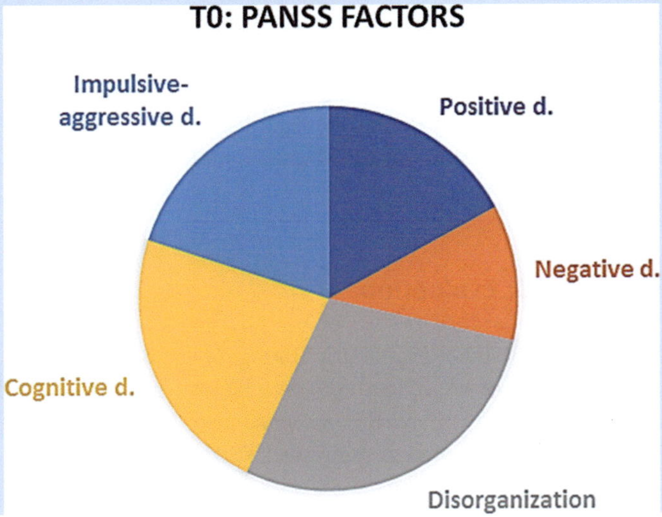

2.2.3 Neuroimaging Assessment

The patient had not been previously scanned with magnetic resonance imaging (MRI) of the brain. At MRI, T1-weighted, T2-weighted, TSE (Turbo Spin Echo), FLAIR (fluid-attenuated inversion recovery), and DWI (diffusion-weighted imaging) scans were acquired. Images were reported as normal (no midline shift, no intracerebral or extra-axial areas of abnormal signal, no evidence of posterior fossa abnormalities, no ventricular enlargements). Liquor and ventricular spaces resulted within the normal age limits (Box 2.3).

Positron emission tomography (PET) images (Box 2.4) revealed, instead, modest glucose cortical hypometabolism in the frontal lobe. Minor cortical metabolic perfusion irregularities were observed bilaterally in the orbitary cortices too. The fixation of the marked glucose was instead preserved in the remaining cerebral and cerebellar structures.

Box 2.3 On the Left and on the Right, Axial Images (T2)

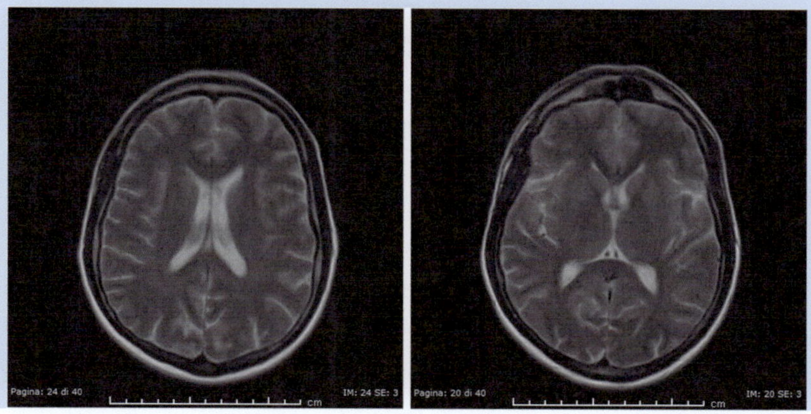

Box 2.4 First PET

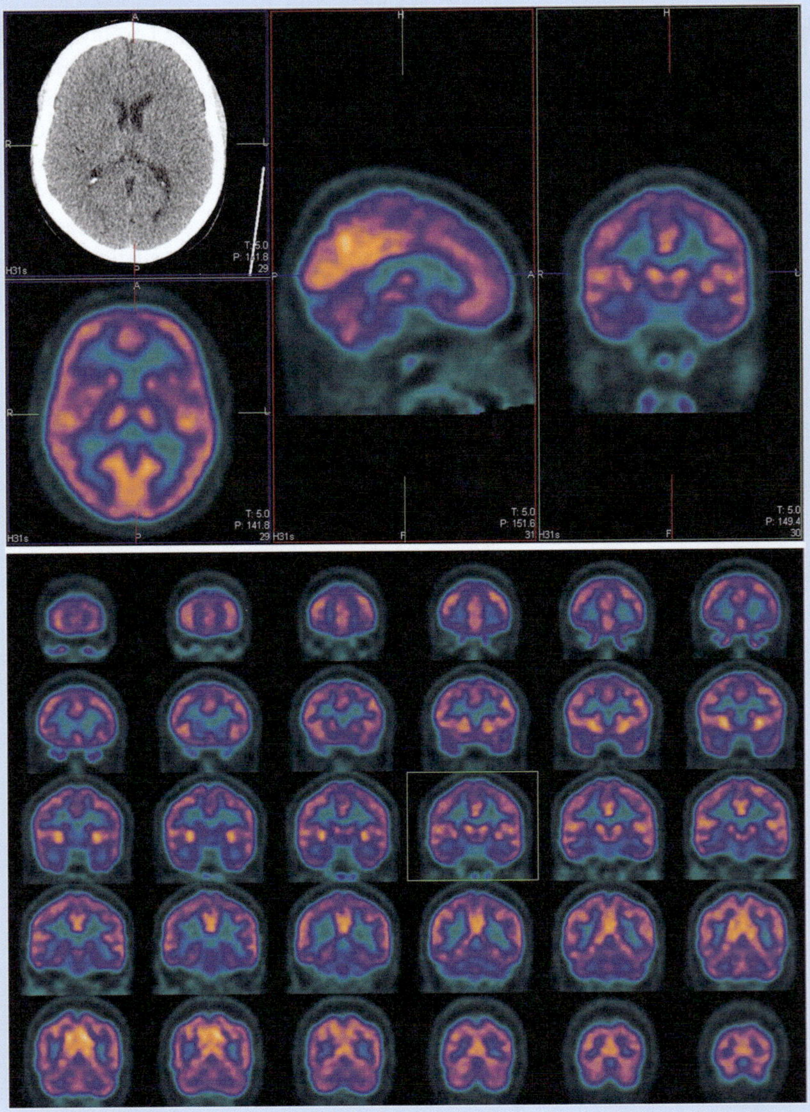

(continued)

Box 2.4 (continued)

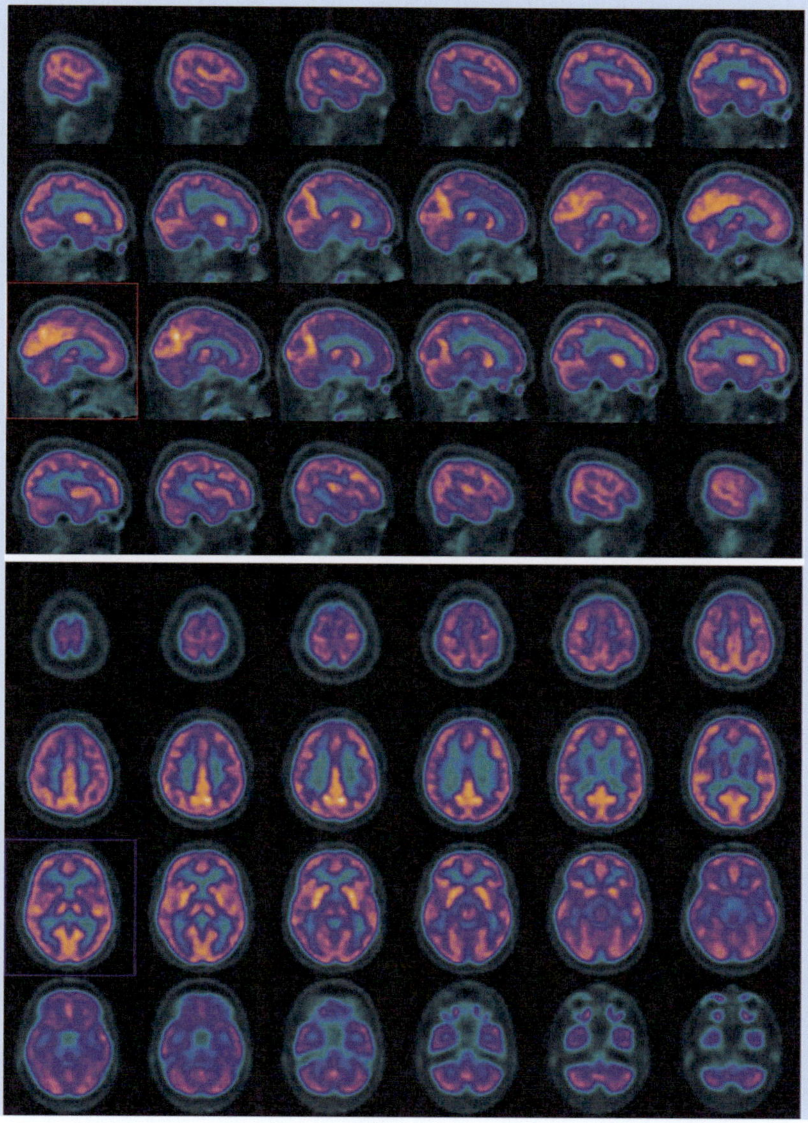

2.2.4 Pharmacological Treatment

Patient was initially treated with 15 mg of aripiprazole, which was subsequently increased (from the second week) to 20 mg/day in accordance with good clinical practice and psychiatric treatment guidelines. After a week of treatment,

Mr. B. denied hearing voices, and objectively, during the interview he appeared markedly less distressed: he showed a sufficient level of attention; he generally didn't lose his train of thought. Nevertheless, when he was in the communal spaces of the ward, in which cameras are more prominent, he used to watch them, but for less time. In such episodes, the patient denied anxiety. He stared at monitors because, he said: "I use to behave in this manner." He began taking off his hat, and he reduced the time spent in assuming bizarre postures. He could not explain why he needed to make those movements nor the relationship between the movements and the communication with "saints." Due to his extremely poor awareness of his illness, therapy with long-acting aripiprazole was chosen, and the patient began a period of rehabilitation in a psychiatric community, given the resolution of the acute clinical condition. But there was persistence of attenuated symptoms such as formal thought disturbance and disorganized sequence of actions during self-care activities. During the final phase of hospitalization, a screening neurocognitive battery of tests was administered. Then, after discharge, the patient underwent a more complete neurocognitive evaluation (Box 2.5).

Box 2.5 First Neurocognitive Evaluation

Test	Score	N.v.	Comment
MMSE (mini mental state examination)	25.60/30	>24.00	Normal
Attentional tests	19.25/60	>31.00	Deficit
Trail making test part A	95.00	<93.00	Deficit
Trail making test part B	330.00	<282.00	Deficit
Raven	23.73/60	>18.60	Normal
Token test (motor function)	24.00/36	>29	Deficit
Boston naming test	43.00/60	>43	*Low score, just above the threshold*
Verbal fluency - phonemic key	7.00	>17	Deficit
Verbal fluency - categorical key	16.00	>25	Deficit
Digit span	5.50	>3.75	Normal
Story recall test – Verbal memory	2.00	>8.00	Deficit
Learning pairs of words – Verbal memory	5.00	>6.50	Deficit
Street completion test	5.00/14	>3.25	Normal
T.O.L. (tower of London)	19.00/36	>20.00	Deficit
FAB (frontal assessment battery)	10.70/18	>13.50	Deficit
Cognitive estimation test	22.97	<18.00	Deficit
Oddity task	9.00	<4.00	Deficit

Two years after discharge, the patient had a severe relapse. He experienced imperative auditory hallucinations; he was extremely anguished and emotionally labile; he manifested aggression toward objects. He attempted suicide by swallowing a hundred pills of alprazolam, but his father, finding the empty blisters, called the emergency services while the patient went into a coma. After the intensive care unit, he was admitted to the psychiatric ward. From the family it emerged that his bizarre thoughts had worsened in the previous 3 weeks. He behaved erratically around the house, and he would talk to strangers, sometimes using neologisms. On certain occasions he also displayed the inability to perform previously learned motor activities (like washing his teeth). Regarding apraxia, at the mental examination, when asked to perform serial step commands, for example, "Take this piece of paper in your left hand, then fold it up, place it in the envelope, and put in the bag," he failed.

Compared with the previous hospitalization, Mr. B. appeared more aggressive. The second hospitalization lasted 3 months. The initial clinical picture consisted of severely disorganized speech. Mr. B. slipped from one topic to the next, which was vaguely connected to the first. When answering a question, the response often had nothing to do with the question at all. He also made up words. His behavior was grossly disorganized too: he would make grimaces, he brushed his teeth without water, and he wore his clothes incorrectly. During hospitalization, a second positron emission tomography was performed (Box 2.6).

He was both pharmacologically treated with atypical antipsychotics (olanzapine, previously aripiprazole) at high doses and typical antipsychotics (haloperidol and chlorpromazine) with poor effect. Of note, Mr. B. also suffered from hyperprolactinemia and gynecomastia caused by the administration of haloperidol. After these attempts, clozapine therapy was started.

After 10 days of treatment, he began showing a more organized behavior: he took the necessary self-care, and during the interview, the episodes of yelling and making strange noises markedly diminished. He also denied hearing voices and desiring to act aggressively against himself or others. He explained that he didn't know why he injured himself; his mood wasn't totally indifferent. For instance, he said he sometimes was sad to see his mother worried. He still appeared bizarre (sometimes making noises, laughing, and talking to himself in a "magic" manner), but he could answer questions, and he didn't appear anguished. He also accepted to spend another rehabilitative period in a highly protective community.

Concerning the second PET scanning (Box 2.6), the tomographic sections showed a reduction in the glucose analog fixation at the level of the posterior portions of the cortical parietal regions. No significant alterations in glucose distribution in the remaining cortical and subcortical structures emerged. In conclusion, there was evidence suggesting changes in the bilateral parietal glucose metabolism. Despite the first exam, in which frontal functional alterations were documented, the second exam exhibited reduction of the glucose analog fixation in parietal cortices.

Box 2.6 Second PET

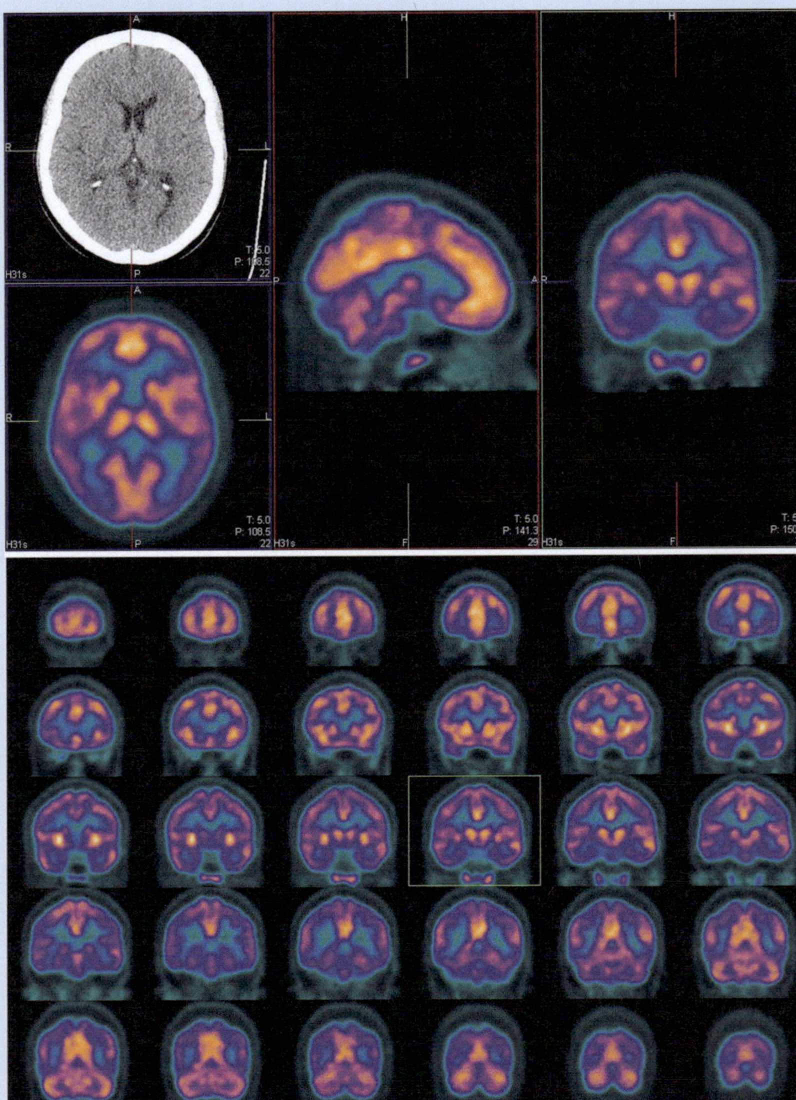

(continued)

Box 2.6 (continued)

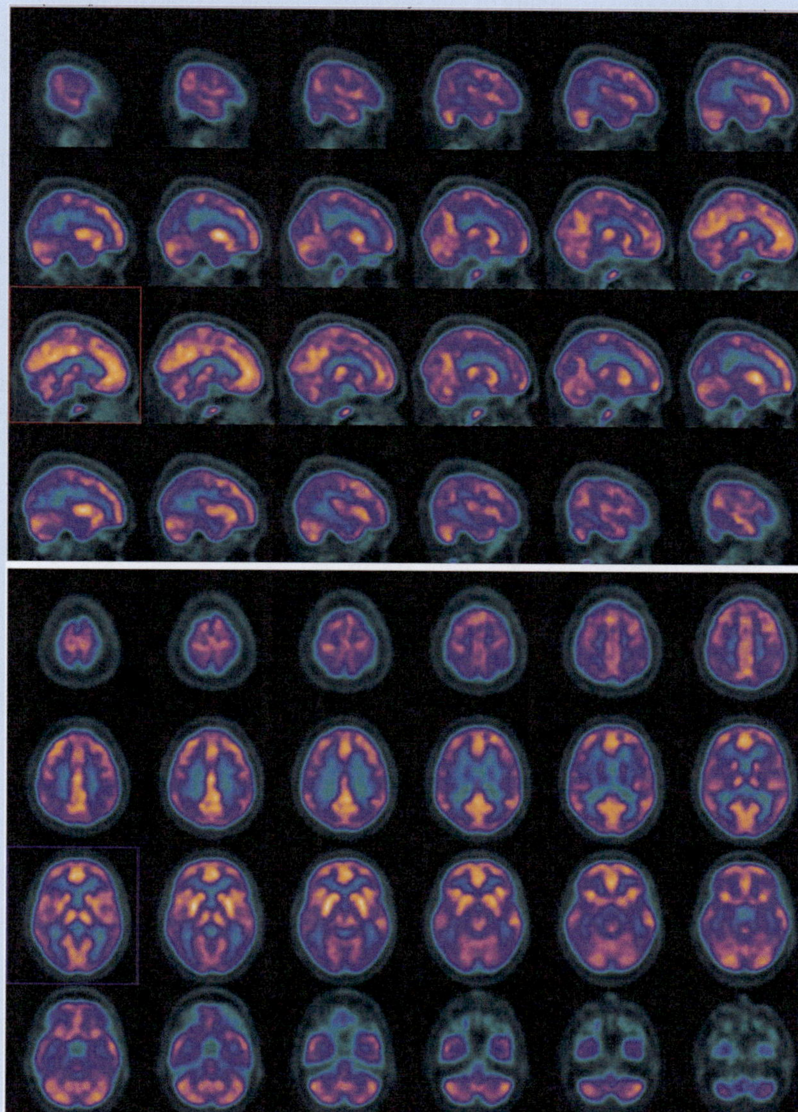

Another PET (Box 2.7) was performed when the patient was rehospitalized because he complained about tiredness, fatigue, and light-headedness; a complete blood count showed that his total WBC count was 2300/cubic mm, and his absolute neutrophil count was 1350/cubic mm, so the clozapine was stopped.

The analysis of PET images highlighted the reduction bilaterally of glucose metabolism in the frontal cortices, lower parietal lobules, and temporal and cerebellar lobes. The subcortical capture of glucose at the basal nuclei and the thalamus was preserved.

After discharge, during the psychosocial rehabilitation program at the community rehabilitation center, another neuropsychological evaluation was made, focused on core neuropsychological deficits of Schizophrenia (Box 2.8).

In November 2016, after 8 years of disease, the patient appeared as a slightly overweight young man, with an overall neat appearance and adequate hygiene. He remained alert and awake throughout the interview, with good eye contact. The orientation was intact to person, place, and time. He had difficulties in abstract thinking; he kept returning to the same limited set of ideas, which were described in a quite circumstantial manner, including some irrelevant details. There was no evidence of loosening of associations, thought blocking, hallucinations or illusions. The insight of disease was extremely poor: the patient didn't realize he was affected by schizophrenia; nonetheless he accepted to take medications. He also cooperated with the staff but had to be stimulated for the exercise of physical and cognitive activities; otherwise he sat passively with his arms crossed in a chair. He took daily therapy with quetiapine 600 mg/day and aripiprazole 5 mg/day (morning administration). He also underwent long-acting therapy with zuclopenthixol depot 300 mg every 2 weeks because of his low adherence to treatment. He has not been hospitalized for 3 years. He is attending a cognitive remediation program.

Box 2.7 Third PET

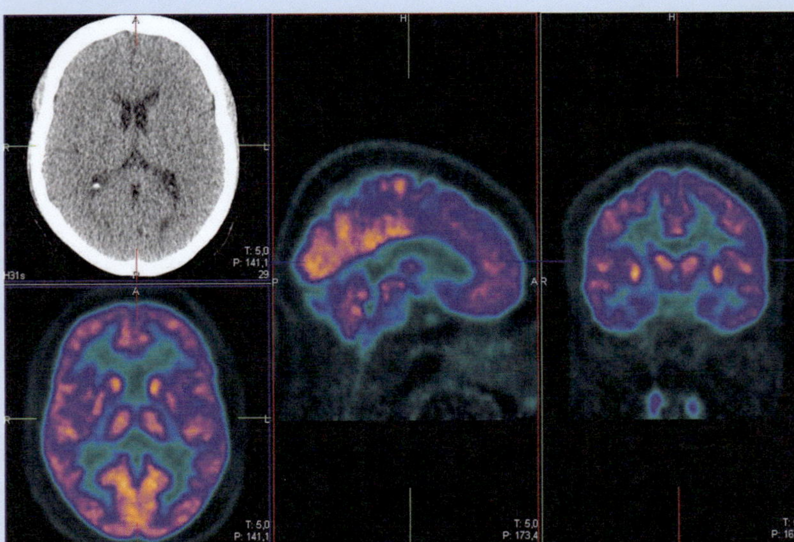

(continued)

Box 2.7 (continued)

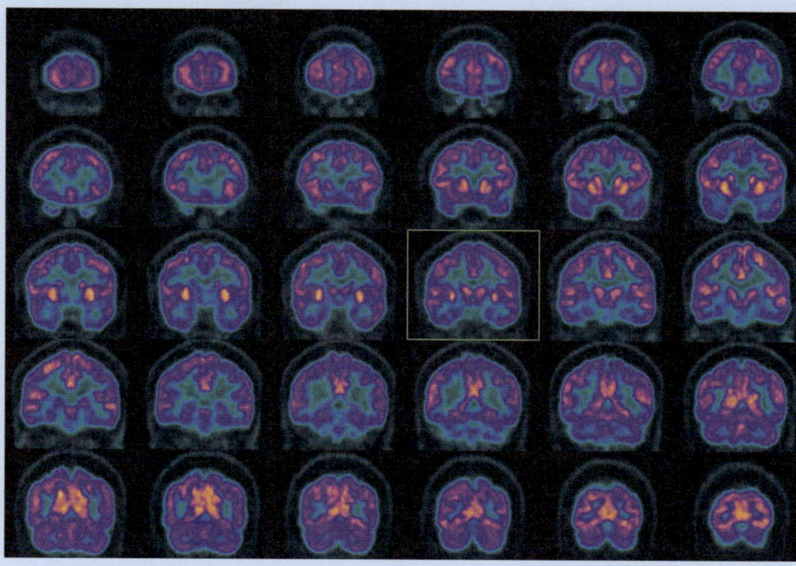

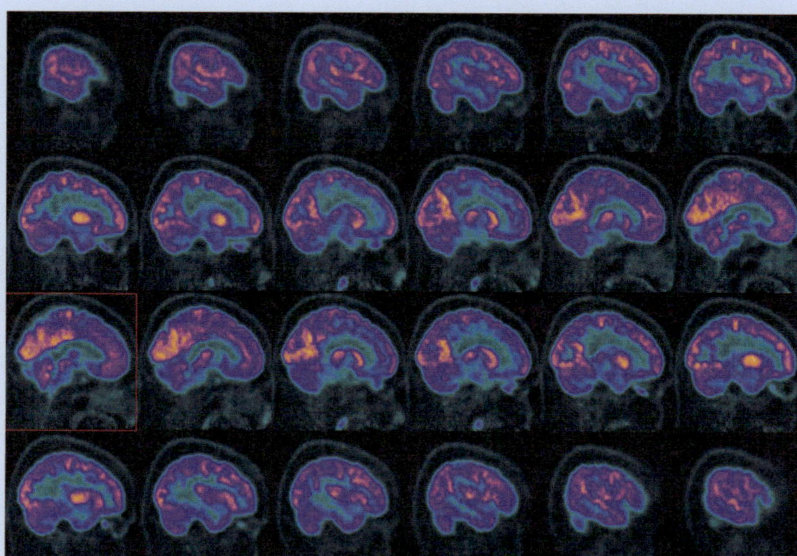

(continued)

Box 2.7 (continued)

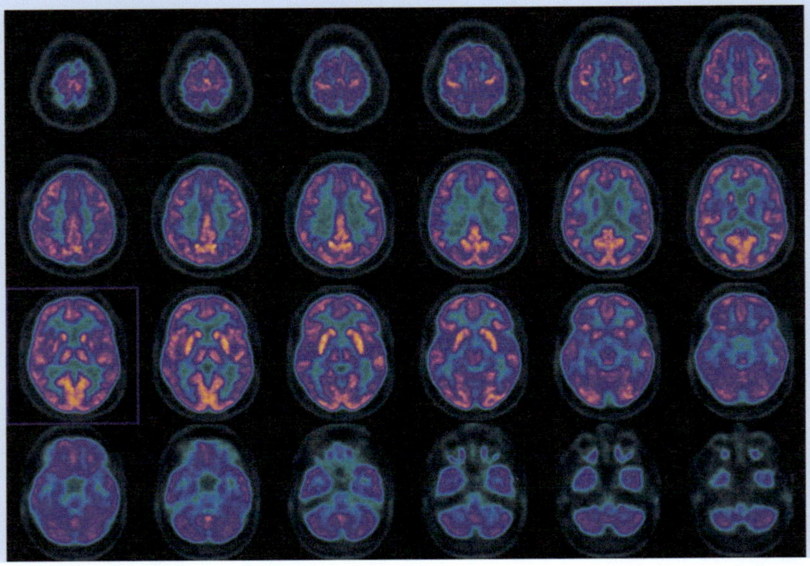

Box 2.8 Second Neurocognitive Evaluation

Test	Cutoff	Score	Comment
BACS (brief assessment cognition schizophrenia)			
Verbal memory	v.n. ≥ 33.01	31.00	Deficit
Working memory	v.n. ≥ 14.93	20.75	Normal
Token task	v.n. ≥ 68.77	25.25	Deficit
Symbol-coding task	v.n. ≥ 40.49	36.00	Deficit
Verbal fluency	v.n. ≥ 31.68	22.25	Deficit
Tower of London	v.n. ≥ 12.37	10.00	Deficit

2.3 Discussion

The diagnostic assessment and treatment of patients with deficit schizophrenia are challenging for clinicians due to the overlap of symptomatological dimensions, with concomitant impaired neurocognition and poor functioning. Consistently with the meta-analysis of Cohen et al. [6] on the psychiatric symptomatology of deficit schizophrenia, Mr. B. exhibited a wide range of psychiatric symptoms, with more severe disorganization symptoms at the onset and, in a second phase of the disease, predominant negative manifestations, with amotivation, flattening, and reduction in speech and activities. Moreover, the fact that the patient had never achieved a minimal functioning, with the disorganized onset, the partial treatment resistance and the severely impaired cognitive evaluation posited a deficit schizophrenia. According to the multidimensional model of schizophrenia, among positive/negative/disorganized symptoms, the latter have been less studied; however, their association with an early-onset, specific neuropsychological characteristics, poor insight, and significant socio-working maladjustment has already been evidenced [30].

Firstly, the patient's neurocognitive assessment revealed deficits related both to frontal lobe function (frontal assessment battery) and executive functions (Tower of London). "Story recall test" and the task of "learning pairs of words" also scored worse, indicating a possible verbal memory dysfunction, which is a consistently reported cognitive deficit in schizophrenia [31]. The Brief Assessment of Cognition in Schizophrenia (second neurocognitive evaluation) assessed a deficit score for verbal memory too. Regarding the cognitive domain of working memory, the performance score was normal. Whereas at the first evaluation, the Trail Making Test, which also assesses spatial working memory, resulted in deficit: this neurocognitive performance difference might reflect the different extent of disorganization at clinical evaluation. The other four cognitive domain tasks (motor function, verbal fluency, speed of processing, executive function) provided a deficit evaluation. It is worth noting that in order to avoid biases from psychopathologic acute condition, the neurocognitive evaluation was made while the patient was in a period of clinical remission. In conclusion, in this case an overall neurocognitive impairment occurs in a patient who displays severe disorganization symptoms with, at the beginning, slightly less severe negative features which, however, persist and worsen.

In this sense, because of the similarity between our results and the literature data about neurocognitive deficit in predominant negative and disorganized schizophrenic dimensions, we speculate that, in our case report, cognitive dysfunction may represent a deficit schizophrenia endophenotype, also in the sense of a measurable state-independent component whose evaluation may be reproducible [32]. A few cognitive deficits are already regarded as a core feature of schizophrenia [33], but better cognitive functioning was found in patients with positive symptoms, compared to deficit schizophrenia [34, 35].

Neuroimaging assessment was also performed. Research has highlighted how psychiatric disorders tend to show specific neuroimaging features and perfusion alterations. The MRI scan (during the first hospitalization) didn't exhibit anatomical cortical region alterations, with decreases having been found more widespread in chronic phase schizophrenia [36]. Moreover, patients with deficit schizophrenia demonstrated disruption of a few white matter tracts [37] compared with non-deficit patients. Regarding positron emission tomography in literature, a relationship emerged between decreased prefrontal cortex glucose metabolism ("hypofrontality") and severity of negative symptoms [38]. Subjects with negative symptoms were found to exhibit lower glucose metabolic rate (positron emission tomography) in the right hemisphere, in particular the temporal and ventral prefrontal cortices [39]. The disorganization cluster was significantly correlated instead with left inferior parietal lobule hypoperfusion [40]. Patients experiencing auditory verbal hallucinations were found to have significantly higher metabolic rates in the left superior and middle temporal cortices, in the bilateral superior medial frontal cortex and in the left caudate nucleus [41]. Mr. B.'s functional neuroimaging data (PET) showed the presence of a frontal reduced uptake at the first exam. Then alterations in glucose uptake in the parietal cortical region emerged in the second exam. In the third, the areas of alteration of uptake were multiple (frontal, parietal, and temporal reduced uptake), with a time interval of about 2.5 years between one PET and the other. These peculiar changes of glucose metabolism may reflect, as already supposed by Sham et al. [42], the neurodevelopmental and, consequently, the neuropathological alterations of deficit schizophrenia.

Despite the fact that disorganization and positive symptoms are traditionally considered more responsive than other symptoms to first-generation antipsychotics, in analyzing a large dataset, Janicak et al. [43] showed that atypical antipsychotics are superior to typical in treatment of the disorganized dimension. As regards the pharmacological management of first episode psychosis, we discussed administration of aripiprazole or risperidone, both available also as long acting. We chose aripiprazole because despite a similar effect on disorganized symptoms compared with risperidone, it has fewer metabolic side effects [44]. Moreover, as demonstrated, treatment with aripiprazole has been correlated with improvement of cognitive skills in young adults with first psychotic episode [45]. Then, after the relapse, one trial with atypical antipsychotic and two trials with typical ones were made. Disorganized dimensions and negative symptoms persisted, so clozapine therapy was administered. Clozapine was discontinued after the manifestation of hematic side effects. Subsequently, the patient benefited from the association of conventional antipsychotic long-acting medication, to ensure adherence to treatment, with quetiapine at medium-high doses. He also took aripiprazole at low doses in order to combat negative symptoms.

At present the patient is attending the psychiatric rehabilitation day center on a psychiatric rehabilitation program. As is known, psychiatric rehabilitation programs for deficit schizophrenic patients involve several procedures among which family interventions, cognitive-behavior therapy, social skills training, vocational rehabilitation, cognitive remediation, and mindfulness. Considering that the proofs of efficacy are still contrasting, it is of primary importance to carry out an accurate assessment of the patient's psychopathology, neurocognition, personal motivational aspects, and social background, in order to organize a suitable rehabilitative option. There is some evidence that cognitive remediation therapy, which targets cognitive deficits with the goal of improving functional outcomes, is promising in schizophrenia [46]. Approaches vary depending on the patient's neurocognitive status and psychiatric symptomatology. Cognitive remediation includes practice exercises with the aim of restoring cognitive functions through neuroplasticity; compensatory trainings are also carried out (computerized exercises, therapist-guided instruction or combined exercises) to circumvent neurocognitive impairment, with certain cognitive programs turning on a specific cognitive domain such as improvement of working memory or social cognition. In the most serious form of schizophrenia, it has been shown that commonly used CRT is not always efficacious [47]. Because of Mr. B.'s neurocognitive evaluation and his functional impairment, a specific cognitive remediation program for implementation of executive functions was developed. Future goals consist of starting a social skills training program to allow Mr. B. to minimize his possible neurocognitive decline and achieve some degree of recovery, especially as regards the ability to manage interpersonal relations, which are crucial in daily routine for living a more productive life.

Key Points

- Disorganized and negative symptoms seem to characterize deficit schizophrenia and tend to be more predominant and represented than psychotic and impulsive-aggressive psychopathological aspects.
- In deficit schizophrenia, a serious neurocognitive impairment has been associated with the negative and disorganized dimension. In this sense, cognitive evaluations could allow a more specific diagnostic assessment and outline targeted rehabilitative treatment. Frequently, structural neuroimaging exams are requested as part of the initial medical work-up in young patients with first-episode psychosis.
- Functional neuroimaging data, as yet to be confirmed, show a number of changes in the disease over the years, with greater anomalies such as the psychiatric picture worsening or becoming chronic. Contrariwise, some studies suggest that conventional structural neuroimaging exams may reveal few abnormalities in young patients with first-episode psychosis.

Self-Assessment Questionnaire

1. Which is the most important clinical aspect to evaluate in follow-up of the patient after clinical stabilization?
   (A) Neurocognition
   (B) Socioeconomic status
   (C) Interpersonal relationships
   (D) Subjective well-being
2. Which is the best rehabilitative option for deficit schizophrenia?
   (A) CBT
   (B) CRT
   (C) Family therapy
   (D) A, B, C
3. What supposed dimension of schizophrenia has been found linked to severe neurocognitive alterations?
   (A) Paranoid
   (B) Aggressive and hostility

(C) Negative and disorganization
(D) Affective

4. Which factors have a significant role in administering long-acting therapy?
   (A) Patient prefers not to take pills
   (B) Non-compliance with pharmacological treatment
   (C) Stabilization of plasma levels
   (D) A + B + C

References

1. Bleuler E. Die Prognose der Dementia Praecox (Schizophreniegruppe). Allg Zeitschrift für Psychiatr und Psych Medizin. 1908;31:436–80.
2. Braff DL, Ryan J, Rissling AJ, et al. Lack of use in the literature from the last 20 years supports dropping traditional schizophrenia subtypes from DSM-5 and ICD-11. Schizophr Bull. 2013;39 (4):751–3.
3. Griswold KS, Del Regno PA, Berger RC. Recognition and differential diagnosis of psychosis in primary care. Am Fam Physician. 2015;91(12):856–63.
4. Salokangas RKR. Symptom dimensions and outcome in schizophrenia. World Psychiatry. 2003;2(3):172–8.
5. Altamura AC, Caletti E, Paoli RA, et al. Correlation between neuropsychological and social cognition measures and symptom dimensions in schizophrenic patients. Psychiatry Res. 2015;230(2):172–80.
6. Cohen AS, Brown LA, Minor KS. The psychiatric symptomatology of deficit schizophrenia: a meta-analysis. Schizophr Res. 2010;118(1–3):122–7.
7. Bozikas VP, Kosmidis MH, Kioperlidou K, et al. Relationship between psychopathology and cognitive functioning in schizophrenia. Compr Psychiatry. 2004;45(5):392–400.
8. Carpenter WT, Heinrichs DW, Wagman AM. Deficit and non deficit forms of schizophrenia: the concept. Am J Psychiatry. 1988;145(5):578–83.
9. Grover S, Kulhara P. Deficit schizophrenia: concept and validity. Indian J Psychiatry. 2008;50:61–6.
10. Kirkpatrick B, Buchanan RW, Ross DE, et al. A separate disease within the syndrome of schizophrenia. Arch Gen Psychiatry. 2001;58(2):165.
11. Keefe RS, Frescka E, Apter SH, et al. Clinical characteristics of Kraepelinian schizophrenia: replication and extension of previous findings. Am J Psychiatry. 1996;153(6):806–11.
12. Jablensky A. The diagnostic concept of schizophrenia: its history, evolution, and future prospects. Dialogues Clin Neurosci. 2010;12(3):271–87.
13. Hovington CL, Lepage M. Neurocognition and neuroimaging of persistent negative symptoms of schizophrenia. Expert Rev Neurother. 2012;12(1):53–69.
14. Olabi B, Ellison-Wright I, AM MI, et al. Are there progressive brain changes in schizophrenia? A meta-analysis of structural magnetic resonance imaging studies. Biol Psychiatry. 2011;70 (1):88–96.
15. Ho B-C, Andreasen NC, Nopoulos P, et al. Progressive structural brain abnormalities and their relationship to clinical outcome: a longitudinal magnetic resonance imaging study early in schizophrenia. Arch Gen Psychiatry. 2003;60(6):585.
16. Bora E, Binnur Akdede B, Alptekin K, et al. Neurocognitive impairment in deficit and non-deficit schizophrenia: a meta-analysis. Psychol Med. 2017;47(14):2401–13.
17. Fervaha G, Agid O, Foussias G, et al. Neurocognitive impairment in the deficit subtype of schizophrenia. Eur Arch Psychiatry Clin Neurosci. 2016;266:397.
18. Morrens M, Hulstijn W, Sabbe B. Psychomotor slowing in schizophrenia. Schizophr Bull. 2007;33(4):1038–53.

19. Bervoets C, Docx L, Sabbe B, et al. The nature of the relationship of psychomotor slowing with negative symptomatology in schizophrenia. Cogn Neuropsychiatry. 2014;19:36–46.
20. Cohen AS, Saperstein AM, Gold JM, et al. Neuropsychology of the deficit syndrome: new data and meta-analysis of findings to date. Schizophr Bull. 2007;33(5):1201–12.
21. Yu M, Tang X, Wang X, et al. Neurocognitive impairments in deficit and non-deficit schizophrenia and their relationships with symptom dimensions and other clinical variables. PLoS One. 2015;10(9):e0138357.
22. APA. Diagnostic and statistical manual of mental disorders. 5th ed. Washington: American Psychiatric Association; 2013.
23. American Psychiatric Association. Diagnostic and statistical manual of mental disorders—DSM-IV-TRH. 4th ed. Arlington: American Psychiatric Publishing; 2000.
24. Ventura J, Thames AD, Wood RC, et al. Disorganization and reality distortion in schizophrenia: a meta-analysis of the relationship between positive symptoms and neurocognitive deficits. Schizophr Res. 2010;121(1):1–14.
25. Sarfati Y, Hardy-Baylé M-C, Brunet E, et al. Investigating theory of mind in schizophrenia: influence of verbalization in disorganized and non-disorganized patients. Schizophr Res. 1999;37(2):183–90.
26. Hardy-Baylé M-C, Sarfati Y, Passerieux C. The cognitive basis of disorganization symptomatology in schizophrenia and its clinical correlates: toward a pathogenetic approach to disorganization. Schizophr Bull. 2003;29(3):459–71.
27. Uhlhaas PJ, Philips W, Mitchell G, et al. Perceptual grouping in disorganized schizophrenia. Psychiatry Res. 2006;145(2):105–11.
28. Takahashi H, Iwase M, Nakahachi T, et al. Spatial working memory deficit correlates with disorganization symptoms and social functioning in schizophrenia. Psychiatry Clin Neurosci. 2005;59(4):453–60.
29. Knight C, Alderman N, Burgess PW. Development of a simplified version of the multiple errands test for use in hospital settings. Neuropsychol Rehabil. 2002;12(3):231–55.
30. Nestsiarovich A, Obyedkov V, Kandratsenka H, et al. Disorganization at the stage of schizophrenia clinical outcome: clinical–biological study. Eur Psychiatry. 2017;42:44–8.
31. Toulopoulou T, Murray RM. Verbal memory deficit in patients with schizophrenia: an important future target for treatment. Expert Rev Neurother. 2004;4(1):43–52.
32. Jablenski A. Subtyping schizophrenia: implications for genetic research. Mol Psychiatry. 2006;11:815–36.
33. Goldberg TE, David A, Gold JM. Neurocognitive deficits in schizophrenia. In: Hirsch SR, Weinberger DR, editors. Schizophrenia. 2nd ed. Cambridge: Blackwell; 2003. p. 168–84.
34. Brazo P, Marie RM, Halbecq BK, Segard L, Delamillieure P, et al. Cognitive patterns in subtypes of schizophrenia. Eur Psychiatry. 2002;17:155–62.
35. Hill SK, Ragland JD, Gur RC, Gur RE. Neuropsychological differences among empirically derived clinical subtypes of schizophrenia. Neuropsychology. 2001;15:492–501.
36. Ellison-Wright I, Glahn DC, Laird AR, et al. The anatomy of first-episode and chronic schizophrenia: an anatomical likelihood estimation meta-analysis. Am J Psychiatry. 2008;165 (8):1015–23.
37. Voineskos AN, Foussias G, Lerch J, et al. Neuroimaging evidence for the deficit subtype of schizophrenia. JAMA Psychiat. 2013;70(5):472–80.
38. Wolkin A, Sanfilipo M, Wolf AP, et al. Negative symptoms and hypofrontality in chronic schizophrenia. Arch Gen Psychiatry. 1992;49(12):959–65.
39. Potkin SG, Alva G, Fleming K, et al. A PET study of the pathophysiology of negative symptoms in schizophrenia. Positron emission tomography. Am J Psychiatry. 2002;159 (2):227–37.
40. Kaplan RD, Szechtman H, Franco S, et al. Three clinical syndromes of schizophrenia in untreated subjects: relation to brain glucose activity measured by positron emission tomography (PET). Schizophr Res. 1993;11(1):47–54.

41. Horga G, Parellada E, Lomeña F, et al. Differential brain glucose metabolic patterns in antipsychotic-naïve first-episode schizophrenia with and without auditory verbal hallucinations. J Psychiatry Neurosci. 2011;36(5):312–21.
42. Sham PC, Castle DJ, Wessely S, et al. Further exploration of a latent class typology of schizophrenia. Schizophr Res. 1996;20(1–2):105–15.
43. Janicak PG, Glick ID, Marder SR, et al. The acute efficacy of aripiprazole across the symptom spectrum of schizophrenia: a pooled post hoc analysis from 5 short-term studies. J Clin Psychiatry. 2009;70:25–35.
44. Robinson DG, Gallego JA, John M, et al. A randomized comparison of Aripiprazole and Risperidone for the acute treatment of first-episode schizophrenia and related disorders: 3-month outcomes. Schizophr Bull. 2015;41(6):1227–36.
45. Yeh C-B, Huang Y-S, Tang C-S, et al. Neurocognitive effects of aripiprazole in adolescents and young adults with schizophrenia. Nord J Psychiatry. 2014;68(3):219–24.
46. Wykes T, Huddy V, Cellard C, et al. A meta-analysis of cognitive remediation for schizophrenia: methodology and effect sizes. Am J Psychiatry. 2011;168(5):472–85.
47. Cella M, Preti A, Edwards C, et al. Cognitive remediation for negative symptoms of schizophrenia: a network meta-analysis. Clin Psychol Rev. 2017;52:43–51.

Psychosis in the Elderly

3

Chiara Rovera and Alessandro Pigoni

Abstract

Psychosis in the elderly represents a frequent and challenging feature, with a prevalence of psychotic symptoms that may reach 10–63% in the hospitalized population. However, both the diagnosis and the treatment of psychotic symptoms in the elder population may present many problems.

In the present chapter, we debate the differential diagnosis between the causes of psychosis in the elderly and how to deal with them. The first cause of psychosis in this population is represented by dementia. Psychiatric symptoms may be present not only in the last phases of neurodegenerative disorders but also in the early stages or at onset, more frequently in specific subtypes of dementia, such as frontotemporal dementia. The second most common cause of psychosis in the geriatric population is depression, while delirium is the third. Delirium, differently from the other described diagnoses, is characterized by an acute change in mental status, disturbances of consciousness, and clouded sensorium and may be caused by several circumstances, ranging from infections to inappropriate medication use.

Considering the background of the present literature, we report the case of a 66-year-old man who was referred to our inpatient clinic for a manic episode with delusions. We investigated the differential diagnostic processes, which encompass a comprehensive clinical evaluation, a very accurate anamnestic interview, blood tests, and eventually brain imaging. Another major issue of concern is treatment, which might be guided by a multidisciplinary endeavor, including pharmacological and non-pharmacological interventions.

C. Rovera (✉) · A. Pigoni
Department of Neurosciences and Mental Health, Fondazione IRCCS Ca' Grande Ospedale Maggiore Policlinico, University of Milan, Milan, Italy

A. C. Altamura, P. Brambilla (eds.), *Clinical Cases in Psychiatry: Integrating Translational Neuroscience Approaches*,
https://doi.org/10.1007/978-3-319-91557-9_3

Keywords
Elderly · Psychosis · Bipolar disorder

3.1 Introduction

Psychosis in the elderly represents nowadays one of the major challenges to clinicians. The elderly are expected to reach 20% of the general population by the year 2030 [1], with an increased concern about problematics in this specific population. Prevalence of psychotic disorders in the elderly ranges widely from 0.2 to 4.75% in community-based samples but may reach 10–63% in the hospitalized population. As the population ages, the interest in psychotic symptoms in aged people is also increasing; however recognition of psychosis is complex due to patients that often deny this kind of symptomatology as a psychological dimension, presenting somatic complaints, anxiety, or cognitive impairments [2].

However, diagnosis and treatment of psychotic symptoms in the elderly population may present many problems, as we will review in this chapter, exemplifying the difficulties clinicians confront with in dealing with such a complex population. Psychosis, indeed, not only represents an acute problem to face but also accounts for poor prognosis, suffering, and social withdrawal [3].

3.1.1 Differential Diagnosis

The diagnosis of psychosis in the aged population needs the collaboration of an interdisciplinary team, with a full clinical, psychiatric, and neurological assessment. The differential diagnosis of elderly patients that present with delusions, hallucinations, and behavioral disturbance should include psychosis related to delirium, psychosis due to general medical conditions, atypical dementia onset, behavioral and psychological symptoms of dementia (BPSD), affective illness, schizophrenia, bipolar disorder (BD) or other primary psychotic disorders, and substance abuse or dependence [4]. Moreover, interactions between different clinical conditions are possible: patients with dementia may have a lower threshold for delirium, and similarly mood alterations may be a prodrome for dementia [5].

Overall, dementia is the most common cause of psychosis in the geriatric population, accounting for almost 40% of the cases [6]. Psychosis is a major aspect of neurodegenerative processes and poses an important health concern for the aged population. Psychiatric symptoms may be present not only in the last phases of neurodegenerative disorders but also in the early stages or at onset [7]. Psychotic symptoms, indeed, are often present in both Alzheimer's disease (AD) and in the behavioral variant of frontotemporal dementia (bvFTD) [8]. As widely reported in the literature, psychosis is one of the most frequent non-cognitive symptoms of AD, with prevalence ranging from 30 to 50% [6]. Moreover, thought disorders seem to be associated with poorer prognosis and greater cognitive impairment [7]. Regarding bvFTD, a large majority of the subjects suffered from delusions, hallucinatory behavior, or suspiciousness during the course of their illness. bvFTD is often associated with heterogeneous presentations, including pure psychiatric symptoms [9], and hallucinations and delusions can even precede the onset of cognitive symptoms and are often recognized as the clinical onset of the disease [10]. From previous studies, a correlation between psychotic symptoms in bvFTD and C9ORF72 hexanucleotide repeat expansion clearly emerges, indicating a possible genetic vulnerability to psychosis in this specific population [7, 11]. Regarding the clinical presentation, delusions of patients affected by dementia are often simple, usually of a paranoid nature, where patients believe that items have been stolen from them or that they have been abandoned [6].

According to literature, the second most common cause of psychosis in the geriatric population is depression, accounting for more than 20% of diagnoses [2]. Subjects with previous diagnoses of mood disorders are at risk of recurrence. On the other hand, late-onset depression (age > 65 years) is also possible, even though it seems to be a different subtype of depression. The prevalence of major depressive disorder (MDD) in the elderly is estimated at 1–3%, although recognition of depressive symptomatology is complex due to patients that often deny depression as a psychological dimension, also presenting somatic complaints, anxiety, or cognitive impairments [12, 13]. It usually shows a peculiar presentation in terms of symptomatology: late age onset depression is more likely to be psychotic and often characterized by delusion focused on somatic concern (hypochondriacal), persecution, guilt, and nihilistic content more frequently than early-onset depression [14]. Moreover, previous literature indicates that late-onset depression is associated with higher prevalence of psychosis [15] and might be a prodrome of further developing dementia, especially AD [16].

Delirium is the third most common cause of psychosis in the elderly [17]. Delirium, differently from the other described diagnoses, is characterized by an acute change in mental status, disturbances of consciousness, and clouded sensorium that might be accompanied by abnormalities in mood, perception, and behavior. Delirium is commonly classified into hyperactive delirium, hypoactive, mixed, and the non-motor subtype. Although hypoactive delirium is usually the most common subtype among hospitalized geriatric patients, psychotic features seem to be more associated with hyperactive delirium subtypes [18]. The geriatric population is particularly at risk of developing delirium, which is described as affecting up to 50% of the elderly hospitalized population, showing a medical and/or multifactorial etiology [19]. A pre-existing medical condition could predispose to delirium or to a psychotic disorder due to a general medical condition. Uncontrolled diabetes, pulmonary or urinary tract infections, and electrolyte alterations are the most common causes of delirium in the elderly. Also misuse or abuse of prescription drugs (such as benzodiazepines) may lead to delirium. Inappropriate medication use in the elderly is the cause of many hospital admissions and may trigger psychotic symptoms (for a more comprehensive review of inappropriate medication use in the elderly, we suggest the Revised Beers Criteria, [20]). Moreover, there might be an intertwining interaction between dementia and delirium, with delirium as a marker of vulnerability to dementia [21].

Other causes of psychotic symptoms in the elderly encompass previously diagnosed psychotic disease, such as schizophrenia, paranoia, and bipolar disorder, and worsening or onset of clinical and neurological conditions, such as brain neoplasia.

Box 3.1 Differential Diagnosis of Psychosis in the Elderly

- *Dementia*: both AD and bvFTD; also in the early stages or at onset
- *Depression*: often with somatic concern (hypochondriacal), persecution, guilt, and nihilistic content
- *Delirium*: acute change in mental status, often due to medical conditions of drug misuse
- *Neurological condition: i.e., brain tumor, brain abscess, etc.*
- Previously diagnosed psychotic disease: i.e., schizophrenia, paranoia, bipolar disorder, etc.

3.1.2 Clinical Evaluation

A comprehensive clinical evaluation is necessary to determine what may be the causes of the psychotic symptoms and to rule out possible etiologies. A wide range workout including psychiatric, neurologic, and medical assessment, as well as blood, urine, brain, cardiac, and imaging studies should be carried out.

A very accurate anamnestic interview should be performed at the admission, with the help of relatives or caregivers. The interview should include previous psychiatric or neurological history, ongoing medical conditions, and pharmacological history.

Clinical acuity of the presentation and intervention impact should guide the evaluation, according to the history and initial examination. Of the three most common psychosis-causing diseases (delirium, depression, and dementia), delirium is usually the most acute and caused by acute and potentially reversible medical conditions. Within a context of frailty and limited reserve capacities, several stressors and intertwining pathological mechanisms may lead to imbalance in the normal homeostasis. Several conditions, therefore, must be assessed. Blood and urine exams, along with vital signs, should be performed as soon as possible, to determine whether an infection has occurred. In elderly patients, urinary tract and pulmonary infections are usually the most common [22], both in hospitalized and in outpatients. Uncontrolled diabetes mellitus and hydro-electrolyte alterations are also common causes of acute delirium and should be assessed as soon as possible, given the fact that all the conditions are reversible with appropriate treatment. Constipation as well as acute and chronic pain represents other reversible causes of delirium that can be promptly treated.

Plasma levels of daily medications prescribed to the patients should be assessed, for measurable drugs. Many psychiatric and nonpsychiatric treatments may be the cause of agitation, confusion, and delirium, if the dosage is above a certain threshold.

A computed tomography (CT) scan should be performed in the case of altered/clouded sensorium and other neurological symptoms of acute onset, to exclude neurological/neurosurgical conditions, such as stroke and intracranial hemorrhage. After the exclusion of the most acute and dangerous causes, a full psychiatric assessment should be performed. Overall, it is of fundamental importance to conduct a full brain imaging assessment, consisting of magnetic resonance imaging (MRI) and positron emission tomography (PET), that may greatly help in the diagnostic processes of dementia [23].

Another major issue of concern is treatment, which might be guided by a multidisciplinary endeavor, including pharmacological and non-pharmacological interventions. The choice and dosage of pharmacological medications should be guided not only by efficacy but also by potential side effects and unwanted interactions with other medications. Drugs' starting doses are usually lower compared to those recommended for younger adults and should be titrated up or down slowly according to clinical response and side effects onset. Overall, treating elderly patients presents more difficulty because of greater sensitivity to drugs and their side effects, higher rates of polypharmacy, and variation in the pharmacokinetic parameters [24]. The drug plasma levels do not correlate with clinical improvement in the elderly population, and this could be explained by an increase in side effects that may aggravate the discomfort felt by the patient [25].

Box 3.2 Exams and Assessment

- Accurate interview with the patient and caregivers
- Physical examination with vital signs
- Blood and urine exams, looking for infections, alteration in electrolytes, uncontrolled metabolic conditions (diabetes)
- Plasma levels of prescribed medications and drugs of abuse
- Brain imaging: CT scan in acute; MRI and PET to assess neurodegenerative disorders (in selected cases)

3.2 Case Presentation

In January 2017, a 66-year-old man was referred to our inpatient clinic for a manic episode with delusions. No previous psychiatric history had been reported before the actual admission. He had no a family history for psychiatric disorder. Neither substance abuse nor medical comorbidity was reported by the patient. At the time of the admission, he was not taking any psychiatric medications except for 10 mg of zolpidem in order to sleep. Regarding other medications, the patients reported taking 10 mg of enalapril for a mild hypertension.

He reported a recent history of stressful events (working pressure) prior to the onset of symptoms. The patient graduated in law and had a good career as a lawyer. However, he had not been working in the last month before the admission.

At the time of evaluation in the emergency room, he showed dysphoric mood, psychomotor acceleration, severe insomnia, and aggressiveness; persecutory and jealousy delusions were also referred by the patient. He firmly refused hospitalization; however, a compulsory hospitalization was necessary.

During his stay in our inpatient unit, blood and urine exams were performed. No significant data emerged from the blood exams, except for a total cholesterol above the normal threshold (chol 235 mg/ml). The urine drug screening was negative for common drugs of abuse. He underwent the Brief Assessment of Cognition in Schizophrenia (BACS), which showed a deficit in the executive functions, frontal efficiency and motor proficiency, and poor performances, just barely above the deficit cutoff, in verbal fluency and working memory, as shown in Table 3.1.

We performed encephalic magnetic resonance imaging (MRI), electroencephalography (EEG), and fluorodeoxyglucose positron emission tomography (FDG-PET). The EEG resulted negative for anomalies, except for unspecific theta waves in the frontotemporal electrodes. The MRI resulted in a moderate atrophy in the parietal lobe and scattered white matter hyperintensities, probably of vascular origins, more prominent in the left hemisphere (Fig. 3.1). The PET did not evidence any significant modification in the normal and symmetric brain metabolism (Fig. 3.2).

A neurologic evaluation was performed and resulted negative for gross alterations. However, the neurologist suggested researching C9ORF72 hexanucleotide repeat expansion and progranulin mutation, which resulted negative as well.

As a pharmacological treatment, he was administered intramuscular aripiprazole 30 mg/d and daily oral valproic acid 1000 mg. The patient presented a good response in terms of reduction of delusions and psychomotor acceleration, as presented in Fig. 3.3. After a few days of treatment, he became more aware of his condition and accepted both the stay in the acute ward and the treatments. However, a full compliance to pharmacological treatment was not completely achieved as the patient

Table 3.1 Neurocognitive evaluation assessed by BACS subtest

Test	Normal score	Score	Comment
Language			
Verbal fluency	≥31.68	33.25	*Low score, just above threshold*
Memory			
Verbal memory	≥33.01	42.50	Normal
Motor proficiency			
Token task	≥68.77	57.50	**Deficit**
Symbol-coding task	≥40.49	30.50	**Deficit**
Frontal proficiency			
Working memory	≥14.93	15.50	*Low score, just above threshold*
Tower of London	≥12.37	10.50	**Deficit**
Frontal proficiency battery	>13.50	9.00	**Deficit**

Italic and Bold emphasis highlights the clinical importance for this kind of results

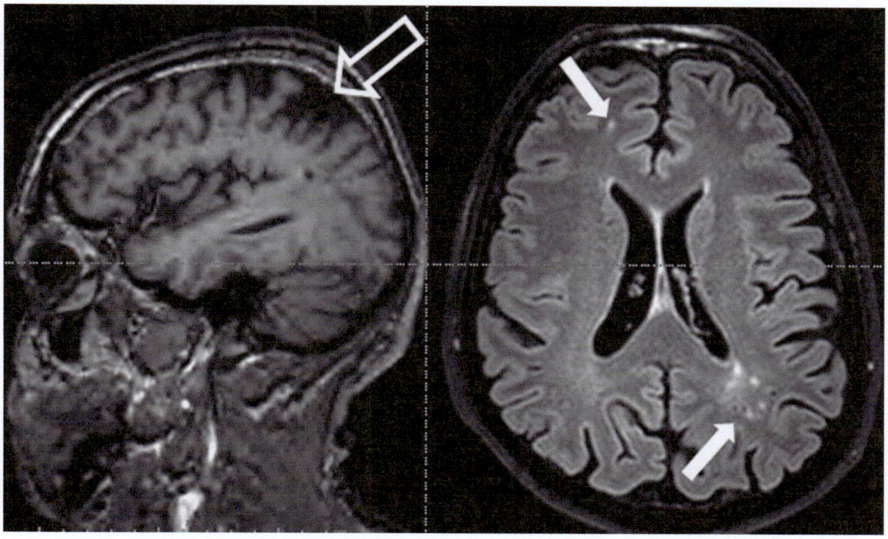

Fig. 3.1 Left panel: T1-weighted sagittal image showing moderate atrophy in the parietal lobe (empty arrow). Transverse Flair image showing white matter hyperintensities more prominent in the left posterior hemisphere, but visible also in the right frontal subcortical regions (full arrows)

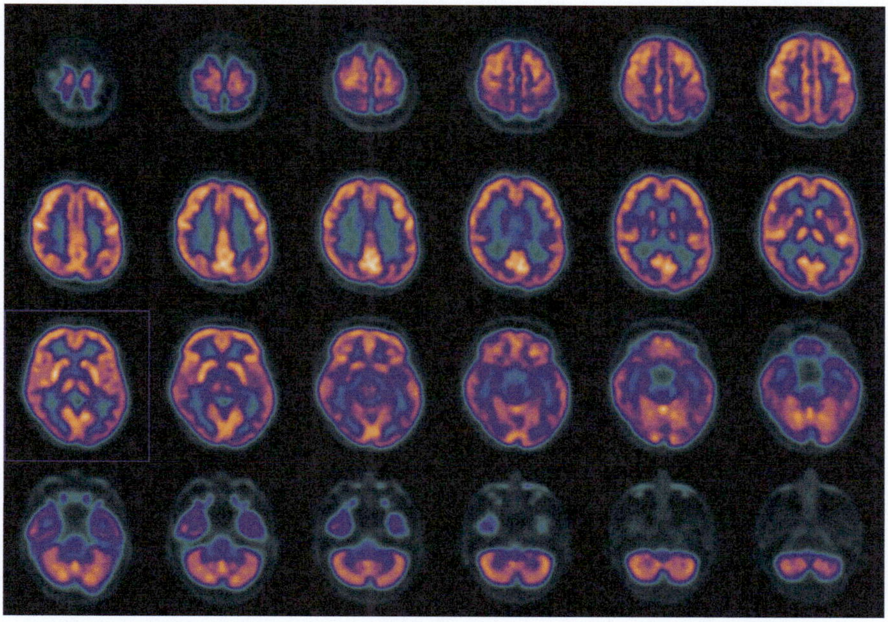

Fig. 3.2 FDG-PET shows no significant alterations of the normal symmetric glucose metabolism of the brain

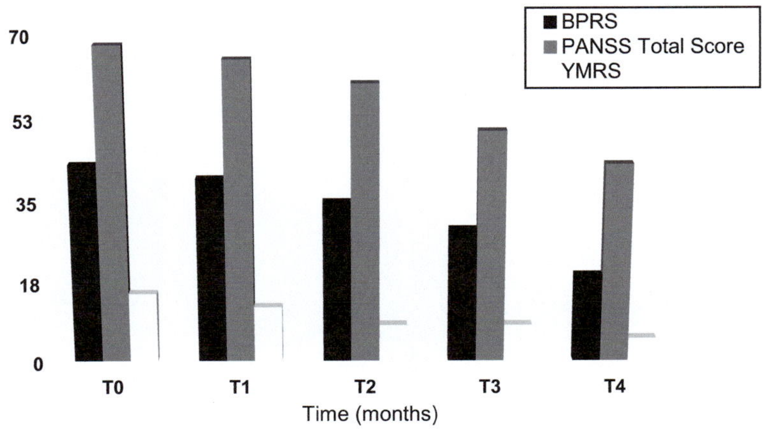

Fig. 3.3 Psychopathological rating scale (mean ± SD) time course

still didn't show full insight into his disease. From the clinical presentation and the response to treatment, a diagnosis of late-onset bipolar disorder type 1 was made. As a consequence of poor compliance, a program to dismiss the patient with a depot therapy was implemented. After discharge, he accepted monthly long-acting injection (LAI) with aripiprazole (400 mg). For the first month of treatment, it was administered an oral supplementation of aripiprazole 20 mg/d. The patient showed a response to treatment after 1 month and reached complete remission after 3 months. Aripiprazole plasma levels ranged from 48 ng/mL to 127 ng/mL. He has not presented any relapses to date (May 2017). The patient did not report any significant adverse effects, and changes in blood examinations did not appear during the follow-up period.

3.3 Literature Review

The presented case regards the difficult approach with late-onset bipolar disorder. Evidence suggest that bipolar disorder might have a trimodal distribution pattern of age of onset, namely, onset in late teens, mid-20s, and early 40s [26, 27]. Differences in clinical presentations between these groups appear nonsignificant. On the other hand, the late-onset subtype has often been recognized as a slightly different entity, overlapping sometimes with neurodegenerative disorders [28]. The onset of mania in later life might be indicative of an underlying medical comorbidity and should be assessed and investigated very carefully.

Neuroimaging studies showed several differences between late-onset bipolar disorder and early-onset bipolar disorder. Volumetric differences in specific frontal and temporal regions and white matter hyperintensities represent the most common findings [29]. Moreover, neurodegenerative disorders share many features with late-onset bipolar disorder, making it difficult to disentangle. Behavioral and psychological symptoms in dementia might resemble the common disinhibition found in acute mania, and both conditions might present psychosis [30]. Apart from the clinical perspective, neurodegeneration and mood disorders seem to share many other aspects. The same inflammation and oxidative stress biomarkers have been identified for both Alzheimer's disease and mood disorders [31]. Moreover, late-onset bipolar disorder seems to share many genetic risk factors with frontotemporal dementia. Indeed, genetic mutation of progranulin and C9ORF72 hexanucleotide repeat expansion, commonly found in frontotemporal dementia, has been associated also with bipolar disorder [32–34], although a large sample study assessing the differences between the two diseases is still lacking.

Although in the presented case, genetic mutations were absent, we cannot rule out the possibility of a neurodegenerative disorder with a subtle psychiatric onset. In this hypothesis, neurocognitive assessment and brain imaging find its place. Our patient presented wide neurocognitive deficits, in several areas, namely, executive functions, working memory, and verbal fluency, which are in contrast with his high educational level and his current profession. The reported difficulties in dealing with working pressure before admission might be the consequence of an altered cognitive state. Moreover, the MRI showed a moderate atrophy in the parietal lobe along with multiple spots of probable vascular origin. The PET, on the other hand, seemed nonsignificant for any ongoing metabolic process alterations. These data, although non-specific for a neurodegenerative diagnosis, underlie a clear, general suffering of the brain. However, a direct relationship between the degree of brain damage and the clinical manifestations is often lacking [35]. Modern research considers the theoretical concept of "cognitive reserve" as a possible explanation for this discrepancy. It refers to structural and dynamic capacities of the brain to buffer against lesions and might be dependent on a naturally existing interindividual variability or modulated by lifestyle factors, such as physical activity, stimulating cognitive activity over a lifetime [36]. Major brain imaging anomalies are common findings in late-onset bipolar disorder [29]. A possible explanation might be that acute, acquired, possibly vascular events may impact on a genetic predisposition.

The choice of a treatment is another great issue in late-onset psychosis. The balance between effectiveness and adverse effects should be the guide. In this case, aripiprazole was the chosen molecule. Specifically, aripiprazole lauroxil (400 mg monthly) showed good effectiveness as mood stabilizer of a psychotic bipolar patient after a manic episode (Fig. 3.3). In addition, the patient did not develop adverse effects or presented metabolic changes. Aripiprazole is considered a safe and effective drug, with low rate of adverse effects [37]. Aripiprazole therefore may be a therapeutic option in cases similar to those presented in the present case (severe symptoms, psychotic features, and poor compliance to pharmacological treatment). In addition, the treatment with first-generation antipsychotics does not seem to significantly change the long-term outcome of psychotic bipolar patients [38]. Second generation antipsychotics have demonstrated stabilizing properties in bipolar disorder, but until now data about the use of atypical antipsychotic LAI are very limited [39]. If future research confirm the efficacy of atypical antipsychotic LAI in the long-term treatment of BD (similarly to schizophrenia), these formulations may not be limited to the treatment of very severe patients, but they might be considered as early as possible in the course of BD. Finally, the stabilizing properties of atypical antipsychotics including aripiprazole would support the change of their obsolete name to "multidimensional stabilizers."

Plasma level determination is advisable during the long-term treatment to optimize the dosage and to investigate a possible correlation with clinical stabilization. These results, although not conclusive, may challenge the current therapeutic ranges proposed for antipsychotic drugs, most likely reflecting oral administration. Aripiprazole LAI showed its efficacy in the clinical stabilization of patients. The absence of severe adverse events demonstrated the tolerability of aripiprazole LAI. For this reason it should be considered an optimal maintenance treatment for bipolar disorder, regardless of the age of onset.

In conclusion, aripiprazole LAI showed its efficacy in the clinical stabilization of patients. The absence of severe adverse events demonstrated the tolerability of aripiprazole LAI, and for this reason it should be considered an optimal maintenance treatment for BP-I also in the elderly.

Key Points

- The elderly are at increased risk for psychosis because of age-related deterioration of cortical areas and neurochemical changes, comorbid physical illnesses, social isolation, sensory deficits, and polypharmacy. The prevalence of psychiatric and neuropsychiatric disorders requiring treatment with an antipsychotic agent is expected to increase dramatically among people aged >64 years.
- The differential diagnosis of elderly patients that present psychotic episode, for the first time, should include psychosis related to delirium, psychosis due to general medical conditions, atypical dementia onset, behavioral and psychological symptoms of dementia, affective illness, schizophrenia, bipolar disorder or other primary psychotic disorders, and substance abuse or dependence.
- Neuroimaging studies showed several differences between late-onset bipolar disorder and early-onset bipolar disorder. Volumetric differences in specific frontal and temporal regions and white matter hyperintensities represent the most common findings, so it is very important to assess the patient with neuroimaging procedures.
- Age-related changes in cognitive function vary considerably across individuals and across cognitive domains, with some cognitive functions appearing more susceptible than others to the effects of aging, and the mapping of cognitive processes onto neural structures constitutes a relatively recent research enterprise driven largely by advances in neuroimaging technology, so it is very important to assess the patient's cognition.
- Plasma level determination is advisable during long-term treatment to optimize the dosage and to investigate a possible correlation with clinical stabilization.

- For patients with psychosis who will not or cannot take oral medications on a regular daily basis or whose other characteristics, such as memory, vision, or auditory impairment, contribute to partial compliance, long-acting injectable antipsychotic medication offers a solution. Older patients are especially at risk of adverse effects associated with traditional antipsychotic agents, such as motor effects, postural hypotension, excessive sedation, and anticholinergic effects because of age-related pharmacokinetic and pharmacodynamic factors, coexisting medical illnesses, and concomitant medications. Therefore, drug dosage recommendations in the elderly are much more conservative than in younger patients.

Self-Assessment Questionnaire

1. What is the best therapeutic option for the stabilization of late-onset bipolar disorder?
   (A) Lithium
   (B) Valproic acid
   (C) Lamotrigine
   (D) Antipsychotic
2. What is the most frequent differential diagnosis that could be confusing in an elderly person presenting psychotic symptoms?
   (A) Dementia
   (B) Unipolar depression
   (C) Delirium
   (D) Brain neoplasia
3. What is the most frequent volumetric difference found between late-onset bipolar disorder and early-onset bipolar disorder?
   (A) Frontal and temporal regions and white matter hyperintensities
   (B) Cerebellum
   (C) Basal ganglia
   (D) Specifically prefrontal cortex
4. Which is the most reliable clinical scale evaluation for psychotic symptoms in bipolar disorder?
   (A) PANSS
   (B) YMRS
   (C) BPRS
   (D) WHO DAS
5. Which is the most important clinical domain among the following for evaluation in the follow-up after the clinical stabilization of the patient?
   (A) Cardiological aspect
   (B) Cognition
   (C) Plasma level
   (D) Endocrinological aspect

References

1. Mintzer J, Targum S. Psychosis in elderly patients: classification and pharmacotherapy. J Geriatr Psychiatry Neurol. 2003;16:199–206.
2. Broadway J, Mintzer J. The many faces of psychosis in the elderly. Curr Opin Psychiatry. 2007;20(6):551–8. [Review]
3. Ostling S, Borjesson-Hanson A, Skoog I. Psychotic symptoms and paranoid ideation in a population-based sample of 95-year-olds. Am J Geriatr Psychiatry. 2007;15:999–1004.
4. Kyomen HH, Whitfield TH. Psychosis in the elderly. Am J Psychiatry. 2009;166(2):146–50.
5. Bennett S, Thomas AJ. Depression and dementia: cause, consequence or coincidence? Maturitas. 2014;79(2):184–90.
6. Murray PS, Kumar S, Demichele-Sweet MAA, Sweet RA. Psychosis in Alzheimer's disease. Biol Psychiatry. 2014;75(7):542–52.
7. Grau-Rivera O, Gelpi E, Carballido-López E, Sánchez-Valle R, López-Villegas MD. Rapidly progressive dementia with psychotic onset in a patient with the C9ORF72 mutation. Clin Neuropathol. 2015;34(5):294–7.
8. Gossink FT, Vijverberg EG, Krudop W, Scheltens P, Stek ML, Pijnenburg YA, Dols A. Psychosis in behavioral variant frontotemporal dementia. Neuropsychiatr Dis Treat. 2017;13:1099–106.
9. Galimberti D, Dell'Osso B, Altamura AC, Scarpini E. Psychiatric symptoms in frontotemporal dementia: epidemiology, phenotypes, and differential diagnosis. Biol Psychiatry. 2015;78 (10):684–92.
10. Hall D, Finger EC. Psychotic symptoms in frontotemporal dementia. Curr Neurol Neurosci Rep. 2015;15(7):46.
11. Shinagawa S, Nakajima S, Plitman E, Graff-Guerrero A, Mimura M, Nakayama K, Miller BL. Psychosis in frontotemporal dementia. J Alzheimers Dis. 2014;42(2):485–99.
12. Mulsant BH, Ganguli M. Epidemiology and diagnosis of depression in late life. J Clin Psychiatry. 1999;20:9–15.
13. Wohlreich MM, Mallinckrodt CH, Watkin JG, et al. Duloxetine for the long-term management of major depressive disorder in patients aged 65 and older: an open-label study. BMC Geriatr. 2004;4:11.
14. Gournellis R, Lykouras L, Fortos A, Oulis P, Roumbos V, Christodoulou GN. Psychotic (delusional) major depression in late life: a clinical study. Int J Geriatr Psychiatry. 2001;16 (11):1085–91.
15. Kessing L. Differences in diagnostic subtypes among patients with late and early onset of a single depressive episode. Int J Geriatr Psychiatry. 2006;21:1127–31.
16. Heser K, Tebarth F, Wiese B, Eisele M, Bickel H, Köhler M, Mösch E, Weyerer S, Werle J, König HH, Leicht H, Pentzek M, Fuchs A, Riedel-Heller SG, Luppa M, Prokein J, Scherer M, Maier W. Wagner M; age CoDe study group. Age of major depression onset, depressive symptoms, and risk for subsequent dementia: results of the German study on ageing, cognition, and dementia in primary care patients (AgeCoDe). Psychol Med. 2013;43(8):1597–610.
17. Kukreja D, Günther U, Popp J. Delirium in the elderly: current problems with increasing geriatric age. Indian J Med Res. 2015;142(6):655–62.
18. Morandi A, Di Santo SG, Cherubini A, Mossello E, Meagher D, Mazzone A, Bianchetti A, Ferrara N, Ferrari A, Musicco M, Trabucchi M, Bellelli G, ISGoD Group. Clinical features associated with delirium motor subtypes in older inpatients: results of a Multicenter study. Am J Geriatr Psychiatry. 2017;25(10):1064–71.
19. Inouye SK, Westendorp RG, Saczynski JS. Delirium in elderly people. Lancet. 2014;383:911–22.
20. By the American Geriatrics Society 2015 Beers Criteria Update Expert Panel. American Geriatrics Society 2015 updated beers criteria for potentially inappropriate medication use in older adults. J Am Geriatr Soc. 2015;63(11):2227–46.
21. Fong TG, Davis D, Growdon ME, Albuquerque A, Inouye SK. The interface between delirium and dementia in elderly adults. Lancet Neurol. 2015;14(8):823–32.

22. Gavazzi G, Delerce E, Cambau E, François P, Corroyer B, de Wazières B, Fougère B, Paccalin M, Gaillat J. Diagnostic criteria for urinary tract infection in hospitalized elderly patients over 75 years of age: a multicenter cross-sectional study. Med Mal Infect. 2013;43 (5):189–94.
23. Barthel H, Schroeter ML, Hoffmann KT, Sabri O. PET/MR in dementia and other neurodegenerative diseases. Semin Nucl Med. 2015;45(3):224–33.
24. Linjakumpu T, Hartikainen S, Klaukka T, et al. Use of medications and polypharmacy are increasing among the elderly. J Clin Epidemiol. 2002;55:809–17.
25. Rovera C, Mauri MC, Bertin E, Di Pace C, Paletta S, Reggiori A, De Gaspari IF, Cattaneo D, Mari D, Altamura AC. Duloxetine in elderly major depression disorder: effectiveness and drug plasma level evaluation. Hum Psychopharmacol. 2016;31(5):349–55.
26. Bellivier F, Golmard JL, Rietschel M, Schulze TG, Malafosse A, Preisig M, McKeon P, Mynett-Johnson L, Henry C, Leboyer M. Age at onset in bipolar I affective disorder: further evidence for three subgroups. Am J Psychiatry. 2003;160(5):999–1001.
27. Carlino AR, Stinnett JL, Kim DR. New onset of bipolar disorder in late life. Psychosomatics. 2013 Jan–Feb;54(1):94–7.
28. Besga A, Gonzalez I, Echeburua E, Savio A, Ayerdi B, Chyzhyk D, Madrigal JL, Leza JC, Graña M, Gonzalez-Pinto AM. Discrimination between Alzheimer's disease and late onset bipolar disorder using multivariate analysis. Front Aging Neurosci. 2015;7:231.
29. Hahn C, Lim HK, Lee CU. Neuroimaging findings in late-onset schizophrenia and bipolar disorder. J Geriatr Psychiatry Neurol. 2014;27(1):56–62.
30. Kales HC, Gitlin LN, Lyketsos CG. Assessment and management of behavioral and psychological symptoms of dementia. BMJ. 2015;350:h369.
31. Besga A, Chyzhyk D, Gonzalez-Ortega I, Echeveste J, Graña-Lecuona M, Graña M, Gonzalez-Pinto A. White matter tract integrity in Alzheimer's disease vs. late onset bipolar disorder and its correlation with systemic inflammation and oxidative stress biomarkers. Front Aging Neurosci. 2017;9:179.
32. Galimberti D, Prunas C, Paoli RA, Dell'Osso B, Fenoglio C, Villa C, Palazzo C, Cigliobianco M, Camuri G, Serpente M, Scarpini E, Altamura AC. Progranulin gene variability influences the risk for bipolar I disorder, but not bipolar II disorder. Bipolar Disord. 2014a;16 (7):769–72.
33. Galimberti D, Reif A, Dell'Osso B, Palazzo C, Villa C, Fenoglio C, Kittel-Schneider S, Leonhard C, Olmes DG, Serpente M, Paoli RA, Altamura AC, Scarpini E. C9ORF72 hexanucleotide repeat expansion as a rare cause of bipolar disorder. Bipolar Disord. 2014b;16 (4):448–9.
34. Rubino E, Vacca A, Gallone S, Govone F, Zucca M, Gai A, Ferrero P, Fenoglio P, Giordana MT, Rainero I. Late onset bipolar disorder and frontotemporal dementia with mutation in progranulin gene: a case report. Amyotroph Lateral Scler Frontotemporal Degener. 2017;18:1–3.
35. Stern Y. What is cognitive reserve? Theory and research application of the reserve concept. J Int Neuropsychol Soc. 2002;8(3):448–60.
36. Cheng ST. Cognitive reserve and the prevention of dementia: the role of physical and cognitive activities. Curr Psychiatry Rep. 2016;18(9):85.
37. Goodwin GM, Abbar M, Schlaepfer TE, Grunze H, Licht RW, Bellivier F, Fountoulakis KN, Altamura AC, Pitchot W, Ågren H, Holsboer-Trachsler E, Vieta E. Aripiprazole in patients with bipolar mania and beyond: an update of practical guidance. Curr Med Res Opin. 2011;27 (12):2285–99.
38. Wu CS, Hsieh MH, Tang CH, Chang CJ. Comparative effectiveness of long-acting injectable risperidone vs. long-acting injectable first-generation antipsychotics in bipolar disorder. J Affect Disord. 2016;197:189–95.
39. Kishi T, Oya K, Iwata N. Long-acting injectable antipsychotics for prevention of relapse in bipolar disorder: a systematic review and meta-analyses of randomized controlled trials. Int J Neuropsychopharmacol. 2016;19(9)

Psychotic Bipolar Disorder

4

Massimiliano Buoli and Alice Caldiroli

Abstract

Psychotic symptoms are considered a clinical predictor of poor outcome in bipolar disorder (BD). The misdiagnosis of psychotic BD (BD-P) with schizophrenia, schizoaffective disorder, or delusional disorder is very common. The identification of biological markers for BD-P would contribute to establish an early diagnosis and to define a proper treatment, to improve outcome of BD-P patients.

A case of a patient suffering from BD-P has been described, focusing on neuroimaging, neuropsychological, and inflammation data that could be considered potential biomarkers of BD-P.

The patient was a 36-year-old female. During her two hospitalizations, inflammatory and metabolic markers were all normal, except for lower plasma levels of vitamin A. After giving her written informed consent, epigenetic tests were performed, showing a significantly different expression of some miRNAs than healthy controls (HC). The magnetic resonance imaging (MRI) showed hypoplasia of the cerebellar vermis and a small left peritrigonal area of hyperintensity in T2 sequence. The only neuropsychological deficit appeared in the motor task. The positron emission tomography (PET) showed a bilateral hypermetabolism in the frontal cortex, cingulate, and striatum.

This case demonstrated that biological markers may be useful to predict outcome of psychotic bipolar patients.

Keywords

Bipolar disorder · Psychotic symptoms · Biomarkers

M. Buoli (✉) · A. Caldiroli
Department of Psychiatry, Fondazione IRCCS Ca' Granda Ospedale Maggiore Policlinico, University of Milan, Milan, Italy

A. C. Altamura, P. Brambilla (eds.), *Clinical Cases in Psychiatry: Integrating Translational Neuroscience Approaches*,
https://doi.org/10.1007/978-3-319-91557-9_4

4.1 Introduction

Bipolar disorder (BD) is a clinical condition that affects about 1–2% of the adult population. It is associated with a high disability and risk of chronicity, particularly if not promptly treated [1]. The identification of potential specific and objective biomarkers could provide an important support to better understand the course of illness, the mechanisms of vulnerability and of disorder expression, finally contributing to the clinical decision-making process.

For example, a recent study has demonstrated the presence of an increased level of a combined inflammation score (IgG class antibodies to the NR2 peptide fragment of the *N*-Methyl-D-Aspartate (NMDA) receptor, IgG class antibodies to gliadin, IgG class antibodies to the Mason-Pfizer monkey virus gag protein, IgM class antibodies to Toxoplasma gondii) in a sample of 57 manic patients, compared to healthy controls (HC) [2]. Moreover, a review analyzed the role of inflammatory cascade and brain-derived neurotrophic factor (BDNF) in cardiovascular diseases among bipolar patients, remarking the importance of BDNF and inflammatory cascade as possible markers of BD [3]. The reason for the elevation of inflammatory markers in individuals with mania remains unknown, but these findings suggest the presence of specific markers in bipolar patients that are not identifiable in the control group. Furthermore, a recent paper describing the work of the Biomarkers Network from the International Society for Bipolar Disorders (ISBD-BIONET) cited neuroimaging and genetic findings in bipolar disorder as important tools to identify bipolar patients (e.g., the loss of gray matter, the altered activation of anterior temporal, the ventral prefrontal and subcortical regions in response to emotional stimuli, the identification of several potential candidate genes associated with increased risk for developing BD) [4]. A review by Schroeter et al. [5] reported that increased serum levels of S100B, a glial growth and differentiation factor, are more frequently present in patients with mood disorders than in HC, particularly in major depressives.

Bipolar patients may experience delusions/hallucinations both during depressive (up to 45% of cases) [6, 7] and manic episodes (up to 88% of cases) [8], and it has been demonstrated that psychotic bipolar patients (BD-P) had better prognosis than schizophrenia (SKZ) patients [9] but worse than non-psychotic bipolar disorder (BD-NP)/unipolar patients [10, 11] in terms of more severity of illness, greater morbidity [12], longer hospitalizations [13], and more frequent relapses [14]. In particular, BD-P patients experiencing mood-incongruent delusions were found to present shorter euthymia and more frequent hearing hallucinations than BD-P patients with mood-congruent delusions [15]. If psychotic symptoms occurred during a depressive episode, the risk of relapse was greater for BD-P with respect to BD-NP [16] than in the case of psychotic mania [12], and the presence of psychotic symptoms in patients with a major depressive episode (MDE) was predictive of non-remission after a standard psychopharmacological treatment [17].

The diagnosis of psychotic BD is not always prompt, and, as a consequence of misdiagnosis (up to 61%) with SKZ, schizoaffective disorder, and delusional disorder, an appropriate treatment of psychotic bipolar patients may be delayed with consequently increased suicidal risk, increased social impairment and higher social costs [18, 19]. Although biological markers for the subset of BD-P have been little investigated, it becomes increasingly clear that the identification of biomarkers for BD-P is important. First, in the light of the high rate of misdiagnosis, biological markers of psychotic BD would help clinicians to establish an early diagnosis; moreover, as longer duration of untreated illness (DUI) in psychotic BD has been related to lower Global Assessment of Functioning (GAF) scores and more hospitalizations [20], biomarker identification would provide the possibility of clinical staging and would help to ameliorate outcome in the long term by recognizing the disorder early and defining a prompt and personalized treatment [21].

4.2 Case Presentation

E. was a 36-year-old woman, graduated in economics, twice hospitalized in our inpatient clinic. She has been described by her relatives as a diligent and scholarly girl, with few friends and the preference for solitary hobbies. Good school performance is reported up to the first year of university, when the patient started to reduce the number of approved exams during each session, however keeping a good overall performance and earning a 3-year degree in 6 years.

Family members reported a depressive episode that occurred during the college period, following the death of the mother in 2008. It was characterized by anhedonia, depressive mood, abandonment of work, and suicidal ideation without concrete projects. The patient turned to a psychotherapist, initially at the Community Mental Health Center affiliated to our Psychiatry Department, and then to a private psychoanalyst with whom she is still in care.

After graduation, the patient undertook apprenticeship at an accountant's for a year and a half. This period is described by relatives as very stressful for the high demands of work. The patient resigned and started working with her father who is also an accountant, but the office was closed down in March 2016 for discontinued activity, and, since then, the patient has been unemployed.

Until this moment, the patient has never taken psychopharmacological therapy, with the exception of a symptomatic treatment (benzodiazepines) prescribed by the general practitioner.

In November 2016 the patient reached our emergency department after a few days of excitement, subtotal insomnia, loosening of associative links, logorrhea, persecutory delusions, and psychomotor agitation symptoms. She was poorly accessible to conversation, agitated, with oscillating mood. Electrocardiogram was performed, and clonazepam 15 drops + promazine 15 drops were administered. She was unaware of the need for care; nevertheless she voluntarily accepted hospitalization, and then she was admitted to our inpatient clinic.

On admission, E. showed euphoric mood, purposeless excessive motor activation, megalomaniac and persecutory delusions, unstable sensitivity, and insomnia. The physical examination and routine blood tests were normal, and the urine drug test was positive for opioids.

At baseline, the Brief Psychiatric Rating Scale (BPRS) [22] was administered with a total score of 43. Young Mania Rating Scale (YMRS) [23] was 27, and Hamilton Scale for Depression (HAM-D) [24] was 17. Her first pharmacological treatment included lithium up to 600 mg/day, haloperidol up to 2 mg/day, and intramuscular olanzapine up to 20 mg/day.

After giving her written informed consent, specific blood tests were carried out, measuring inflammatory markers (total lymphocyte count, B lymphocyte count, T lymphocyte count, C-reactive protein, serum albumin, serum bilirubin, serum uric acid, vitamins A and E) and performing a genetic test (miRNA-wide analyses), showing a significantly different expression of some miRNAs than HC. Of note, E. presented an overexpression of hsa-miR-579-3p, hsa-miR-4454 + hsa-miR-7975, and hsa-let-7a-5p. All inflammatory markers were normal, except for vitamin A which was 0.27 mg/L (normal range 0.30–0.70 mg/L).

During hospitalization, she underwent a magnetic resonance imaging (MRI) showing a mild hypoplasia of the cerebellar vermis (Fig. 4.1).

Furthermore, the patient underwent neuropsychological examination showing normal scores in language, verbal memory, working memory, frontal efficiency test, and a deficit in motor functioning (Table 4.1).

After 1 week of hospitalization, the BPRS score decreased to 27, MRS was 11, and HAM-D was 9. The patient was collaborating and calm, and mood improvement was observed. Speaking was fluid, and associative links were more frequently maintained. Delusions gradually disappeared, and the patient became aware of the need to continue treatment and visits. She was discharged with olanzapine 20 mg/day and lithium 900 mg/day, and the indication for follow-up at our Day Hospital Unit and at the Community Mental Health Center affiliated with our department.

Box 4.1 First Hospitalization

- Admitted with a possible manic episode with psychotic symptoms
- Urine drug test positive for opioids
- Vitamin A (antioxidant) lower than normal range
- miRNA
- MRI: mild hypoplasia cerebellar vermis
- BPRS 43 → 27; HAM-D 17 → 9; MRS 17 → 11
- Discharged with olanzapine 20 mg/day + lithium 900 mg/day

She attended our Day Hospital on three occasions, where the psychiatrist did not change the treatment with olanzapine and lithium. She was cooperative and appropriate to the context. Initially abundantly logorrheic and easily distracted, her speech gradually returned to normality, and associative links were maintained. The sleep rhythm was preserved.

In December 2016 she attended the Community Mental Health Center affiliated with our department. Olanzapine was suspended for hypersomnia, and a depressive nuance appeared, without evolving into an MDE. Lamotrigine was prescribed in order to stabilize the subthreshold depressive symptoms. In May 2017 the patient attended the last psychiatric visit at the Community Mental Health Center, and she was treated with lithium 600 mg/day and lamotrigine 200 mg/day.

E. maintained a good psychopathological compensation until the end of May, when she was accompanied to the emergency department by public forces for

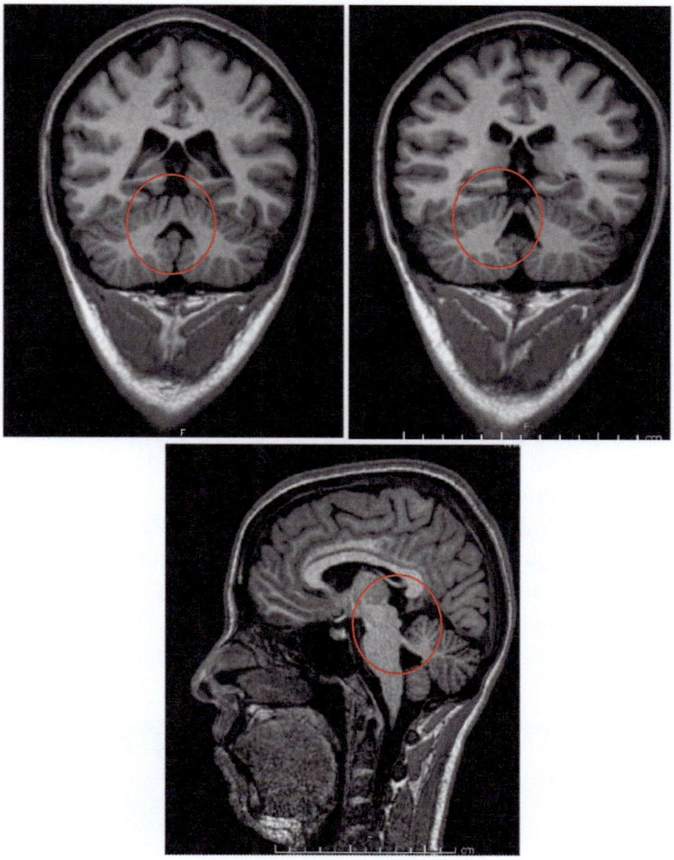

Fig. 4.1 T1-weighted sagittal and coronal images showing mild hypoplasia of the cerebellar vermis

Table 4.1 Neurocognitive evaluation (BACS subtest and Beck Cognitive Insight Scale)

Test	Normal range	Score	Equivalent score	Comment
Language				
Verbal fluency (*subtest BACS*)	v.n. $\geq$ 31.68	45.25	2	Normal
Memory				
Verbal memory (*subtest BACS*)	v.n. $\geq$ 33.01	50.00	4	Normal
Motor function				
Token task (*subtest BACS*)	v.n. $\geq$ 68.77	67.00	0	***Deficit***
Symbol-coding task (*subtest BACS*)	v.n. $\geq$ 40.49	58.75	4	Normal
Frontal efficiency tests				
Working memory (*subtest BACS*)	v.n. $\geq$ 14.93	23.25	4	Normal
Tower of London test (*subtest BACS*)	v.n. $\geq$ 12.37	22.00	4	Normal

	Score	Comment
Beck cognitive insight scale	8	The patient has insight

behavioral abnormalities in a library. Symptoms such as insomnia, excessive expenses, accelerated thinking with loose associative links, and megalomaniac delusions were reported. She was hospitalized for 2 days in another hospital and treated with lithium up to 900 mg/day, levomepromazine up to 50 mg/day, and olanzapine up to 20 mg/day. She was fatuous and hilarious, overall manageable; the ideation was abundant with loose links and delusions without emotional participation. Hallucinations were not detectable. She showed attention and concentration deficit. The hypnotic profile was restored and, after two days, she was moved to our inpatients clinic.

During hospitalization, treatment with lithium 900 mg/day and intramuscular aripiprazole up to 30 mg/day was prescribed in anticipation of long-acting treatment.

Lithium plasma levels were 0.91 mEq/L with lithium 900 mg/day. Augmentation therapy with folic acid was prescribed because routine blood tests showed mild normocytic anemia with folate deficiency.

E. showed a gradual return to euthymia. Associative links were more frequently maintained and megalomaniac delusions resolved. After 1 week of hospitalization, E. underwent long-acting intramuscular therapy with aripiprazole 400 mg which was well tolerated by the patient. Furthermore, as diagnostic completion, E. underwent another MRI which showed a small left peritrigonal area of hyperintensity in T2 sequence (Fig. 4.2) and a positron emission tomography (PET) showing bilateral hypermetabolism in the frontal cortex, cingulate, and striatum (Fig. 4.3).

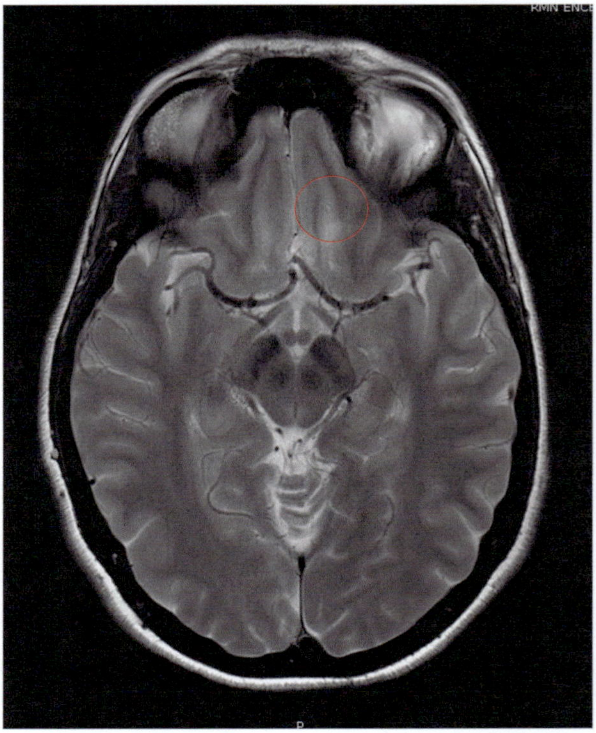

Fig. 4.2 T2-weighted image showing a mild left peritrigonal hyperintensity

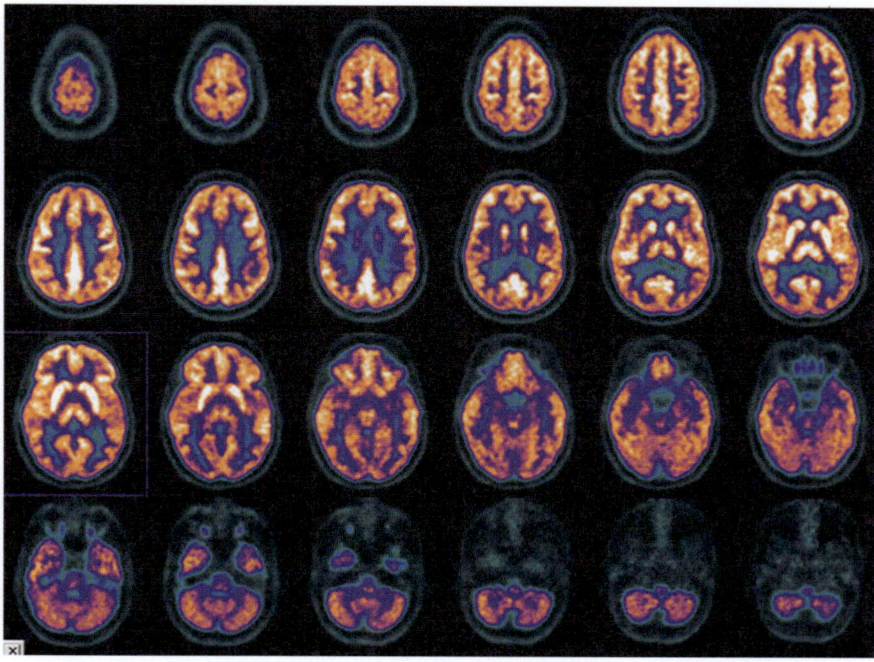

Fig. 4.3 FDG-PET shows bilateral hypermetabolism in the frontal cortex, cingulate, and striatum

After 1 week of hospitalization, the patient was discharged with lithium 900 mg/day, aripiprazole 15 mg/day, folic acid 5 mg/day, and indication to be followed-up at our Day Hospital Unit for the long-acting treatment.

Box 4.2 Second Hospitalization

- Admitted with a possible manic episode with psychotic features
- Treated with lithium 900 mg/day and aripiprazole 30 mg/day
- Lithium plasma levels = 0.91 mEq/L
- MRI: small left peritrigonal hyperintensity
- PET: bilateral hypermetabolism in the frontal cortex, cingulate, and striatum
- Discharged with depot treatment with Aripiprazole Maintena 400 mg + aripiprazole 15 mg/day + lithium 900 mg/day

Lithium plasma levels were 0.9 mEq/L while taking lithium 900 mg/day. At the Day Hospital, she showed a good psychopathological compensation, although she complained about asthenia. Aripiprazole was suspended, and she continued the stabilizing treatment with lithium 600 mg/day and weekly controls. After 1 month, lithium plasma levels decreased to 0.39 mEq/L, and the treatment was increased to lithium 750 mg/day.

Currently, depot treatment with Aripiprazole Maintena 300 mg is administered monthly, and she is taking lithium 750 mg/day, presenting a good compensation and a good psychosocial functioning.

4.3 Review of the Literature

Biological markers of psychotic bipolar disorder have been little investigated. A recent systematic review showed that most of the studies in literature are concordant in recognizing BD-P as a subset with specific biological abnormalities, with a different degree of severity with respect to SKZ and BD-NP [25].

From a genetic point of view, the most studied genes are NRG1, 5HTTLPR(s), COMT, and DAOA, with inconclusive results. The finding concerning the positive association between mood-incongruent BD-P and one NRG1 polymorphism (NRG241930G) has been replicated [26, 27], as well as the association between the psychotic dimension in BD and two chromosome regions (13q32 and 16p12) [28–30].

Studies of infectious, neuroendocrine, and inflammatory markers demonstrated that BD-P patients had higher levels of kynurenic acid (KYNA) [31] and lower serum levels of dehydroepiandrosterone sulfate (DHEAS) and progesterone than BD-NP [32] and increased anti-*Saccharomyces cerevisiae* antibodies (ASCA) levels than HC [33, 34], but these data have not been replicated yet.

Neurophysiological studies showed SKZ patients and BD-P patients sharing abnormal low-frequency activity [35]. Four papers reported reduced P50 suppression in BD-P compared to BD-NP/HC [36–39], while two studies did not find any differences [40, 41]. BD-P had higher M20 wave lateralization than BD-NP [42], higher P85 ratio [43] and an increased beta-gamma activity after P200 response window compared to HC [44], but these data need further studies to be confirmed.

Most of the functional neuroimaging studies have not been replicated. Data from structural neuroimaging showed that SKZ patients had lower gray matter volumes than BD-P [45–47]. Sensitive differences were found between BD-P and BD-NP, with contrasting results: the absence of major reduction of gray matter volumes has been replicated [48–50], as well as the lower DLPFC volume [51, 52] and the ventricular enlargement [53, 54]. These preliminary findings have to be replicated in further studies.

Finally, in neuropsychological studies, SKZ and BD-P resulted impaired in emotion processing and ToM compared to HC [55–57]. Lower scores of BD-P than BD-NP have been found in some cognitive domains (executive functioning, verbal and logical memory, working memory, verbal, and semantic fluency) by different authors [58–61]. However, three papers reported no differences between BD-P and BD-NP in terms of cognitive performance [62–64].

In the presented case, E. underwent genetic and neuroimaging examination and dosages of the inflammatory pattern. Her MRI showed a small left peritrigonal hyperintensity, which is in line with the previous findings that reported significantly more white matter hyperintensities in BD-P than in HC. The inflammatory markers were all normal, except for vitamin A (retinol), which was lower than the normal values. The inflammatory pattern may vary according to the different phases of the disorder. Of note, it has been demonstrated that vitamin A deficiency contributed to the oxidative damage in developing rat brain [65] and that a supplementation therapy with nonenzymatic antioxidants (such as vitamin A) significantly reduced anxiety and depression in a psychiatric population [66]. Moreover, Chowdhury et al. [67] found significantly lower concentrations of antioxidants (vitamins A, E, and C) in BD patients than in HC.

From a clinical point of view, the Bipolar-Schizophrenia Network on Intermediate Phenotypes (B-SNIP) demonstrated several overlaps between SKZ and BD-P in terms of symptomatology, psychosocial functioning, and family characteristics [68]; furthermore, the presence of overlapping biological abnormalities with SKZ, but with a lesser degree of severity, suggested that BD-P is an intermediate phenotype between psychotic and mood disorders and part of the "psychotic continuum" [69]. Of note, a growing body of data in literature indicates that different biological markers are associated with the psychotic dimension in BD, but further research is needed to detect specific biomarkers of BD-P. Taken together, these findings are in favor of the hypothesis of the presence of a continuum among psychotic disorders, suggesting that a dimensional model may be more reliable than a categorical one.

Box 4.3 Biomarkers of BD-P

- *Genetics*: NRG1 polymorphism (NRG241930G) and two chromosome regions (13q32 and 16p12)
- *Neuro-immuno-endocrinology*: ↑KYNA, ↑ASCA, ↓ DEAHS and progesterone
- *Neurophysiology*: ↓ P50 suppression, ↑ P85 ratio vs HC, ↑ β-γ activity after P200 response window, ↑ M20 wave lateralization
- *MRI*: ↓ DLPFC volume (mood incongruent), ↓ DLPFC and insula volumes, ↓ left vlPFC, dmPFC, left temporal pole volumes, ↑ **white matter hyperintensity**
- *Neuropsychology*:↓ executive functions, ↓ verbal and logical memory, ↓ (spatial) working memory, ↓ verbal and semantic fluency, ↓ verbal learning, ↓ visual attention, ↓ emotional processing

Key Points

- BD is a chronic disorder associated with high disability and morbidity. Some biological abnormalities have been reported in BD as an attractive basis for the discovery of promising biomarkers, such as the loss of gray matter, the altered activation of anterior temporal, ventral prefrontal and subcortical regions in response to emotional stimuli, and increased level of a combined inflammation score during mania.

- Psychotic bipolar disorder seems to be a specific subset, characterized by oscillating mood and the presence of psychotic features. Patients suffering from BD-P have worse prognosis than BD-NP in terms of more severity of illness, greater morbidity, longer hospitalizations, and more frequent relapses.
- The rate of misdiagnosis of BD-P is high, up to 61%. BD-P is more often misdiagnosed with schizophrenia, schizoaffective disorder, and delusional disorder, thus delaying the prescription of an appropriate treatment, with consequently increased suicidal risk, increased social impairment, and higher social costs.
- Several findings in the field of mood disorders could provide the background for the identification of potential specific and objective biomarkers, which could be an important aid to better understand the course of illness, the mechanisms of vulnerability and of disease expression. Biological markers of psychotic BPAD would help clinicians to establish an early diagnosis and to ameliorate prognostic evaluations.
- To date, only few studies about potential biological markers of BD-P have been replicated. Among these are the association between BD-P and NRG1, 5HTTLPR(s), COMT, and DAOA genes; data about BD-P and 16p12/13q regions; increased IL-1 and IL-6 plasma levels in both BD-P and SKZ; higher lateralization of M20 wave and higher P85 ratio in BD-P than HC; reduced white matter integrity than HC; and more severe cognitive impairment than BD in the same domains and an intermediate cognitive performance between SKZ and HC.

Self-Assessment

1. What is the prevalence of bipolar disorder?
   (a) 5%
   (b) 20%
   (c) 80%
   (d) 1–2%
   (e) 10%
2. Which abnormality has been found in manic BD patients?
   (a) Increased BDNF levels
   (b) Increased inflammatory markers
   (c) Decreased inflammatory markers
   (d) Increased serum antioxidants levels
   (e) Decreased BDNF levels
3. Psychotic bipolar patients present:
   (a) Worse prognosis than unipolar and BD-NP patients
   (b) Better prognosis than unipolar and BD-NP patients
   (c) Worse prognosis than SKZ patients
   (d) Better prognosis than schizoaffective patients
   (e) We do not have data about BD-P patients' prognosis
4. Identification of biological markers for BD-P:
   (a) Would hinder the diagnostic and therapeutic process

(b) Could help patients to decide independently which treatment is the best
(c) Would help clinicians to promptly diagnose the disorder and to properly identify the treatment
(d) Could help clinicians to decide independently which treatment is the best
(e) Would hinder the clinical staging processing, worsening outcome in the long term

5. Vitamin A (retinol) is:
(a) The forerunner of beta-carotene
(b) An enzymatic antioxidant
(c) Fundamental for metabolism of lipids and proteins
(d) A cholesterol derivative
(e) A nonenzymatic antioxidant

References

1. Fagiolini A, Forgione R, Maccari M, et al. Prevalence, chronicity, burden and borders of bipolar disorder. J Affect Disord. 2013;148:161–9.
2. Dickerson F, Stallings C, Origoni A, et al. A combined marker of inflammation in individuals with mania. PLoS One. 2013;e73520:8.
3. Goldstein BI, Young LT. Toward clinically applicable biomarkers in bipolar disorder: focus on BDNF, inflammatory markers, and endothelial function. Curr Psychiatry Rep. 2013;15:425.
4. Frey BN, Andreazza AC, Houenou J, et al. Biomarkers in bipolar disorder: a positional paper from the International Society for Bipolar Disorders Biomarkers Task Force. Aust N Z J Psychiatry. 2013;47:321–32.
5. Schroeter ML, Sacher J, Steiner J, et al. Serum S100B represents a new biomarker for mood disorders. Curr Drug Targets. 2013;14:1237–48.
6. Brugue E, Colom F, Sanchez-Moreno J, et al. Depression subtypes in bipolar I and II disorders. Psychopathology. 2008;41:111–4.
7. Mantere O, Suominen K, Leppämäki S, et al. The clinical characteristics of DSM-IV bipolar I and II disorders: baseline findings from the Jorvi bipolar study (JoBS). Bipolar Disord. 2004;6: 395–405.
8. Keck PE Jr, McElroy SL, Havens JR, et al. Psychosis in bipolar disorder: phenomenology and impact on morbidity and course of illness. Compr. Psychiatry. 2003;44:263–9.
9. Zenisek R, Thaler N, Ringdahl E, et al. C-29Intellectual ability and functional outcome in bipolar disorder and schizophrenia. Arch Clin Neuropsychol. 2014;29:582–3.
10. Østergaard SD, Bertelsen A, Nielsen J, et al. The association between psychotic mania, psychotic depression and mixed affective episodes among 14.529 patients with bipolar disorder. J Affect Disord. 2013;147:44–50.
11. Vythilingam M, Chen J, Bremner JD, et al. Psychotic depression and mortality. Am J Psychiatry. 2003;160:574–6.
12. Coryell W, Leon AC, Turvey C, et al. The significance of psychotic features in manic episodes: a report from the NIMH collaborative study. J Affect Disord. 2001;67:79–88.
13. Kessing LV. Subtypes of manic episodes according to ICD-10: prediction of time to remission and risk of relapse. J Affect Disord. 2004;81:279–85.
14. Tohen M, Zarate CA Jr, Hennen J, et al. The McLean-Harvard first-episode mania study: prediction of recovery and first recurrence. Am J Psychiatry. 2003;160:2099–107.
15. Azorin JM, Akiskal H, Hantouche E. The mood-instability hypothesis in the origin of mood-congruent versus mood-incongruent psychotic distinction in mania: validation in a French National Study of 1090 patients. J Affect Disord. 2006;96:215–23.

16. Kessing LV. Subtypes of depressive episodes according to ICD-10: prediction of risk of relapse and suicide. Psychopathology. 2003;36:285–91.
17. Buoli M, Caldiroli A, Altamura AC. Psychotic versus non-psychotic major depressive disorder: a comparative naturalistic study. Asian J Psychiatr. 2013;6:333–7.
18. Altamura AC, Dell'Osso B, Berlin HA, et al. Duration of untreated illness and suicide in bipolar disorder: a naturalistic study. Eur Arch Psychiatry Clin Neurosci. 2010;260:385–91.
19. Birnbaum HG, Shi L, Dial E, et al. Economic consequences of not recognizing bipolar disorder patients: across-sectional descriptive analysis. J Clin Psychiatry. 2003;64:1201–9.
20. Altamura AC, Buoli M, Caldiroli A, et al. Misdiagnosis, duration of untreated illness (DUI) and outcome in bipolar patients with psychotic symptoms: a naturalistic study. J Affect Disord. 2015;182:70–5.
21. Lin A, Reniers RL, Wood SJ. Clinical staging in severe mental disorder: evidence from neurocognition and neuroimaging. Br J Psychiatry Suppl. 2013;54:s11–7.
22. Overall JE, Gorham DR. The brief psychiatric rating scale. Psychol Rep. 1962;10:799–812.
23. Young RC, Biggs JT, Ziegler VE, Meyer DA. A rating scale for mania: reliability, validity and sensitivity. Br J Psychiatry. 1978;133:429–35.
24. Hamilton M. A rating scale for depression. J Neurol Neurosurg Psychiatry. 1960;23:56.
25. Buoli M, Caldiroli A, Cumerlato Melter C, et al. Biological aspects and candidate biomarkers for psychotic bipolar disorder: a systematic review. Psychiatry Clin Neurosci. 2016;70:227–44.
26. Goes FS, Willour VL, Zandi PP, et al. Bipolar disorder Phenome group; NIMH genetics initiative bipolar disorder consortium. Family-based association study of Neuregulin 1 with psychotic bipolar disorder. Am J Med Genet B Neuropsychiatr Genet. 2009;150B:693–702.
27. Prata DP, Breen G, Osborne S, et al. An association study of the neuregulin 1 gene, bipolar affective disorder and psychosis. Psychiatr Genet. 2009;19:113–6.
28. Cheng R, Juo SH, Loth JE, et al. Genome-wide linkage scan in a large bipolar disorder sample from the National Institute of Mental Health genetics initiative suggests putative loci for bipolar disorder, psychosis, suicide, and panic disorder. Mol Psychiatry. 2006;11:252–60.
29. Goes FS, Zandi PP, Miao K, et al. Mood-incongruent psychotic features in bipolar disorder: familial aggregation and suggestive linkage to 2p11-q14 and 13q21-33. Am J Psychiatry. 2007; 164:236–47.
30. Kremeyer B, García J, Müller H, et al. Genome-wide linkage scan of bipolar disorder in a Colombian population isolate replicates loci on chromosomes 7p21-22, 1p31, 16p12 and 21q21-22 and identifies a novel locus on chromosome 12q. Hum Hered. 2010;70:255–68.
31. Olsson SK, Sellgren C, Engberg G, et al. Cerebrospinal fluid kynurenic acid is associated with manic and psychotic features in patients with bipolar I disorder. Bipolar Disord. 2012;14: 719–26.
32. Johansson AG, Nikamo P, Schalling M, Landén M. Polymorphisms in AKR1C4 and HSD3B2 and differences in serum DHEAS and progesterone are associated with paranoid ideation during mania or hypomania in bipolar disorder. Eur Neuropsychopharmacol. 2012;22:632–40.
33. Severance EG, Dupont D, Dickerson FB, et al. Immune activation by casein dietary antigens in bipolar disorder. Bipolar Disord. 2010;12:834–42.
34. Severance EG, Gressitt KL, Yang S, et al. Seroreactive marker for inflammatory bowel disease and associations with antibodies to dietary proteins in bipolar disorder. Bipolar Disord. 2014; 16:230–40.
35. Narayanan B, O'Neil K, Berwise C, et al. Resting state electroencephalogram oscillatory abnormalities in schizophrenia and psychotic bipolar patients and their relatives from the bipolar and schizophrenia network on intermediate phenotypes study. Biol Psychiatry. 2014; 76:456–65.
36. Cabranes JA, Ancín I, Santos JL, et al. P50 sensory gating is a trait marker of the bipolar spectrum. Eur Neuropsychopharmacol. 2013;23:721–7.
37. Olincy A, Martin L. Diminished suppression of the P50 auditory evoked potential in bipolar disorder subjects with a history of psychosis. Am J Psychiatry. 2005;162:43–9.

38. Sánchez-Morla EM, García-Jiménez MA, Barabash A, et al. P50 sensory gating deficit is a common marker of vulnerability to bipolar disorder and schizophrenia. Acta Psychiatr Scand. 2008;117:313–8.
39. Schulze KK, Hall MH, McDonald C, et al. P50 auditory evoked potential suppression in bipolar disorder patients with psychotic features and their unaffected relatives. Biol Psychiatry. 2007; 62:121–8.
40. Carroll CA, Kieffaber PD, Vohs JL, et al. Contributions of spectral frequency analyses to the study of P50 ERP amplitude and suppression in bipolar disorder with or without a history of psychosis. Bipolar Disord. 2008;10:776–87.
41. Patterson JV, Sandman CA, Ring A, et al. An initial report of a new biological marker for bipolar disorder: P85 evoked brain potential. Bipolar Disord. 2009;11:596–609.
42. Reite M, Teale P, Rojas DC, et al. Bipolar disorder: anomalous brain asymmetry associated with psychosis. Am J Psychiatry. 1999;156:1159–63.
43. Patterson JV, Sandman CA, Jin Y, et al. Gating of a novel brain potential is associated with perceptual anomalies in bipolar disorder. Bipolar Disord. 2013;15:314–25.
44. Hamm JP, Ethridge LE, Shapiro JR, et al. Spatiotemporal and frequency domain analysis of auditory paired stimuli processing in schizophrenia and bipolar disorder with psychosis. Psychophysiology. 2012;49:522–30.
45. Mahon PB, Lee DS, Trinh H, et al. Morphometry of the amygdala in schizophrenia and psychotic bipolar disorder. Schizophr Res. 2015;164:199–202.
46. Nenadic I, Maitra R, Langbein K, et al. Brain structure in schizophrenia vs. psychotic bipolar I disorder: a VBM study. Schizophr Res. 2015;165:212–9.
47. Ratnanather JT, Poynton CB, Pisano DV, et al. Morphometry of superior temporal gyrus and planum temporale in schizophrenia and psychotic bipolar disorder. Schizophr Res. 2013;150: 476–83.
48. Ivleva EI, Bidesi AS, Keshavan MS, et al. Gray matter volume as an intermediate phenotype for psychosis: bipolar-schizophrenia network on intermediate phenotypes (B-SNIP). Am J Psychiatry. 2013;170:1285–96.
49. McDonald C, Bullmore E, Sham P, et al. Regional volume deviations of brain structure in schizophrenia and psychotic bipolar disorder: computational morphometry study. Br J Psychiatry. 2005;186:369–77.
50. Yüksel C, McCarthy J, Shinn A, et al. Gray matter volume in schizophrenia and bipolar disorder with psychotic features. Schizophr Res. 2012;138:177–82.
51. Radaelli D, Poletti S, Gorni I, et al. Neural correlates of delusion in bipolar depression. Psychiatry Res. 2014;221:1–5.
52. Tost H, Ruf M, Schmäl C, et al. Prefrontal-temporal gray matter deficits in bipolar disorder patients with persecutory delusions. J Affect Disord. 2010;120:54–61.
53. Edmiston EE, Wang F, Kalmar JH, et al. Lateral ventricle volume and psychotic features in adolescents and adults with bipolar disorder. Psychiatry Res. 2011;194:400–2.
54. Strasser HC, Lilyestrom J, Ashby ER, et al. Hippocampal and ventricular volumes in psychotic and nonpsychotic bipolar patients compared with schizophrenia patients and community control subjects: a pilot study. Biol Psychiatry. 2005;57:633–9.
55. Daros AR, Ruocco AC, Reilly JL, et al. Facial emotion recognition in first-episode schizophrenia and bipolar disorder with psychosis. Schizophr Res. 2014;153:32–7.
56. Ruocco AC, Reilly JL, Rubin LH, et al. Emotion recognition deficits in schizophrenia-spectrum disorders and psychotic bipolar disorder: findings from the bipolar-schizophrenia network on intermediate phenotypes (B-SNIP) study. Schizophr Res. 2014;158:105–12.
57. Thaler NS, Allen DN, Sutton GP, et al. Differential impairment of social cognition factors in bipolar disorder with and without psychotic features and schizophrenia. J Psychiatr Res. 2013; 47:2004–10.
58. Aminoff SR, Hellvin T, Laerber TV, et al. Neurocognitive features in subgroups of bipolar disorder. Bipolar Disord. 2013;15:272–83.

59. Frydecka D, Eissa AM, Hewedi DH, et al. Impairments of working memory in schizophrenia and bipolar disorder: the effect of history of psychotic symptoms and different aspects of cognitive task demands. Front Behav Neurosci. 2014;8:416.
60. Simonsen C, Sundet K, Vaskinn A, et al. Neurocognitive dysfunction in bipolar and schizophrenia spectrum disorders depends on history of psychosis rather than diagnostic group. Schizophr Bull. 2011;37:73–83.
61. Zanelli J, Reichenberg A, Morgan K, et al. Specific and generalized neuropsychological deficits: a comparison of patients with various first-episode psychosis presentations. Am J Psychiatry. 2010;167:78–85.
62. Ancín I, Cabranes JA, Santos JL, et al. Executive deficits: a continuum schizophrenia-bipolar disorder or specific to schizophrenia? J Psychiatr Res. 2013;47:1564–71.
63. Savitz J, van der Merwe L, Stein DJ, et al. Neuropsychological status of bipolar I disorder: impact of psychosis. Br J Psychiatry. 2009;194:243–51.
64. Selva G, Salazar J, Balanzá-Martínez V, et al. Bipolar I patients with and without a history of psychotic symptoms: do they differ in their cognitive functioning? J Psychiatr Res. 2007;41: 265–72.
65. Banala RR, Karnati PR. Vitamin a deficiency: an oxidative stress marker in sodium fluoride (NaF) induced oxidative damage in developing rat brain. Int J Dev Neurosci. 2015;47:298–303.
66. Gautam M, Agrawal M, Gautam M, et al. Role of antioxidants in generalized anxiety disorder and depression. Indian J Psychiatry. 2012;54:244–7.
67. Chowdhury MI, Hasan M, Islam MS, et al. Elevated serum MDA and depleted non-enzymatic antioxidants, macro-minerals and trace elements are associated with bipolar disorder. J Trace Elem Med Biol. 2017;39:162–8.
68. Tamminga CA, Ivleva EI, Keshavan MS, et al. Clinical phenotypes of psychosis in the bipolar-schizophrenia network on intermediate phenotypes (B-SNIP). Am J Psychiatry. 2013;170: 1263–74.
69. Altamura AC, Serati M, Albano A, et al. An epidemiologic and clinical overview of medical and psychopathological comorbidities in major psychoses. Eur Arch Psychiatry Clin Neurosci. 2011;261:489–508.

Bipolar Disorder and Borderline Personality Disorder

5

A. Fiorentini, L. Cremaschi, and C. Prunas

Abstract

Bipolar disorder (BD) and borderline personality disorder (BPD) are two disabling psychiatric conditions that are associated with a high rate of comorbidity and mortality (approximately 10% of BPD patients suffer from BD I disorder and another 10% show a comorbid BD II disorder). Up to date, the relationship between BD and BPD is controversial, and the debate on whether or not BPD should be considered within a broader bipolar spectrum is still ongoing. According to the literature, the underestimation of comorbidity rates between these disorders could lead to a significant delay in diagnosis and planning of appropriate treatments. In addition, this would lead to an increased risk of exposure to improper medications, a poorer outcome and a higher risk of complications. In this chapter we present a clinical case of a patient who suffers from BD and BPD. The discussion of this case report draws attention to the challenges associated with the diagnosis and treatment of a wide population of psychiatric patients, whose symptoms and clinical history appear to be somehow overlapping between BD and BPD, considered as distinct disorders according to the current nosography.

Keywords

Bipolar disorder · Borderline personality disorder · Comorbidity · Differential diagnosis

A. Fiorentini (✉) · L. Cremaschi · C. Prunas
Department of Neurosciences and Mental Health, Fondazione IRCCS Ca' Granda Ospedale Maggiore Policlinico, University of Milan, Milan, Italy

A. C. Altamura, P. Brambilla (eds.), *Clinical Cases in Psychiatry: Integrating Translational Neuroscience Approaches*,
https://doi.org/10.1007/978-3-319-91557-9_5

5.1 Introduction

Bipolar disorder (BD) is a disabling mood disorder with a chronic recurrent course, associated with a high rate of comorbidity and mortality [1]. It is characterized by unusual shifts in mood, energy, activity levels, and in the ability to carry out day-to-day tasks. According to DSM-5, the diagnosis of BD type I disorder requires criteria for a manic episode to be met, while for BD type II disorder, criteria for a hypomanic episode and a major depressive episode should be fulfilled.

Borderline personality disorder (BPD), with a lifetime prevalence of around 3% [2], is one of the most frequently diagnosed personality disorders, affecting up to 20% of all psychiatric inpatients and thus representing a relevant socioeconomic burden [3]. A personality disorder is characterized by an enduring and inflexible pattern of behavior, thoughts, and feelings that seriously affect the person's psychosocial functioning. Regarding BPD, its DSM-5 diagnostic criteria include a "significant impairment in self (identity or self-direction) and interpersonal functioning (empathy or intimacy)" as well as pathological personality traits in the following domains: negative affectivity (emotional lability, anxiousness, separation insecurity, depressivity), disinhibition (impulsivity and risk taking), and antagonism (hostility).

To date, the relationship between BD and BPD is controversial. The debate on whether or not BPD should be inscribed within the **bipolar spectrum** (hypomania, cyclothymia, switching on antidepressants, atypical depression and mixed states) is still ongoing, drawing heterogeneous conclusions. In fact, some authors raise the question on whether BPD should be considered an affective disorder or a personality disorder [4, 5]. For instance, Akiskal and his group included BPD within the bipolar spectrum. If considered—on the contrary—as separate disorders, the differential diagnosis, especially in the case of BD II, may not be straightforward. Discriminating between BD and BPD is often complicated [6, 7], but becomes crucial, especially in order to evaluate the comorbidity rates between the two disorders [8], as their co-occurrence significantly affects patients' outcome.

The growing interest in this topic may be explained by the **high comorbidity rates** of these two conditions. For instance, it has been estimated that approximately 10% of BPD patients also suffer from BD I disorder and another 10% show a comorbid BD II disorder. Nevertheless, in the vast majority of cases (80–90%), each disorder is diagnosed in the absence of the other. Among BP II patients, 20% received a diagnosis of BPD, whereas it is only 10% among BD I patients [8]. According to NESARC (National Epidemiologic Survey on Alcohol and Related Conditions) data, the lifetime prevalence of BPD is 29% among BD I patients and 24% among BD II ones. A recent meta-analysis by Fornaro and colleagues pointed out a 21.6% prevalence rate of BPD among 5273 BD subjects [9]. They established that 18.5% of people with BPD have a comorbid BD diagnosis and, interestingly, they found higher rates of comorbid BPD in BD II participants (37.7%) and in North American studies (26.2%). They also documented that male gender and increasing age are predictors of a lower prevalence of comorbid BPD in BD patients. This is in line with previous data reporting higher rates of female patients and younger mean age at onset in comorbid BD and BPD cases vs. BD samples without comorbid BPD [10, 11].

Some **clinical factors** are relevant when considering the **comorbidity** between BPD and BD (see Box 5.1). First of all, it is worth noting that BD patients with BPD have an earlier age at BD onset [10, 12–15] and, consequently, a longer duration of bipolar illness. Furthermore, they show higher suicidality rates compared with BD alone [15–20]. Interestingly, among BPD criteria, only impulsivity and unstable relationships were found to independently predict a past history of attempted suicide [21]. The enhanced suicidal risk associated with BPD was reported to be due to features either shared (impulsivity) or not shared (intense/unstable interpersonal relationship) with severe mood disorders [21]. Increased suicide risk was found only among females and among subjects with a genetic risk for suicide [22, 23], being also associated with high-lethality suicidal behavior [24, 25]. McDermid and coworkers reported a more frequent history of childhood trauma in BD with BPD compared to those without BPD [10]. Moreover, BD with comorbid BPD shows a higher rate of other comorbid psychiatric disorders, in particular a greater prevalence of substance abuse [10, 14, 15, 19, 26–28]. Regarding the clinical course, DB with BPD is related to an increased number of mood episodes over time and more major depressive episodes [10].

Box 5.1 Characteristics of BD in Comorbidity with BPD

- Earlier age of BD onset
- Higher suicidality rates and high-lethality suicidal behavior
- History of childhood trauma
- Higher rates of other comorbid psychiatric disorders
- Higher prevalence of substance abuse
- Increased number of mood episodes
- Increased number of major depressive episodes

The two disorders share some **core symptoms,** such as affective instability and impulsivity. Affective instability is defined by DSM-5 as "a marked reactivity of mood (e.g., intense episodic dysphoria, irritability or anxiety, usually lasting a few hours and only rarely more than a few days)". It represents a DSM-5 criterion for BPD, but not for BD, although being commonly observed also among these patients [29]. This is especially true when considering currently depressed BD type II cases [30], soft bipolar atypical forms of depressions [31], "ultrarapid" [32] and "stably unstable" bipolar cases [33]. It has also been proposed to include within the bipolar spectrum most of the forms of affective instability [34]. Impulsivity is another central feature of BPD, closely linked to mood lability. It is often seen as sexual impulsivity in both BPD and BD, although it can also be physical, aggressive, financial [7], or binge eating-related [35, 36]. Nevertheless, impulsivity in DB patients is more episodic, while in BPD it is more pervasive [37] (see Table 5.1).

Table 5.1 Clinical features of BPD and BD

	BPD	DB
Age of onset	Early adulthood	Between 20 and 25
Epidemiology	3% of general population, 10% of outpatients, 20% of inpatients	1–2% lifetime risk
Positive family history		×
Symptoms		
Effort to avoid real or imagined abandonment	×	
Pattern of unstable and intense interpersonal relationships	×	As complication of poor control of the disorder
Identity disturbance	×	× (especially in cyclothymic disorder)
Impulsivity	×	× (mania and mixed features)
Recurrent suicidal behavior	×	× (during acute episode)
Affective instability	×	× (during acute episode)
Chronic feelings of emptiness	×	
Intense anger, difficulties in controlling anger	×	× (episodic, during mania/ mixed features)
Transient stress-related paranoid ideation or severe dissociative symptoms	×	× (during psychotic episodes)
Past or present history of manic episodes		× (BD I)
Hypomanic episodes		× (BD II)
Grandiosity		×
Depressive episodes	× (chronic low mood)	× (as distinct periods of low mood)

Taken all together, it is clear that the underestimation of BD comorbidity rates in BPD cases could lead to a significant delay in diagnosis and planning of an appropriate treatment. In addition, this would lead to an increased risk of exposure to improper medications, a poorer outcome, and a higher risk of complications.

5.2 Case Presentation

Mrs. C., a 29-year-old woman, first came to our clinical attention in 2015, when she was admitted to the Psychiatry Department for depressive symptoms in BPD, including depressed mood, anhedonia, anergy, sleep disturbance, suicidal thoughts, and hyporexia. These symptoms worsened in 3 weeks prior to the visit, being responsible for a significant functional impairment.

For instance, in the examination room, she reported that her mood was low, she no longer enjoyed social gatherings and spent most of the day lying in her bed, with a significant tiredness and hopelessness. When asked about suicidal thoughts, she first denied, but she then confessed to have been thinking about hurting herself, so that "people would understand her suffering." She lost 3 kg in the previous 3 weeks from hyporexia.

From the physical point of view, she did not report any particular disease and she did not take any regular medication. In relation to past psychiatric history, she reported three past episodes similar to the present one, but affirmed that she usually feels "sad and empty". She reported one previous hospitalization for depression. She confessed to having recurrent self-mutilating behaviors, consisting in superficial cutting of the forearms. When asked about manic or hypomanic symptoms, she could recognize two distinct periods, which lasted approximately 1 month, characterized by grandiosity, reduced need for sleep, increased goal-directed activities, and excessive involvement in dangerous situations. She tried many different antidepressants, with only partial benefits, as they made her irritable and interfered with her sleep.

Considering the abovementioned symptomatology and the high risk of self-injurious behaviors, she was then hospitalized in our Psychiatric Department. During hospitalization, in order to perform an appropriate differential diagnosis, blood (general routine and thyroid function) and urine (drug screening) exams were performed, showing no significant alterations. She underwent the Brief Assessment of Cognition in Schizophrenia (BACS), which did not show any significant alterations (see Table 5.2). A cerebral magnetic resonance imaging (MRI) and fluorodeoxyglucose positron emission tomography (FDG-PET) were also collected and did not report any alteration (see Figs. 5.1 and 5.2).

Furthermore, she underwent a clinical assessment, based on the administration of structured diagnostic interviews (SCID-I and SCID-II) and psychometric scales. She fulfilled criteria for:

- Bipolar disorder type II, current depressive episode
- Borderline personality disorder
- Episodic alcohol abuse
- Anorexia nervosa, with purging behavior

Table 5.2 Assessment of cognition in schizophrenia (BACS)

Test	Normal score	Score	Comment
Language			
Verbal fluency	v.n. $\geq$ 31.68	54.25	Normal
Memory			
Verbal memory	v.n. $\geq$ 33.01	49.25	Normal
Motor proficiency			
Token task	v.n. $\geq$ 68.77	49.00	Deficit
Symbol-coding task	v.n. $\geq$ 40.49	49.25	Normal
Frontal proficiency			
Working memory	v.n. $\geq$ 14.93	21.00	Normal
Tower of London	v.n. $\geq$ 12.37	14.75	Normal

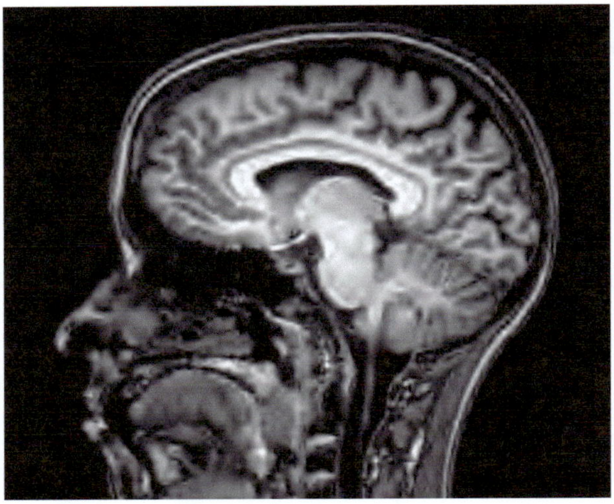

Fig. 5.1 Encephalic magnetic resonance imaging (MRI)

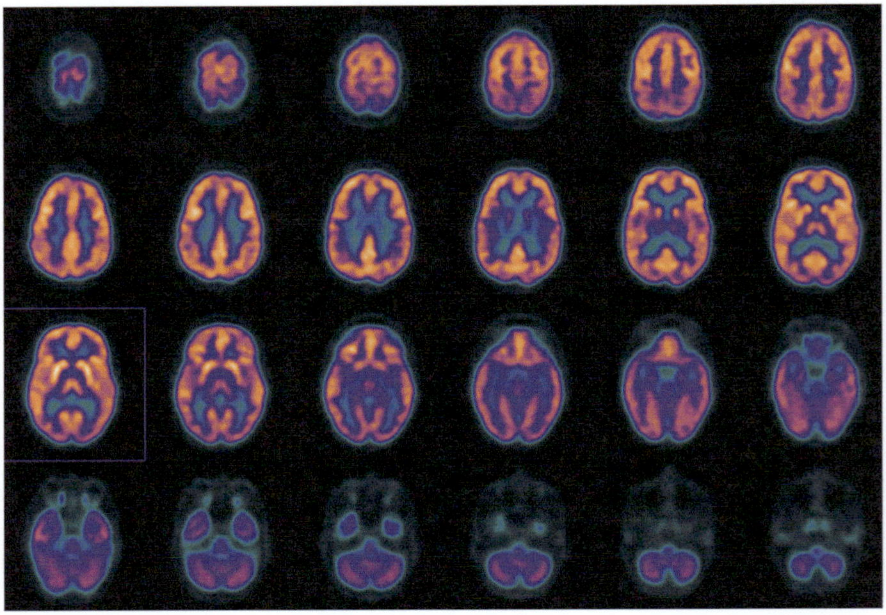

Fig. 5.2 Fluorodeoxyglucose positron emission tomography (FDG-PET)

Criteria for BD II and BPD are presented in Boxes 5.2 and 5.3, respectively.

Box 5.2 DSM-5 Criteria for BD II

- Criteria have been met for at least one hypomanic episode and at least one major depressive episode.
- There has never been a manic episode.
- The occurrence of the hypomanic episode(s) and major depressive episode (s) is not better explained by schizoaffective disorder, schizophrenia, schizophreniform disorder, delusional disorder, or other specified or unspecified schizophrenia spectrum and other psychotic disorders.
- The symptoms of depression or the unpredictability caused by frequent alteration between periods of depression and hypomania causes clinically significant distress or impairment in social, occupational, or other important areas of functioning.

Box 5.3 DSM-5 Criteria for BPD

A pervasive pattern of instability of interpersonal relationships, self-image, and affects and marked impulsivity beginning by early adulthood and present in a variety of contexts, as indicated by >5 (five) of the following:

1. Frantic efforts to avoid real or imagined abandonment.
2. A pattern of unstable and intense interpersonal relationships characterized by alternating between extremes of idealization and devaluation.
3. Identity disturbance: markedly and persistently unstable self-image or sense of self.
4. Impulsivity in at least two areas that are potentially self-damaging (e.g., excessive spending, substances of abuse, sex, reckless driving, binge eating). Note: Do not include suicidal or self-mutilating behavior covered in Criterion 5.
5. Recurrent suicidal behavior, gestures, or threats, or self-mutilating behavior.
6. Affective instability due to a marked reactivity of mood (e.g., intense episodic dysphoria, irritability, or anxiety usually lasting a few hours and only rarely more than a few days).
7. Chronic feelings of emptiness.
8. Inappropriate, intense anger, or difficulty controlling anger (e.g., frequent displays of temper tantrums, constant anger, and recurring fights).
9. Transient, stress-related paranoid ideation or severe dissociative symptoms.

The patient was discharged after 3 weeks on a pharmacological treatment with aripiprazole, lamotrigine, vortioxetine, and quetiapine, and addressed to our day hospital service for clinical stabilization.

Clinical History

- Family history for suicide (maternal grandfather) and generalized anxiety disorder (mother)
- No history of complications of pregnancy or delivery
- No history of neurodevelopmental disorders
- Good scholar and social functioning
- No other medical conditions

The patient reported that her symptoms first appeared at the age of 12. At that time, her parents got divorced, and she literally stopped eating, losing 10 kilos in a 6-month period and reaching a BMI of 15. In approximately 1 year's time, she recovered without taking any specific treatment but continued having restricting eating behaviors during the following years, although not fulfilling the criteria for anorexia nervosa.

The BD onset is collocated at 17, when she experienced her first depressive episode. This was characterized by depressed mood, reduced interest in most of the usual activities, weight loss, feeling of worthlessness, and difficulties in concentrating. She did not seek any medical attention, and the episode remitted approximately within 1 month. She got her high school diploma and was then admitted to the university (economics course).

When she was 19, she experienced her second depressive episode and, at this point, she received her first pharmacological treatment with antidepressants. Such episode resolved over a 2-month period, making her able to return to her university studies. One year later, she interrupted the psychopharmacological treatment. At the age of 22, during a period of study abroad in Germany, the patient experienced her first hypomanic episode, characterized by a persistently elevated mood, grandiosity, decreased need for sleep, and increase in goal-directed and potentially dangerous activities. She did not seek any kind of help and the episode spontaneously remitted in few weeks.

She came back to Italy and, at 24 years old, she had her first hospitalization in a psychiatric ward for a third depressive episode, complicated with a suicide attempt (she cut her forearm). During the hospitalization, hypomanic symptoms were not recognized by the clinicians. Hence, she was dismissed with a diagnosis of recurrent major depressive disorder and an antidepressant therapy. She also fulfilled the criteria for borderline personality disorder. In fact, she showed a pervasive pattern of instability of interpersonal relationships, self-image and affects, and marked impulsivity. She often reported unstable and intense interpersonal relationships, characterized by alternating between extremes of idealization and devaluation; recurrent suicidal behavior, gestures, or threats and self-mutilating behavior; affective instability due to a marked reactivity of mood; and chronic feeling of emptiness and inappropriate and intense anger.

After the first hospitalization, the patient did not fulfill the criteria for either a depressive or a hypomanic phase. She experienced frequent mood swings, alterations in her eating behavior, episodic alcohol abuse, and recurrent self-mutilating behaviors, together with a pattern of instability concerning her interpersonal relationships. Nevertheless, she was able to complete her university studies and start working as a researcher in the university where she graduated.

At the age of 26, the second episode of hypomania occurred. This was similar to the first one she experienced and, also in this case, she did not seek any medical help.

In 2015, when she was 27, she went through four different mood episodes, responsible for four hospitalizations: two major depressive episodes and two major depressive episodes with mixed features. The mixed episodes were characterized by depressive symptoms (e.g., depressed mood, psychomotor agitation, diminished ability to concentrate, recurrent thoughts of death) and hypomanic ones (e.g., decreased need for sleep, increased goal directed, and potentially dangerous activities). The psychopharmacological treatment was changed from aripiprazole, lamotrigine, duloxetine, and quetiapine to carbolithium, aripiprazole, and quetiapine. She also started a CBT-oriented psychotherapy.

With this combination of treatments, she has been asymptomatic since 2015. She was able to find a job and a stable relationship.

In conclusion, within an 11-year period since the first episode occurred, the patient experienced nine episodes of illness: five major depressive episodes, two hypomanic episodes, and two major depressive episodes with mixed features (see Fig. 5.3). The BD onset was at the age of 17; the first appropriate treatment was introduced at 27, with a duration of untreated illness of 10 years.

5.3 Literature Review

The above-described case represents an example of coexisting BD and BPD, with the latter being the first diagnosis and BD being diagnosed later on during the course of illness. This aspect may be related to the overlapping symptomatology that frequently delays the time of an appropriate diagnosis.

By reviewing the existing literature on BD and BPD, Ghaemi and colleagues tried to identify similarities and differences between the two disorders in relation to some relevant nosological validators (Table 5.3). Although similar in terms of some variables such as mood lability and impulsivity, they seem to significantly differ with respect to others, including:

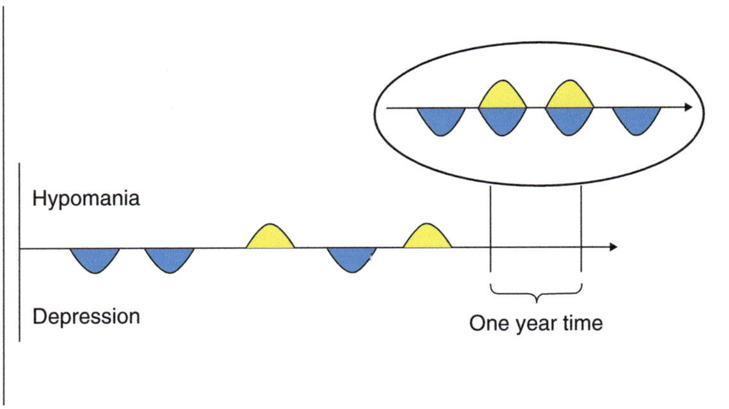

Fig. 5.3 Timeline of BD phases

- Past sexual abuse: higher prevalence in BPD vs. BD.
- Parasuicidal self-harm: twofold increased relative risk in BPD vs. BD.
- Genetics and neurobiology: BD almost completely genetic with several neurobiological abnormalities vs. BPD mostly environmental in causation with fewer neurobiological alterations. According to a recent genome-wide association study of BPD, a genetic overlap with BD, major depressive disorder, and schizophrenia emerged, suggesting the existence of an etiological overlap with the major psychoses [38].
- Treatment response: in BD the approach is based on an appropriate psychopharmacological treatment, with psychotherapies being adjunctive, whereas BPD requires psychotherapies as central to its management, with psychotropics being adjunctive [7] (Table 5.3).

Considering the frequent co-occurrence of BD and BPD, which, as shown in our case, may highly impact the course of illness and its outcome [39], the mainstay of the treatment approach in BPD is based on integrated psycho-/pharmacotherapy. Controlled treatment trials provide initial evidence on the efficacy of pharmacological interventions in the treatment of personality disorder signs and symptoms. A dimensional perspective is encouraged, in order to target symptoms' domains: for instance, BPD patients with affective instability or transient depression may benefit from the use of SSRIs, mood stabilizers, and/or atypical antipsychotics, the latter being useful also in the case of psychotic-like symptoms. SSRIs and mood stabilizers may help also in cases of impulsivity or aggression, whereas antidepressants in general are indicated in patients with anxiety.

Moreover, it is worth considering that several studies showed a higher risk of new-onset BD in BPD patients compared to subjects with other personality disorders, in particular 1.1–16.9 times for BD I and 2.1–9.5 times for BD II [10, 40–42].

Table 5.3 Diagnostic validators: bipolar illness vs. borderline personality

	Borderline personality	Bipolar illness
Symptoms	Recurrent para suicidal self-mutilating behaviour	Manic symptoms or episodes
Genetics	Strong environmental heritability	Very strong genetic heritability
Course	High prevalence of sexual abuse	Low prevalence of sexual abuse
Treatment	Psychotherapies necessary, medications adjunctive	Medications necessary, psychotherapies adjunctive
Neurobiology	Non specific	Amygdalar enlargement, hippocampal atrophy

In addition, the clinical history of the patient including various suicide attempts leads to some considerations about the enhanced risk of suicide attempts in cases of comorbidity between BD and BPD. The suicide risk in BD is among the highest within the psychiatric conditions. In fact, approximately one-fourth of BD I patients and one-fifth of BD II subjects attempt suicide in their lifetime [43], with a pooled suicide rate of 164 per 100,000 persons-years and accounting for 3.4–14% of all suicide deaths. The prevalence of attempted suicide in BD II and I does not appear to be significantly different, although BD II patients seem to use significantly more violent and lethal methods than those with BD I [44]. Depressive onset in BD has been associated with a greater rate of lifetime suicide attempts, with a 2.4-fold risk [45].

Moreover, according to the APA, up to 10% of individuals meeting criteria for BPD eventually commit suicide. Tsanas and colleagues reported suicide rates in BPD ranging between 6 and 8% as well as non-suicidal self-injurious behaviors in up to 90% of patients [46].

This prevalence rate may be even higher considering patients showing the two comorbid conditions. For instance, the comorbidity with cluster B/borderline personality disorder is among the variables significantly associated with suicide attempts in BD.

The present clinical case also sheds light on the issue regarding the prescription of antidepressants in BD patients, as in this patient their use probably induced an acceleration of cycling.

Most of the international guidelines in the treatment of BD (APA, WFSBP, CANMAT, NICE) discourage the prescription of antidepressant monotherapy in depressed patients, especially BD I, because of the enhanced risk of relapses and recurrences as well as suicide attempts, with a significant impact on clinical outcome [47, 48].

Some evidence supports their use in combination with an antimanic agent and for a limited period of time, within the resolution of the acute phase. They should be avoided in case of rapid cycling, previous cycle acceleration, or switching to manic phases during an antidepressant treatment with a concomitant mood stabilizer [49, 50].

Nevertheless, the risk/benefit ratio often leans toward the prescription of antidepressants, apparently widely used in BD [51–53], also during hypomanic/manic phases. Their prevalence rate seems to be higher in European countries [54] compared with American ones [55–59].

The use of antidepressants may be due, on one hand, to the misdiagnosis of BD as unipolar depression, reported in approximately 40% of cases [60] and responsible for a significant extension of the untreated illness period.

On the other hand, when BD is properly diagnosed, the antidepressant prescription may be related to the fact that depressive symptoms tend to last longer compared with hypomanic/manic ones [61, 62]. The administration of antidepressants may also be explained by the frequent comorbidity with anxiety disorders, even considering that anxious symptoms seem to last more than those of any other phase [63].

In literature, some authors reported that euthymic patients with residual symptoms, comorbid psychiatric disorders, and unsatisfactory global functioning may present an insufficient response to lithium [64], thus potentially explaining the tendency to choose antidepressants.

In addition, given that BD patients (especially II) are frequently less aware of elevation mood and other activation symptoms—often mild and difficult to recognize—they don't come to clinical attention and antidepressant treatment is continued beyond the necessary period.

5.4 Conclusions

The present case report draws attention to the challenges associated with the diagnosis and treatment of a wide population of psychiatric patients, whose symptoms and clinical history appear to be intermediate—or somehow overlapping—between BD and BPD, considered as distinct disorders according to the current nosography.

To date, evidence coming from the available literature does not allow us to resolve the debate regarding the nosological distinction between these two disorders. In particular, it's still uncertain if, on one hand, they should be considered as separate disorders that may, in some cases, simultaneously occur, or, on the other hand, they may take place within a continuum, either from an etiological or symptomatological point of view.

In the latter case, some features may point to the clinical picture as being intermediate, rather than related to distinctive phases of illness and, hence, to different disorders. The absence of clear manic lifetime episodes in our patient may support such a perspective, as the prevalence of mixed symptoms over the course of illness made the clinical presentation closer to the typical BPD picture. Furthermore, in her history, some of the intrinsic BPD characteristics, such as self-cutting, binge eating, substance abuse, and explosive anger, occurred almost exclusively during depressive or mixed phases. For instance, when the patient achieved an adequate mood stabilization, no such pathological personality features—by definition stable in BPD—emerged.

In support of the continuity between the two disorders, it is worth further underlining that recent genetic studies documented a significant overlap between these two conditions in pathogenetic terms. In our case, the familial anamnesis could not help us in potential investigation of the different genetic pools involved in the two disorders.

Finally, the present case report underlines some relevant therapeutic implications: antidepressant use, although hypothetically explained by the prevalence of depressive episodes, may have raised the emotional component, thus leading more easily to impulsive behaviors, more rapid cycles, and higher suicidality risk. The introduction of lithium—following the antidepressants' interruption—represented a decisive turning point in the clinical history of our patient. The achieved mood stability smoothed the pathological personality alterations, till they gradually disappeared. The psychotherapeutic intervention aimed to provide education about her everyday lifestyle, in particular to introduce changes in dietary and physical activity habits, to promote abstinence from substance and alcohol use, and to enhance the circadian rhythm integrity, during both depressive and hypomanic phases. This helped the patient to reduce her mood instability and related consequences on the behavioral levels. Among the other therapeutic focuses, the interpersonal relationships and self-esteem, which at first appeared to be significantly impaired, thus led to precipitate critical phases of illness, with repercussions on behavior and mood.

Key Points

- BD and BPD have high comorbidity rates.
- BD and BPD share some core symptoms, such as affective instability and impulsivity, and there is an ongoing debate on whether BPD should be considered within the bipolar spectrum.
- When BD occurs with BPD, it has particular clinical features that are important to consider, such as earlier age of onset, higher suicidality rates, high-lethality suicidal behavior, and increased number of mood episodes.
- The underestimation of comorbidity rates could lead to a significant delay in diagnosis and planning of appropriate treatments.
- Antidepressant use may raise the emotional component of BPD patients, thus leading more easily to impulsive behaviors, more rapid cycles, and higher suicidality risk.

Self-Assessment Questionnaire

1. What is the estimated prevalence of BPD among BD subjects?
   (A) 10% of BD II and 20% of BD I patients
   (B) 20% of BD II and 10% of BD I patients
   (C) 5% of BD II and I patients
   (D) 0.1% of BD II and I patients
2. Which of the following is a core symptom of both BD and BPD?

(A) Depressed mood
(B) Hypomanic symptoms
(C) Affective instability
(D) Anxiety

3. Which condition could be a risk factor for rapid cycling?
   (A) Alcohol abuse
   (B) History of childhood trauma
   (C) Antidepressant treatment
   (D) History of suicide attempts
4. Compared to BD without BPD, BD in comorbidity with BPD has a higher risk of:
   (A) Heroin abuse
   (B) PTSD
   (C) Suicide attempts
   (D) Agoraphobia
5. What is the most efficacious treatment approach for BD?
   (A) Antipsychotics
   (B) Mood stabilizers
   (C) Psychotherapy
   (D) Antidepressants

References

1. Grande I, Berk M, Birmaher B, Vieta E. Bipolar disorder. Lancet. 2016;387(10027):1561–72.
2. Tomko RL, Trull TJ, Wood PK, Sher KJ. Characteristics of borderline personality disorder in a community sample: comorbidity, treatment utilization, and general functioning. J Personal Disord. 2014;28:734–50.
3. Lieb K, Zanarini MC, Schmahl C, et al. Borderline personality disorder. Lancet. 2004;364:453–61.
4. Galione J, Zimmerman M. A comparison of depressed patients with and without borderline personality disorder: implications for interpreting studies of the validity of the bipolar spectrum. J Personal Disord. 2010;24:76372.
5. Parker G. Is borderline personality disorder a mood disorder? Br J Psychiatry. 2014;204:252–3.
6. Barroilhet S, Vohringer PA, Ghaemi SN. Borderline versus bipolar: differences matter. Acta Psychiatr Scand. 2013;128:385–6.
7. Ghaemi SN, Dalley S, Catania C, Barroilhet S. Bipolar OR borderline: a clinical overview. Acta Psychiatr Scand. 2014;130(2):99–108.
8. Zimmerman M, Morgan TA. Problematic boundaries in the diagnosis of bipolar disorder: the interface with borderline personality disorder. Curr Psychiatry Rep. 2013;15:422.
9. Fornaro M, Orsolini L, Marini S, De Berardi D, et al. The prevalence and predictors of bipolar and borderline personality disorders comorbidity: systematic review and meta-analysis. J Affect Disord. 2016;195:105–18.
10. McDermid J, Sareen J, El-Gabalawy R, et al. Comorbidity of bipolar disorder and borderline personality disorder: findings from the National Epidemiologic Survey on alcohol and related conditions. Compr Psychiatry. 2015;58:18–28.
11. Sajatovic M, Blow FC, Ignacio RV. Psychiatric comorbidity in older adults with bipolar disorder. Int J Geriatr Psychiatry. 2006;21:582–7.
12. Goldberg JF, Garno JL. Age at onset of bipolar disorder and risk for comorbid borderline personality disorder. Bipolar Disord. 2009;11:205–8.

13. Moor S, Crowe M, Luty S, et al. Effects of comorbidity and early age of onset in young people with bipolar disorder on self-harming behavior and suicide attempts. J Affect Disord. 2012;2012(136):1212–5.
14. Neves FS, Malloy-Diniz LF, Correa H. Suicidal behavior in bipolar disorder: what is the influence of psychiatric comorbidities? J Clin Psychiatry. 2009;70:13–8.
15. Perugi G, Angst J, Azorin JM, BRIDGE Study Group, et al. The bipolar-borderline personality disorders connection in major depressive patients. Acta Psychiatr Scand. 2013;128:376–83.
16. Carballo JJ, Harkavy-Friedman J, Burke AK, et al. Family history of suicidal behavior and early traumatic experiences: additive effect on suicidality and course of bipolar illness? J Affect Disord. 2008;109:57–63.
17. Carpiniello B, Lai L, Pirarba S, et al. Impulsivity and aggressiveness in bipolar disorder with co-morbid borderline personality disorder. Psychiatry Res. 2011;188:40–4.
18. Garno JL, Goldberg JF, Ramirez PM, Ritzler BA. Bipolar disorder with comorbid cluster B personality disorder features: impact on suicidality. J Clin Psychiatry. 2005;66:339–45.
19. Swann AC, Lijffijt M, Lane SD, Steinberg JL, Moeller FG. Antisocial personality disorder and borderline symptoms are differentially related to impulsivity and course of illness in bipolar disorder. J Affect Disord. 2013;148:384–90.
20. Yen S, Frazier E, Hower H, et al. Borderline personality disorder in transition age youth with bipolar disorder. Acta Psychiatr Scand. 2015;132(4):270–80. https://doi.org/10.1111/acps.12415
21. Zeng R, Cohen LJ, Tanis T, et al. Assessing the contribution of borderline personality disorder and features to suicide risk in psychiatric inpatients with bipolar disorder, major depression and schizoaffective disorder. Psychiatry Res. 2015;226:361–7.
22. Neves FS, Malloy-Diniz LF, Romano-Silva MA, et al. Is the serotonin transporter polymorphism (5-HTTLPR) a potential marker for suicidal behavior in bipolar disorder patients? J Affect Disord. 2010;125:98–102.
23. Oquendo MA, Bongiovi-Garcia ME, Galfalvy H, et al. Sex differences in clinical predictors of suicidal acts after major depression: a prospective study. Am J Psychiatry. 2007;164:134–41.
24. Joyce PR, Light KJ, Rowe SL, Cloninger CR, Kennedy MA. Self-mutilation and suicide attempts: relationships to bipolar disorder, borderline personality disorder, temperament and character. Aust NZ J Psychiatry. 2010;44:250–7.
25. Wilson ST, Stanley B, Oquendo MA, et al. Comparing impulsiveness, hostility, and depression in borderline personality disorder and bipolar II disorder. J Clin Psychiatry. 2007;68:1533–9.
26. Hidalgo-Mazzei D, Walsh E, Rosenstein L, Zimmerman M. Comorbid bipolar disorder and borderline personality disorder and substance use disorder. J Nerv Ment Dis. 2015;203:54–7.
27. Perroud N, Cordera P, Zimmermann J, et al. Comorbidity between attention deficit hyperactivity disorder (ADHD) and bipolar disorder in a specialized mood disorders outpatient clinic. J Affect Disord. 2014;168:161–6.
28. Preston GA, Marchant BK, Reimherr FW, et al. Borderline personality disorder in patients with bipolar disorder and response to lamotrigine. J Affect Disord. 2004;79:297–303.
29. Goodwin FK, Jamison KR. Manic-depressive illness: bipolar disorders and recurrent depression. Oxford: Oxford University Press; 2007.
30. Henry C, Mitropoulou V, New AS, et al. Affective instability and impulsivity in borderline personality and bipolar II disorders: similarities and differences. J Psychiatr Res. 2001;35:307–12.
31. Perugi G, Fornaro M, Akiskal HS. Are atypical depression, borderline personality disorder and bipolar II disorder overlapping manifestations of a common cyclothymic diathesis? World Psychiatry. 2011;10:45–51.
32. Mackinnon DF, Pies R. Affective instability as rapid cycling: theoretical and clinical implications for borderline personality and bipolar spectrum disorders. Bipolar Disord. 2006;8:1–14.
33. Akiskal HS. The temperamental borders of affective disorders. Acta Psychiatr Scand Suppl. 1994;379:32–7.

34. Angst J, Gamma A. A new bipolar spectrum concept: a brief review. Bipolar Disord. 2002;4:11–4.
35. Nagata T, Yamada H, Teo AR, et al. Using the mood disorder questionnaire and bipolar spectrum diagnostic scale to detect bipolar disorder and borderline personality disorder among eating disorder patients. BMC Psychiatry. 2013;13:69.
36. Perugi G, Akiskal HS. The soft bipolar spectrum redefined: focus on the cyclothymic, anxious-sensitive, impulse-dyscontrol, and binge-eating connection in bipolar II and related conditions. Psychiatr Clin North Am. 2002;25:713–37.
37. Koenigsberg H. Affective instability: toward an integration of neuroscience and psychological perspectives. J Personal Disord. 2010;24:60–82.
38. Witt SH, et al. Genome-wide association study of borderline personality disorder reveals genetic overlap with bipolar disorder, major depression and schizophrenia. Transl Psychiatry. 2017;7(6):e1155.
39. Frias A, et al. Anxious adult attachment may mediate the relationship between childhood emotional abuse and borderline personality disorder. Personal Ment Health. 2016;10 (4):274–84.
40. Grant BF, et al. Sociodemographic and psychopathologic predictors of first incidence of DSM-IV substance use, mood and anxiety disorders: results from the wave 2 national epidemiologic survey on alcohol and related conditions. Mol Psychiatry. 2009;14:1051–66.
41. Gunderson JG, et al. Interactions of borderline personality disorder and mood disorders over 10 years. J Clin Psychiatry. 2006;75:829–34.
42. Zanarini MC, Frankenburg FR, Vujanovic AA, Hennen J, Reich DB, Silk KR. Axis II comorbidity of borderline personality disorder: description of a 6-year course and prediction to time-to-remission. Acta Psychiatr Scand. 2004;110:416–20.
43. Merikangas KR, et al. Prevalence and correlates of bipolar spectrum disorder in the world mental health survey initiative. Arch Gen Psychiatry. 2011;68(3):241–51.
44. Novick DM, et al. Suicide attempts in BD I and II disorder: a review and meta-analysis of the evidence. Bipolar Disord. 2010;12(1):1–9.
45. Cremaschi L, et al. Onset polarity in BD: a strong association between first depressive episode and suicide attempts. J Affect Disord. 2017;209:182–7.
46. Tsanas A, Saunders KE, Bilderbeck AC, Palmius N, Osipov M, Clifford GD, et al. Daily longitudinal self-monitoring of mood variability in bipolar disorder and borderline personality disorder. J Affect Disord. 2016;205:225–33.
47. Altamura AC, Percudani M. The use of antidepressants for long-term treatment of recurrent depression: rationale, current methodologies, and future directions. J Clin Psychiatry. 1993;54 (Suppl):29–37. discussion 38; Review
48. Ghaemi SN, Hsu DJ, Soldani F, Goodwin FK. Antidepressants in bipolar disorder: the case for caution. Bipolar Disord. 2003;5(6):421–33.
49. Nivoli AM, Colom F, Murru A, Pacchiarotti I, Castro-Loli P, González-Pinto A, Fountoulakis KN, Vieta E. New treatment guidelines for acute bipolar depression: a systematic review. J Affect Disord. 2011;129(1–3):14–26.
50. Post RM. Bipolar depression dilemma: continue antidepressants after remission or not? Curr Psychiatr Ther. 2004;3(7):40–9.
51. Grande I, de Arce R, Jiménez-Arriero MÁ, Lorenzo FG, Valverde JI, Balanzá-Martínez V, Zaragoza S, Cobaleda S, Vieta E, SIN-DEPRES Group. Patterns of pharmacological maintenance treatment in a community mental health services bipolar disorder cohort study (SIN-DEPRES). Int J Neuropsychopharmacol. 2013;16(3):513–23.
52. Kessing LV, Vradi E, Andersen PK. Nationwide and population-based prescription patterns in bipolar disorder. Bipolar Disord. 2016;18(2):174–82.
53. Post RM, Leverich GS, Kupka R, Keck PE Jr, McElroy SL, Altshuler LL, Frye MA, Rowe M, Grunze H, Suppes T, Nolen WA. Clinical correlates of sustained response to individual drugs used in naturalistic treatment of patients with bipolar disorder. Compr Psychiatry. 2016;66:146–56.

54. Frangou S, Raymont V, Bettany D. The Maudsley bipolar disorder project. A survey of psychotropic prescribing patterns in bipolar I disorder. Bipolar Disord. 2002;4(6):378–85.
55. Jarema M. Atypical antipsychotics in the treatment of mood disorders. Curr Opin Psychiatry. 2007;20(1):23–29 [J Psychopharmacol 26(5):603–17].
56. Depp C, Ojeda VD, Mastin W, Unützer J, Gilmer TP. Trends in use of antipsychotics and mood stabilizers among Medicaid beneficiaries with bipolar disorder, 2001–2004. Psychiatr Serv. 2008;59(10):1169–74.
57. Ghaemi SN, Hsu DJ, Thase ME, Wisniewski SR, Nierenberg AA, Miyahara S, Sachs G. Pharmacological treatment patterns at study entry for the first 500 STEP-BD participants. Psychiatr Serv. 2006;57(5):660–5.
58. Russo P, Smith MW, Dirani R, Namjoshi M, Tohen M. Pharmacotherapy patterns in the treatment of bipolar disorder. Bipolar Disord. 2002;4(6):366–77.
59. Baldessarini RJ, Leahy L, Arcona S, Gause D, Zhang W, Hennen J. Patterns of psychotropic drug prescription for U.S. patients with diagnoses of bipolar disorders. Psychiatr Serv. 2007;58 (1):85–91.
60. Ghaemi SN, Sachs GS, Chiou AM, Pandurangi AK, Goodwin K. Is bipolar disorder still underdiagnosed? Are antidepressants overutilized? J Affect Disord. 1999;52(1–3):135–44.
61. Undurraga J, Baldessarini RJ, Valentí M, Pacchiarotti I, Tondo L, Vázquez G, Vieta E. Bipolar depression: clinical correlates of receiving antidepressants. J Affect Disord. 2012;139(1):89–93.
62. Tondo L, Baldessarini RJ, Vázquez G, Lepri B, Visioli C. Clinical responses to antidepressants among 1036 acutely depressed patients with bipolar or unipolar major affective disorders. Acta Psychiatr Scand. 2013;127(5):355–64.
63. McIntyre RS, Rosenbluth M, Ramasubbu R, Bond DJ, Taylor VH, Beaulieu S, Schaffer A, Canadian Network for Mood and Anxiety Treatments (CANMAT) Task Force. Managing medical and psychiatric comorbidity in individuals with major depressive disorder and bipolar disorder. Ann Clin Psychiatry. 2012;24(2):163–9.
64. Bowden CL. Novel treatments for bipolar disorder. Expert Opin Investig Drugs. 2001;10 (4):661–71.

Pseudodementia: A Case Report on the Connection Between Dementia and Bipolar Spectrum Disorders

6

Valentina Ciappolino and Giulia Orsenigo

Abstract

Bipolar disorder (BD) is a major affective disorder marked by recurrent/cyclical episodes of mania/hypomania and depression, and it is considered one of the major causes of disability worldwide, having a detrimental effect on the cognitive, social, and occupational functioning of the individual. BD may have a late onset. The elderly can indeed present two different patterns of mixed affective-cognitive disturbances, with several overlapping features: depressive pseudodementia (symptoms of dementia in depression, reversible) and pseudodepressive dementia (symptoms of depression in dementia, irreversible). Interestingly, the similarities in several features between BD and dementia suggest a potential overlooked continuum between these disorders. Of note, this report describes the case of a 71-year-old male, presenting a quickly cognitive deficits onset and a change from his prior pattern of behavior. Up to date, there are no clear guidelines available to diagnose and treat late-onset bipolar illness presenting as pseudodementia; thus it is necessary to keep in mind the existence of this clinical presentation which requires a follow-up of symptoms over time.

Keywords

Bipolar disorder · Pseudodementia · Cognitive functions · Neuroimaging studies

V. Ciappolino (✉) · G. Orsenigo
Department of Neurosciences and Mental Health, Fondazione IRCCS Ca' Granda Ospedale Maggiore Policlinico, University of Milan, Milan, Italy

A. C. Altamura, P. Brambilla (eds.), *Clinical Cases in Psychiatry: Integrating Translational Neuroscience Approaches*,
https://doi.org/10.1007/978-3-319-91557-9_6

6.1 Introduction

6.1.1 Cognitive Functions

Cognitive functioning can be defined as "the ability of an individual to perform various mental activities most closely associated with learning and problem solving. Examples include verbal, spatial, psychomotor, and processing speed ability." These abilities are supported by specific neural networks, as evidenced by neuroimaging studies [1, 2]. Cognitive and practical competencies are the key to enhancing individual and community well-being and to living a life of purpose [3, 4].

COGNITIVE FUNCTIONS

ATTENTION	is the function that allows to isolate pertinent and relevant information about a problem to be resolved, considering infinite incoming information both from within and out of us, it is the ability to sustain concentration on a particular object, action, or thought, and ability to manage competing demands in our environment		
MEMORY	is the function that receives from the learning systems, it sorts, stores and recovers every sort of information; short-term / working memory (limited storage) and long-term memory (unlimited storage)		
PERCEPTION	is the function that allows recognition and interpretation of sensory stimuli (smell, touch, hearing, etc.)		
REASONING	is the function responsible for logic processes, from which the most important is the language		
ORIENTATION	to space, to time, to relationships with oneself and with others		
LANGUAGE	is a competence to manage logical and symbolic systems, allowing us to translate sounds into words and generate verbal output		
INTELLIGENCE	is a function that allows to solve complex problems		
MOTOR SKILLS	is the ability to mobilize our muscles and bodies, and ability to manipulate objects		
VISUAL AND SPATIAL PROCESSING	is the ability to process incoming visual stimuli, to understand spatial relationship between objects, and to visualize images and scenarios		
EXECUTIVE FUNCTIONS	are the abilities that enable goal-oriented behavior, such as the ability to plan, and execute a goal.	Flexibility	the capacity for quickly switching to the appropriate mental mode
		Theory of mind	insight into other people's inner world, their plans, their likes and dislikes
		Anticipation	prediction based on pattern recognition
		Problem-solving	defining the problem in the right way to then generate solutions and pick the right one
		Decision making	the ability to make decisions based on problem-solving, on incomplete information and on emotions (ours and others')
		Working Memory	the capacity to hold and manipulate information "on-line" in real time
		Emotional self-regulation	the ability to identify and manage one's own emotions for good performance
		Sequencing	the ability to break down complex actions into manageable units and prioritize them in the right order
		Inhibition	the ability to withstand distraction, and internal urges

As mentioned above, if these cognitive functions are impaired, they have a major negative impact on psychosocial functioning, as they are responsible for it [5]. Therefore it is important to quickly identify the causes of these possible cognitive deficits. In particular it's necessary to distinguish between cognitive deficits associated to mood disorders, which are often reversible, or cognitive deficits linked to dementias (e.g., Alzheimer's disease), which are not reversible, so that it is possible to intervene quickly and limit negative consequences.

6.1.2 Bipolar Disorder and Dementia

Bipolar disorder (BD) is a major affective disorder marked by recurrent/cyclical episodes of mania/hypomania and depression, as described in the *Diagnostic and Statistical Manual of Mental Disorders*, 5th edition [6], and it is considered one of the major causes of disability worldwide [7]. The subtypes of BD include bipolar disorder I (BD-I) and bipolar disorder II (BD-II). Patients with BD-I experience at least one manic episode, which can be preceded or followed by hypomanic or major depressive episodes, whereas BD-II is marked by at least one hypomanic episode, at least one major depressive episode, and the absence of manic episodes. To satisfy a clinical diagnosis of BD, the abnormal mood episodes should also have a detrimental effect on the social and occupational functioning of the individual, as pointed in DSM 5. Late-life BD or geriatric BD [8–10] usually refers to patients older than 60 years who receive a diagnosis of BD. However, some authorities use an age cut-off of 50, 55, or 65 years. BD has an estimated lifetime prevalence of 0.6% for type I, 0.4% for type II, and 1.4% for subthreshold type [11, 12].

Several cross-sectional studies have suggested the presence of cognitive dysfunction in bipolar disorder also during euthymic states [13–17], related to chronicity or to a longer duration of illness, especially if untreated, or to a high number of episodes [18] that seem to have a negative impact on cognitive function, regardless of the acute phase of illness [15, 19–24]. Emerging scientific data suggest that BD is not only an illness of mood but that it affects multiple domains impacting overall functioning, considering it to be a multisystem condition in which cognitive impairment is one of the most important aspects. In fact we found in the recent literature several commonalities in pathophysiological processes of BD and dementia [25].

Dementia is an acquired and persistent impairment of intellectual faculties, which occurs in the absence of an altered level of consciousness [26], affecting several cognitive domains (memory, executive functioning, language comprehension) and which is sufficiently severe to impair competence in daily living, occupation, or social interaction [27, 28]. It is a major public health concern and is becoming increasingly burdensome as the population ages [29, 30]; dementia contributes more than any other disorder to disability and has a major financial and psychosocial impact on patients and their family, as well as on society in general [30, 31]. The incidence of dementia has doubled with every 5.9 year increase in age, from 3.1/1000 person years (pyr) at age 60–64 to 175/1000 pyr at age 95+. The total number of new cases of dementia each year worldwide was then estimated to be nearly 7.7 million, implying one new case every 4.1 seconds [32–34].

Bipolar disorder and dementia in the elderly can easily overlap, often in comorbidity. In these patients, depressive symptoms can be secondary to an underlying psycho-organic process, configuring a sort of depressive or maniacal "affective coverage" of mental deterioration syndromes [35].

While the association between major depression and both cognitive decline and dementia has been proposed and investigated for many years [36], that between bipolar disorder and dementia has been less investigated and is still uncertain

[37]. However, the risk of developing depressive and manic episodes seems to increase after a patient is diagnosed with dementia [25], particularly in the form of frontotemporal dementia (FTD) [38–40].

For many years bipolar disorders and dementia have been regarded as distinct and unrelated clinical entities. In fact, in contrast to unipolar depression, which can conceivably start at any age, bipolar spectrum disorders, for their onset, which is usually premature, have been neglected in the differential diagnosis of dementia. Nevertheless, despite the fact that new onset of bipolar disorder becomes less common with age, some studies have suggested that a range of 6–8% of patients with bipolar illness may have a late-life onset and this may still constitute about 5–8% of all annual inpatient psychiatric admissions. Furthermore, bipolar disorder may account for as much as 20% of mood conditions in the elderly [10, 25].

Although the first episode of full-blown classic mania in the elderly is rather uncommon, it may be of paramount importance to rule out physical illness [10], mood instability, mixed states, and "atypical" depressions that may emerge from the sixth decade of life along with cognitive decline in previously healthy individuals. In a review on this topic, it has been suggested that the intersection of dementia and BD might give rise to a mixed–manic state, which has been considered as a putative late-onset bipolar spectrum disorder type VI [9, 41], beyond the *Akiskal and Pinto* [42] framework of type I (mania and depression), type II (cyclothymia and hypomania), type III (depression plus drug-induced hypomania), type IV (late-onset depression superimposed on hyperthymic temperament), and type V (cyclic mixed depressions). This presentation of bipolar illness seems to have a lower association with family history, and patients generally have a higher premorbid psychosocial functioning as compared to early-onset illness before 65 years. However, even euthymic elderly people with late-onset bipolar illness, when compared to age-matched controls, seemed to do worse in domains of cognition, including language and executive functioning [10].

Both disorders have a multidimensional nature, so that mood and cognition can be considered fundamental aspects both in bipolar disorder and dementia.

In fact, the elderly can present two different patterns of mixed affective-cognitive disturbances [43]:

- Depressive pseudodementia (symptoms of dementia in depression)
- Pseudodepressive dementia (symptoms of depression in dementia)

The first term refers to a primary mood disturbance accompanied by a cognitive disturbance, while the second refers to a primary dementing illness that is accompanied by a secondary major depression, which may be due to the early insight into their failing capacities and consequent reaction to this loss of abilities with depression [44, 45]. When severe, the first type has also been called **pseudodementia**. The creation of the term pseudodementia (Pseudodemenz) is usually credited to Wernicke [46]. However, the exact circumstances and the debates that accompanied the emergence of this term have never been conclusive, and the references are not accurate. A Greek physician of the first century after Christ, Rufus of Ephesus, had already spoken of a kind of *melancholic pseudodementia*; after him another physician in Parma, Pompeo Sacco (1634–1718), spoke instead of *melancholy with pseudodementia*; lastly Ganser (1898) dealt with the parody of dementia in detainees awaiting execution. It was indeed Leslie Gordon Kiloh, in a study of 1961, who argued that the concept of pseudodementia does not belong to any nosological system, is purely descriptive, and does not imply an accurate diagnosis. Initially the term was introduced, as mentioned above, into German psychiatry: firstly, when Stertz (1910) described pseudodementia as a psychogenic illness that could occur after any trauma; secondly, Bonhoeffer's approach based on psychogenic reactions (1911); and lastly, the views of Alzheimer and his student Schuppius whose understanding was open to psychoanalytic concepts (1914) [46, 47].

6.1.3 Bipolar Disorder and Dementia: Differential Diagnosis

Several overlapping features make it very difficult to differentiate between bipolar disorder and dementia [20, 22], demonstrating the subtleties in the diagnosis.

Symptoms of bipolar disorder that may present as symptoms of a dementia process
Decreased need for sleep or a disrupted sleep process
Increased distractibility and impaired attention span
Irritability or aggression
Psychomotor agitation or retardation
Paranoia and other psychotic symptoms such as auditory hallucinations
Inappropriate guilt (which may be delusional)
Fatigue or loss of energy
Significant weight loss
Racing thoughts/ impaired thought process
Diminished ability to think or concentrate
Sadness
Diminished interest and pleasure in activities

DEMENTIA	PSEUDODEMENTIA
Insidious onset	Sudden onset
Slow progression	Rapid progression
Lack of insight	Presence of insight
Confabulations	Memory disorders
Tendency to diminish disability	Tendency to emphasize disability
Behavior corresponding to the severity of illness	Behavior often not corresponding to the severity of illness
Lack of answers	General responses (i.e. "I do not know")
Worsening at night	No night changes
Incongruity of affect	Depressed mood
Few vegetative symptoms	Frequent vegetative symptoms
Infrequent psychiatric history	Frequent psychiatric history
Low suicide risk	High suicide risk

Clinical studies [44, 48, 49] have sought to identify symptoms and signs that could be used in the differential diagnosis of reversible and irreversible dementia. However, the clinical presentation of reversible dementia is variable and depends on the age of the patient, the underlying psychiatric disorder, and the setting in which the patient is treated. Early observations suggested that patients with reversible dementia express excessive complaints and distress about their cognitive loss and often claim that they are unable to find the correct answers during a mental status examination, whereas those with dementing illnesses usually make an attempt, and a "near-miss" answer is common. Furthermore, the depressive syndrome of elderly patients with reversible dementia is usually severe. These patients suffer from more intense motor retardation, hopelessness, helplessness, and anxiety than cognitively unimpaired elderly depressed patients and are more likely to demonstrate delusions. The dementia syndrome, on the other hand, tends to be of mild severity and consists principally of impairment in attention, free recall, motor speed, spontaneous elaboration of detail, word fluency, and syntactic complexity. Despite these observations, clinical examination alone does not permit reliable identification of depressed patients in whom the dementia syndrome will subside after effective antidepressant treatment.

Cognitive Symptoms in Pseudodementia
Deficits in short and long term memory
Deficits in concentration, distraction
Absence of disturbance of orientation
Absence of confabulation, aphasia, agnosia
Identification of symptoms beginning
Rapid progression
Whiny behavior
Insight

Similarly, biological studies, including the blood level of progranulin [50], the presence of C9ORF72, dexamethasone suppression test, investigation of platelet monoamine oxidase activity, brain imaging, and EEG and sleep EEG studies, do not appear to help in the differential diagnosis of reversible dementia [49].

Behavioral and psychological symptoms like agitation, overactivity without agitation, aggression, apathy, affective lability, dysphoria, euphoria, disinhibition, impaired self-regulation, suicidal behavior, psychosis, dealing with money, spending needlessly, donating money to strangers, inappropriate erotic behavior and moral transgressions, and lack of insight are symptoms which in the elderly can be attributed to cognitive impairment rather than to the effect of altered mood on reasoning [25, 28, 51]. In fact, more often than not, affective and related behavioral disturbances in the dementia population are either considered as behavioral complications or attributed to depression and treated with antidepressants. It has been hypothesized that these symptoms may be secondary to a neurodegenerative process but might also be an exacerbation from previous affective temperament. The major difference, however, is that pseudodementia is a reversible cognitive impairment [49]. Pseudodementia is, in fact, a cognitive and functional impairment mimicking a neurodegenerative disease but believed to be caused by neuropsychiatric symptoms and, in contrast to primary neurodegenerative disease, potentially reversible with management of psychiatric symptoms [44, 52].

6.1.4 Other Causes of "Reversible Dementia"

There are other causes of reversible dementia that need to be considered and excluded before making a diagnosis of pseudodementia [29, 33, 48, 53].

NEUROCOGNITIVE DISORDERS				
Neurocognitive disorders –> DSM5 criteria (APA, 2013)				
Alzheimer's disease				
Most frequent form				
Vascular abnormalities	Lewy body dementia	Frontotemporal dementia	Parkinson's disease	Huntington's disease
≈ 5%				
Causes of "Reversible Dementia" Syndrome				
1. All causes of subacute delirium (especially drugs, metabolic disturbance and infection)				
2. Chronic metabolic disturbances (hypo- and hyperthyroidism, hypo- and hyperparathyroidism, vitamin B12 and possibly folate deficiency, hypo- and hypernatriemia, hypercalcemia, metabolic encephalopathy caused by hepatic or renal failure)				
3. Chronic infectious diseases (Whipple's disease, neurosyphilis, HIV, neuroborreliosis)				
4. Disorders of rheologic nature (plasmocytoma, coagulopathies, polycythemia)				
5. Structural abnormalities (tumor, subdural hematoma, normal pressure hydrocephalus)				
6. Profound depression ("pseudo-dementia") - 0.9-5%				
How to differentiate these different forms?				
Basic laboratory work-up that should include a complete blood count, measurement of electrolytes, serum calcium, renal and liver function, serum glutamic pyruvic transaminase, serum B-12, measurements of thyroid function (TSH), Venereal Disease Research Laboratory (VDRL)		EEG	CT scan, magnetic resonance imaging (MRI) and PET (Positron Emission Tomography) of the brain, that are recommended if there is a short or atypical history, or where focal neurological symptoms or signs are present	

6.1.5 Prognosis of Pseudodementia

The differential prognosis of reversible dementias, particularly depressive pseudodementia, requires a timely diagnosis and adequate treatment of depression, which can represent the difference between recovery and the persistence of symptoms [43]. The diagnosis and treatment of this type of depression can be even more complex if it is part of a bipolar disorder rather than a unipolar depression [25].

However, it is to be considered that cognitive impairment is frequently part of the depressive syndrome and of bipolar disorder, with significant cross-diagnostic overlap, but also cognitive-specific alterations within diagnostic categories, often related to demographic and cognitive variables rather than diagnosis [54]. In any case, often, this cognitive impairment, which is more severe and pervasive during relapses of illness, is also persistent in the euthymic state, especially in the elderly [13, 33, 45]. Nonetheless, only a few elderly bipolar or unipolar depressed patients develop severe cognitive dysfunctions, and one reason may be that these patients have an underlying dementing disorder. This is supported by the observation that many patients with reversible dementia do not achieve complete cognitive recovery even when their intellectual function improves following remission of depression [20]. Cognitive areas which seem to be more impaired during relapses of illness are attention, learning and memory, psychomotor functioning, and frontal executive functions, particularly working memory, inhibitory control, mental flexibility, and information processing speed [13, 16, 22, 23, 55–57]. It is unclear which of these cognitive dysfunctions remain in remission states. Some studies have suggested that verbal learning and memory, attention, and executive functions may also be impaired in euthymic patients, but results are still contradictory [20, 23].

The question arises as to whether a patient's depression may represent a prodrome of a later dementing process and depression is therefore a risk factor for dementia or whether depressive symptoms are an early sign of dementia pathology [19, 36].

Furthermore, a history of depression or bipolar disorder seems to increase the risk of a person developing dementia [19, 27, 36, 45, 58–60] and in some cases doubles it [45, 53]. In general, bipolar disorder is regarded as a more severe mood disorder than major depression [27], related to a higher risk of developing dementia, especially in middle age, [37] and to a greater number of prior affective episodes [19].

Moreover, in literature, some authors conclude that reversible cognitive impairment, in late-life moderate to severe depression, appears to be a strong predictor of later-onset dementia [49, 61, 62]. In fact, even when cognitive impairment improves with treatment of the depressive symptoms, patients are still at a substantially greater risk for dementia than demographically similar people without depression [36].

However, the relationship between depression or bipolar disorder and risk for later development of dementia is still unclear [60].

Some authors, in fact, have suggested that depression with reversible cognitive impairment may be an early phase of dementia rather than a risk factor for it. One hypothesis postulates that people with higher cognitive reserves are able to compensate enough at baseline so that their cognition stays marginal but out of the clinical arena. Any additional insult to their brain, however (such as a physical illness or depression), overwhelms their already strained cognitive reserves, and these people's cognition crosses the threshold of clinical impairment [10]. If the physical illness, depression, or other brain insult improves, these patients' cognitive abilities may return but not to baseline. Thus, brain insults such as depression obviate the patients' decreased cognitive reserves. Therefore, depressive symptoms can be the first signs of dementia or can complicate it. For this reason, some authors recommend that patients with depressive pseudodementia have a full dementia screening, comprehensive cognitive testing, and ongoing monitoring of their cognitive function [33, 49, 62].

In any case, when patients present changes from a prior pattern of behavior, clinicians need to consider promptly a possible causative neurodegenerative disease. They must remain alert to the possibility of dementia masquerading as another psychiatric condition, such as bipolar disorder or a psychotic disorder, but bizarre behaviors arising in late life should be thoroughly investigated as possible symptoms of dementia [28].

The similarities in several features between bipolar disorder and dementia suggest an overlooked continuum between these disorders [63].

6.1.5.1 Case Presentation

Family and physiological medical history
71-year-old Caucasian male
Firstborn of four children (one brother, two sisters)
Mother died at age 88 with Alzheimer's
Father died at age 72 for acute myocardial infarction; he was diabetic

(continued)

He lives with his wife and a son of 40 years old
Familiarity with mental illness (sister and nephew affected by affective disorders)
Eutocic delivery at the end of pregnancy, breastfeeding supplementing with infant formula, somatopsychic development reported in the standard
High school diploma (technical expert)
He worked till the age of 67 in the petrochemical industry (Russia, China, United Arab Emirates); currently retired
Smoking, about one packet of cigarettes/day; alcohol, only in social contexts
No known allergies

6.1.5.2 Pathological History

- Surgical intervention of left temporal craniotomy with carotid aneurysmatic clipping, following a trauma, in 1987, when the patient was 41 years old. Currently in therapy with Aggrenox
- Diabetes mellitus type 2 in therapy with Metformin 500 mg × 3/day
- Dyslipidemia in therapy with torvast 20 mg/day

6.1.5.3 Psychopathological History

He had no other psychiatric or substance abuse history, except for a "breakdown" about 30 years ago, during a working trip to Russia. He was described by his family and colleagues, just the week before departure, as very unsettled, easily irritable, and loquacious and with a short sleep duration. These traits are compatible with a *hyperthymic temperament*, the definition of which is characterized by four or more of the following attributes, which are not episode-bound and constitute part of the habitual long-term functioning of the individual [64]:

- Upbeat and exuberant
- Articulate and jocular
- Overoptimistic and carefree
- Overconfident and boastful
- High energy level, full of plans and improvident activities
- Versatile with broad interests
- Overinvolved and meddlesome
- Uninhibited and risk-taking
- Habitual short sleeper (less than 6 hours/night)

During the trip, lasting 1 week, he had to work many hours, almost without ever stopping, and on the third day, he had an anxiety attack, characterized principally by cardiorespiratory symptoms (chest discomfort, palpitations, and dyspnea). He had never had these symptoms before, so he was convinced it was a heart attack and consequently went to the emergency room of a hospital. After a reportedly normal cardiology workup, the patient received a benzodiazepine and then was discharged from the hospital with "take as needed" alprazolam.

6.1.5.4 Recent Psychopathological History

In recent months his family has described him as very confused, with fluctuating orientation levels. Cognitive deficits soon arose such as grossly disorganized speech and thinking and severe memory loss that had not occurred prior to a few weeks earlier. He was also emotionally unstable and frequently felt guilty toward his wife. Again, as in the past, he appeared very restless, easily irritable, talkative, and sleepless (only 2 to 3 hours a night). He also had paranoid symptoms with family members but mostly with strangers. Lastly he displayed marked impairment in his activities of daily living as well.

As a short test to assess cognitive function, the Mini-Mental State Examination (MMSE) was used, as a measure of global cognition. His MMSE score varied between 17 and 24 points (out of 30). The fluctuations seemed to be related to poor motivation to finish testing or being paranoid at times, rather than due to orientation in time, place, or person.

We decided to use neuropsychology testing to get a better picture of the patient's cognitive deficits. Results of the first testing showed a significant impairment in attention span, memory, language, frontal assessment battery, and executive function, suggesting a possible neurodegenerative condition such as dementia.

TEST	NORMAL SCORE	SCORE	COMMENT
GLOBAL COGNITIVE FUNCTION			
Mini Mental State Examination	v.n.>24	23.86/30	**Deficit**
Clock Drawing Test	v.n.>3	4/5	Normal
ATTENTION AND EXECUTIVE FUNCTION			
Attentive matrices	v.n.>31	16.75/60	**Deficit**
Trail Making Test			
(Part A)	v.n.<93	n.a.	n.a.
(Part B)	v.n.<282	n.a.	n.a.
(Part B-A)	v.n.<186	n.a.	n.a.
DEDUCTIVE REASONING			
Raven's Coloured Progressive Matrices	v.n.>18.60	15/36	**Deficit**
MEMORY			
Digit Span	v.n.>3.75	3.75	Low score, just above threshold
Short stories	v.n.>8.0	18	Normal
FRONTAL ASSESSMENT BATTERY			
Frontal assessment battery	v.n.>13.50	4.90	**Deficit**
Cognitive ability tests	v.n.<18.00	19.97	**Deficit**
Oddities	v.n.<0.4	5	**Deficit**
LANGUAGE			
Phonemic verbal fluency task	v.n.>17	5	**Deficit**
Categorical verbal fluency task	v.n.>25	14	**Deficit**

ADLs activities of daily living = 5/6.
IADL instrumental activities of daily living scale = 3/5.

He maintains a partial personal autonomy with difficulties in managing daily activities, so that he needs his wife's help.

In order to differentiate between a neurodegenerative and a psychiatric cause of this cognitive decline, we conducted several examinations through laboratory investigations and neuroimaging studies.

Routine chemistries were found to be within normal limits. A broad range of lab tests were ordered to rule out common and uncommon causes of cognitive impairments in the elderly such as HIV, neurosyphilis, vitamin B12 and folate deficiency, hyperparathyroidism, and neoplastic and paraneoplastic syndromes.

ECG: normal sinus rhythm, 74 beats per minute, and normal sinoatrial, atrioventricular, and intraventricular conduction, QTc 422 ms.

EEG obtained in awake and drowsy states was normal.

Cranial CT scan with contrast dye: prominent cerebral sulci and ventriculomegaly, bilateral ischemic lesions in frontal regions, and periventricular leukoaraiosis, likely of vascular degenerative significance. There are no apparent new alterations of parenchymal density at both supra- and subtentorial levels. No intra- or extra-axial mass injuries.

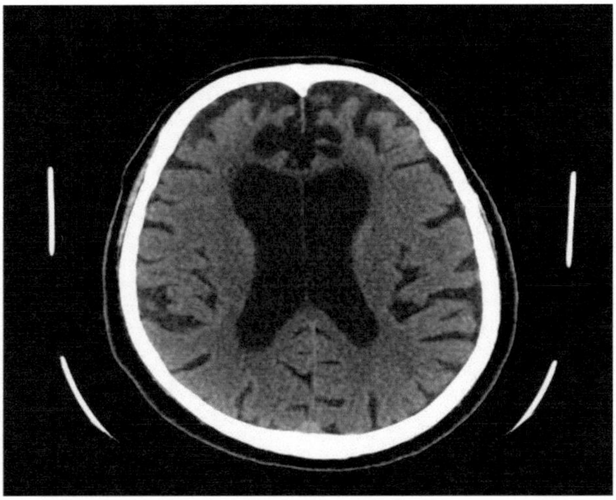

We did not carry out a **MRI**, despite the advantages over CT scan (ability to change the contrast of the images, giving higher detail and ability to change the imaging plane without moving the patient, producing images in any plane), because the patient had a clip, after the surgical intervention of left temporal craniotomy in 1987, that was not compatible with electromagnetic field of MRI.

PET (positron emission tomography): area of hypocaptation of radiopharmaceuticals in the medial frontal region, near the cortex of cingulate gyrus, referable to prior neurosurgical intervention. In spite of a moderate extension of cerebral sulci and lateral ventricles, the captation of radiopharmaceuticals appears standard on the cortex, supra- and subtentorial levels, basal ganglia, and thalamus.

Conclusions: non-suggestive presentation of neurodegenerative pathology like frontotemporal dementia or Alzheimer's disease.

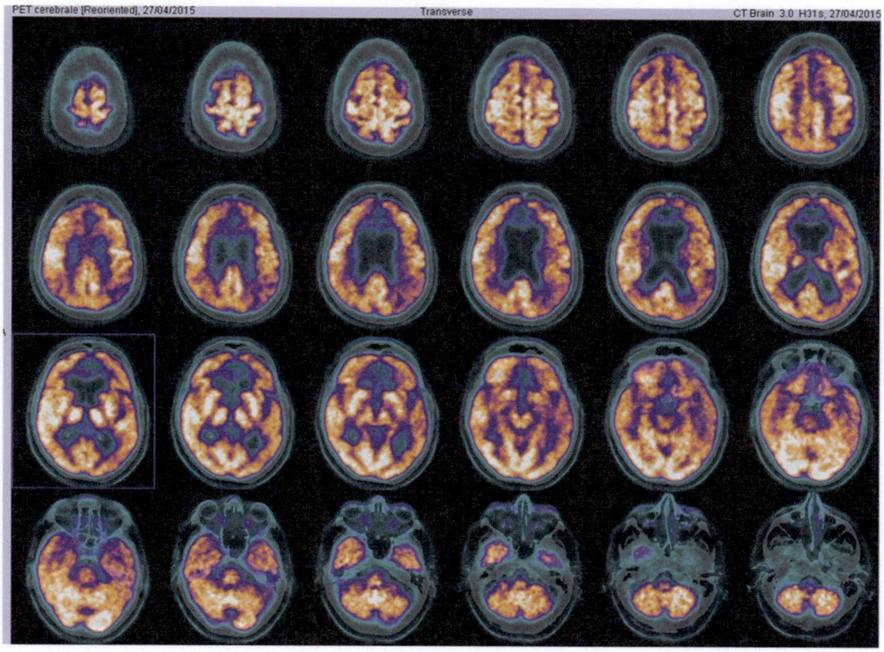

For the prevalence of persecutory delusions, he was initially treated with a trial of haloperidol up to 4 mg and olanzapine up to 10 mg for 2 weeks but developed extra pyramidal side effects (especially tremor) from these medications, so they were interrupted. For dysphoria with mood deflection associated with easy irritability and emotional lability, antipsychotic drugs were replaced with intravenous trazodone 25 mg associated with selegiline 5 mg and valproate 1000 mg. However, there was a worsening of the clinical picture, with increased tremor in the upper limbs, confusion, disorientation, difficulty in carrying out daily activities, difficulty in speaking and appearance of aphasia, deficit in short-term memory, psychomotor retardation with a slowing down of psychomotor movements, and mood deflection. The neurologist concluded for a drug-induced parkinsonism, probably due to valproate.

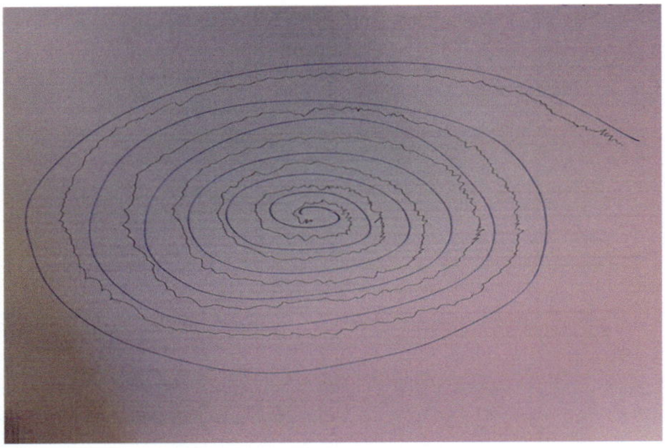

So we stopped all therapy, replacing it with sertraline 50 mg, with the dose gradually titrated up to 150 mg a day.

About 2 months later, there was a progressive remission of psychopathological symptoms and an improvement in cognition, first language attention and memory, and then frontal assessment battery and executive function. There is evidence suggesting that cognitive dysfunction is a primary sign in depression rather than an indirect behavioral consequence of the affective symptoms of depression. However, improvement particularly in memory appears to correlate with amelioration of depressive symptoms and signs [65].

In fact, repeat neuropsychological testing was obtained at the time of discharge and showed significant improvement compared with admission, with a remission of the dementia syndrome, which is defined as a clinical state that no longer meets the DSM-5 criteria for dementia and a Mini-Mental State score greater than 23.

TEST	NORMAL SCORE	SCORE	COMMENT
GLOBAL COGNITIVE FUNCTION			
Mini Mental State Examination	v.n.>23.8	28.86/30	Normal
ATTENTION AND EXECUTIVE FUNCTION			
Attentive matrices	v.n.>31	45.25/60	Normal
Trail Making Test			
(Part A)	v.n.<93	60	Normal
(Part B)	v.n.<282	88	Normal
DEDUCTIVE REASONING			
Raven's Coloured Progressive Matrices	v.n.>18.60	24.8/36	Normal
TAL ASSESSMENT BATTERY - WORKING MEMORY			
Frontal assessment battery	v.n.>11.59	15.43/18	Normal
Weigl's Test	v.n.>8.1	13.3/15	Normal
Digit Span Backward	v.n.>2.65	6.34	Normal
Digit Span	v.n.>4.26	6.27	Normal
Corsi Span	v.n.>3.46	6.24	Normal
ANTEROGRADE EPISODIC MEMORY			
Free and Cued Selective Reminding Test (imm.)	v.n.>19.6	28.27/36	Normal
Free and Cued Selective Reminding Test (diff.)	v.n.>6.7	9.45/12	Normal
VISUOSPATIAL DIMENSION			
Clock Drawing Test	v.n.>6	10/10	Normal
PRAXIA			
Constructional apraxia (REY complex Figure Test)	v.n.>28.8	29/36	Normal
LANGUAGE			
Phonemic verbal fluency task	v.n.>17.35	30.5	Normal
Categorical verbal fluency task	v.n.>25	52	Normal
Sartori's denomination test	v.n.>50.36	58/64	Normal

Antidepressant therapy was therefore suspended for subsequent anxiety elevation with sleep disturbances and replaced with quetiapine 200 mg a day, for stabilizing mood.

Fortunately, the patient had a complete remission of symptoms as his diagnosis turned out to be a lifelong undiagnosed bipolar illness (with possible hypomanic episodes in the past), or a possible late-life onset bipolar disorder, most recent episode mixed versus depressed with psychotic features, considering that psychiatric diagnoses after a traumatic brain injury, like that of our patient in 1987, may vary over time, with their prevalence declining at approximately 2 years post-injury and then gradually increasing in subsequent years [66].

Currently, the patient maintains a relatively good clinical stability, sometimes with mood variability, characterized by alterations in sleep pattern and increased activity (night shopping on the internet and sometimes complaints to online sales companies), when it is necessary to increase the therapy with quetiapine up to 400 mg; this alternates with periods of greatest apathy, treated with a low dose of sertraline, up to 50 mg. Moreover, he also maintains a good cognitive functioning, going to exhibitions, theaters, and reading many books.

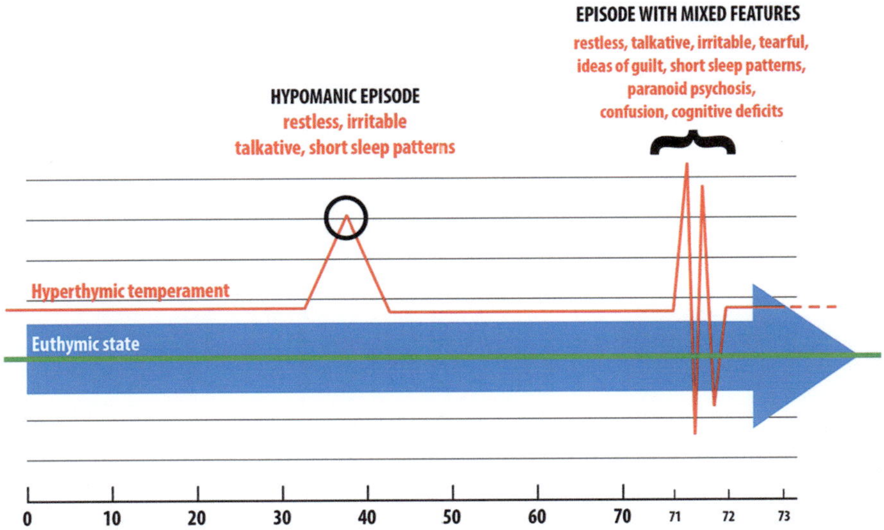

6.2 Discussion

This case highlights the phenomenon of late-life onset bipolar disorder presenting as pseudodementia.

With regard to the cognitive deficits, which in our case seemed to completely regress with the remission of acute symptoms, there are several studies that show how some cognitive alterations, particularly executive functions, encoding or consolidation of information, as well as retrieval and working memory, are impaired in bipolar disorder, which indicates that fronto-subcortical or mesolimbic circuitry and temporo-hippocampal structures are involved in this disease and may reflect underlying dysfunction in the structural or functional neuroanatomy of the prefrontal cortex [21].

Precisely disturbances in frontal lobe functions on neuropsychological testing support the abnormalities demonstrated with functional neuroimaging (PET, SPECT) in BD [67, 68]. These neuropsychological disturbances correlated with reduced blood flow in the mesial prefrontal cortex [67, 68]. Moreover, positron emission tomography (PET) studies showed reduced glucose metabolism in frontal and temporal regions, in the insula, in the basal ganglia, and in prefrontal areas of patients diagnosed with bipolar depression [49, 69, 70]. Even if more commonly PET imaging studies have reported resting prefrontal hypometabolism in medial, dorsolateral, and subgenual regions of the prefrontal cortex in patients with bipolar depression and mania too, some studies have revealed hypermetabolism in anterior cingulate, parahippocampal cortex, and medial temporal structures. However, most of these studies have reported regions of either hypo- or hypermetabolism, though one study captured simultaneous prefrontal hypometabolism and limbic hypermetabolism, which have been associated with cognitive deficits. In conclusion, abnormal cerebral blood flow patterns may be a trait marker of depression and mania in BD [70].

Regarding magnetic resonance imaging studies [27, 71], data showed a reduced volume in the frontotemporal regions, prefrontal cortex [16, 72], parietal lobe [73], and hippocampus [21, 71], which has been associated with the degree of cognitive impairment [74], so the severity of dementia increased in line with decreasing hippocampal volume. Reports on basal ganglia volumes in BD compared to healthy subjects are conflicting because some studies found an increased and others a decreased volume, particularly in the putamen and caudate [27, 49]. Moreover white matter abnormalities in BD patients have been reported to be widespread, especially in the inferior fronto-occipital fasciculus, in the cingulum, in the internal capsule, and in the posterior brain regions. Lastly, alterations have also been found at the level of connectivity, seen with resting-state functional MRI, between frontal and mesolimbic areas. In particular the medial prefrontal cortex is the major locus of shared abnormality, with bipolar disorder being characterized by reduced default-mode network connectivity to the hippocampus and fusiform gyrus as well as increased connectivity with the primary visual cortex [71].

Studies are needed to examine whether more sensitive assessment of neuropsychological dysfunction can reliably identify patients who proceed to develop irreversible dementia. Potential neuropsychological predictors may include disturbances in semantic encoding and retrieval, naming, visuospatial tasks, and "automatic" cognitive processes that do not require effort. Neuropsychological abnormalities in these areas have distinguished geriatric depressed patients from mildly demented patients or patients with bipolar disorder [27, 49].

Our case shows that the distinction between "organic" cognitive symptoms (thought of as symptoms due to structural or neurodegenerative, irreversible conditions) and "functional" symptoms (thought of as symptoms due to reversible conditions, usually psychiatric in nature) can be challenging when using our current diagnostic nosology.

The patient's cognitive presentation and neuroimaging results were unclear, with a generalized cerebral atrophy, more focused on the frontal area, possibly consistent with a neurodegenerative condition such as vascular dementia. However, the time-course of symptoms and cognitive and neuropsychological symptoms remission, as a result of treatment for affective symptoms, made us hypothesize that the patient's disordered thought process (i.e., due to a primary mood or psychotic disorder) might be more central to his presentation.

However, there are no clear guidelines available to treat this late-life onset bipolar illness presenting as pseudodementia. In addition to a thorough medical workup and psychiatric assessment, a firm understanding of baseline psychosocial functioning and timeline of events is very important.

It is reported that mood stabilizers in the form of anticonvulsants are better tolerated in the elderly with bipolar illness than lithium and that acute mania responds well to valproate and atypical antipsychotics [25, 75]. Lithium, however, may have neuroprotective abilities and may reduce the risk of developing dementia. It is difficult to decide whether lithium has this protective effect against dementia

[76], whether it is due to increased neurogenesis, to mood-stabilizing abilities that prevent recurrences of affective episodes, or to other treatment-related factors, such as the inhibition of glycogen synthase kinase 3, which is a key enzyme in the metabolism of amyloid precursor protein and in the phosphorylation of the tau protein involved in the pathogenesis of Alzheimer's disease. However, continued treatment with lithium suppresses the harmful effect on cognition associated with affective episodes and thus decreases the rate of developing dementia to the same rate as that among the general population [77].

In our case, besides affective symptoms, our patient also had paranoia, so we did not try a traditional mood stabilizer first but straight away tried an antipsychotic therapy.

As to the treatment of pseudodementia associated to a late onset of BD with only antidepressants and/or acetylcholinesterases, some authors have demonstrated that patients seem to worsen or aggravate symptoms by unmasking mania [25]. For this reason, treatment with these drugs in patients with pseudodementia should be closely monitored, with early detection of symptoms so as to prevent a maniacal shift [75].

There is a lack of literature about a treatment of this condition, and most of what is currently available is deduced from case series and mixed-age population studies [10].

In conclusion, it is not possible to make a clear distinction between "functional" and "organic" conditions based on our current diagnostic tests, since dementia and severe mood disorder have significant overlapping features even on MRI and neuropsychological testing, presenting similarly, based on the neural circuits they affect, and consequently the real diagnoses of most "organic" conditions are made only by autopsy. In this direction a detailed clinical history could be very useful in making a differential diagnosis. Then in the treatment of this condition, we have to consider that some types of drugs (antidepressants and acetylcholinesterases) can carry additional risks of paradoxical disinhibition and memory loss.

Lastly, at present, we can only make a follow-up of these patients, because if we cannot establish the progression of their symptoms over time, then we must carry out ongoing monitoring of their cognitive function for the next few years.

Key Points

- Bipolar disorder is not only an illness of mood but that it affects multiple domains impacting overall functioning.
- Cognitive impairment is one of the most important aspect of bipolar disorder.
- A late life onset of bipolar disorder could occur as pseudodementia even if a bipolar disorder and dementia, in the elderly, can easily overlap, often in comorbidity.
- Importance of differential diagnosis between dementia and pseudodementia because of several overlapping features between bipolar disorder and dementia.

- The major difference, however, is that pseudodementia is a reversible cognitive impairment.
- Pseudodementia is a cognitive and functional impairment mimicking a neurodegenerative disease potentially reversible with management of psychiatric symptoms.

Assessment Questionnaire

	A	B	C	D
COGNITIVE DYSFUNCTIONS ARE PRESENT BOTH IN DEMENTIA BUT ALSO IN BIPOLAR DISORDER, PARTICULARLY SEVERAL CROSS SELECTIONAL STUDIES SHOWED:	BD PATIENT PRESENTED COGNITIVE DYSFUNCTION ONLY IN DEPRESSIVE PHASE	BD PATIENT PRESENTED COGNITIVE DYSFUNCTION ONLY IN MANIC PHASE	BD PATIENT PRESENTED COGNITIVE DYSFUNCTION BOTH IN MANIC AND DEPRESSIVE PHASE	BD PATIENT PRESENTED COGNITIVE DYSFUNCTION NOT ONLY IN MANIC AND DEPRESSIVE PHASES BUT ALSO IN EUTHYMIC PHASE
MANY STUDIES DEMONSTRATED THAT BIPOLAR AND DEMENTIA CAN HAVING SEVERAL COMMONALITIES SUCH AS:	MANIC SYMPTHOMS	DEPRESSIVE SYMPTHOMS	COGNITIVE IMPAIRMENT	ALL OF THESE SYMPTHOMS
PSEUDODEPRESSIVE DEMENTIA CAN BE DEFINED AS:	A PRIMARY DEMENTING ILLNESS THAT IS ACCOMPAINED BY A SECONDARY MAJOR DEPRESSION	A PRIMARY MOOD DISTURBANCE ACCOMPAINED BY A SECONDARY COGNITIVE DISTURBANCE	A SUBTYPE OF DEMENTIA	NONE OF THESE DEFINITIONS
SEVERAL OVERLAPPING FEATURES MAKE VERY DIFFICULT TO DISTINGUISH BIPOLAR DISORDER FROM DEMENTIA, AMONG SYMPTHOMS THAT ARE DIFFERENT:	INSIGHT IN DEMENTIA IS PRESENT	PSEUDODEMENTIA HAS A SLOW PROGRESSION	IN PSEUDODEMENTIA PATIENTS TEND TO EMPHASIZE DISABILITY	ALL OF ANSWERS ARE CORRECT
FOR DIFFERENTIAL DIAGNOSIS BETWEEN REVERSIBLE OR NOT REVERSIBLE DEMENTIA WHICH OF THESE EXAMINATIONS ARE USEFUL BUT NOT FUNDAMENTAL:	EEG	BRAIN MRI	DEXAMETHASONE SUPPRESSION TEST	ALL OF ANSWERS ARE CORRECT

References

1. Fernández M, Mollinedo-Gajate I, Peñagarikano O. Neural circuits for social cognition: implications for autism. Neuroscience. 2017. https://doi.org/10.1016/j.neuroscience.2017.07.013.
2. Pulvermüller F. Neural reuse of action perception circuits for language, concepts and communication. Prog Neurobiol. 2018;160:1–44. https://doi.org/10.1016/j.pneurobio.2017.07.001.
3. Saraçlı Ö, Akca ASD, Atasoy N, et al. The relationship between quality of life and cognitive functions, anxiety and depression among hospitalized elderly patients. Clin Psychopharmacol Neurosci. 2015;13(2):194–200. https://doi.org/10.9758/cpn.2015.13.2.194.
4. L'esame BA, Dell'adulto N. Applicazioni Cliniche E Forensi. 2nd ed. Firenze: Giunti Organizzazioni Speciali ed; 2013.
5. Savva GM, Zaccai J, Matthews FE, et al. Prevalence, correlates and course of behavioral and psychological symptoms of dementia in the population. Br J Psychiatry. 2009;194(3):212–9. https://doi.org/10.1192/bjp.bp.108.049619.
6. American Psychiatric Association. Diagnostic and statistical manual of mental disorders (DSM–5). Washington: American Psychiatric Association; 2013.

7. Organization WH. The global burden of disease: 2004 update. Geneva: World Health Organization; 2008.
8. Sajatovic M, Chen P. Geriatric bipolar disorder. Psychiatr Clin North Am 2011;34(2):319–333, vii. doi:https://doi.org/10.1016/j.psc.2011.02.007.
9. Azorin J-M, Kaladjian A, Adida M, Fakra E. Late-onset bipolar illness: the geriatric bipolar type VI. CNS Neurosci Ther. 2012;18(3):208–13. https://doi.org/10.1111/j.1755-5949.2011.00255.x.
10. Banga A, Gyurmey T, Matuskey D, Connor DF, Kaplan RF, Steffens DC. Late-life onset bipolar disorder presenting as a case of pseudo-dementia: a case discussion and review of literature. Yale J Biol Med. 2013;86(2):235–44. http://www.ncbi.nlm.nih.gov/pubmed/23766743. Accessed 31 Aug 2017
11. Merikangas KR, Jin R, He J-P, et al. Prevalence and correlates of bipolar Spectrum disorder in the world mental health survey initiative. Arch Gen Psychiatry. 2011;68(3):241. https://doi.org/10.1001/archgenpsychiatry.2011.12.
12. Bortolato B, Köhler CA, Evangelou E, et al. Systematic assessment of environmental risk factors for bipolar disorder: an umbrella review of systematic reviews and meta-analyses. Bipolar Disord. 2017;19(2):84–96. https://doi.org/10.1111/bdi.12490.
13. Bearden CE, Hoffman KM, Cannon TD. The neuropsychology and neuroanatomy of bipolar affective disorder: a critical review. Bipolar Disord 2001;3(3):106–150 (discussion 151-3). http://www.ncbi.nlm.nih.gov/pubmed/11465675. Accessed 31 Aug 2017.
14. Murphy FC, Sahakian BJ. Neuropsychology of bipolar disorder. Br J Psychiatry Suppl. 2001;41:s120–7. http://www.ncbi.nlm.nih.gov/pubmed/11450171. Accessed 31 Aug 2017
15. Savitz J, Solms M, Ramesar R. Neuropsychological dysfunction in bipolar affective disorder: a critical opinion. Bipolar Disord. 2005;7(3):216–35. https://doi.org/10.1111/j.1399-5618.2005.00203.x.
16. Robinson LJ, Nicol Ferrier I. Evolution of cognitive impairment in bipolar disorder: a systematic review of cross-sectional evidence. Bipolar Disord. 2006;8(2):103–16. https://doi.org/10.1111/j.1399-5618.2006.00277.x.
17. Dell'Osso B, Cinnante C, Di Giorgio A, et al. Altered prefrontal cortex activity during working memory task in bipolar disorder: a functional magnetic resonance imaging study in euthymic bipolar I and II patients. J Affect Disord. 2015;184:116–22. https://doi.org/10.1016/j.jad.2015.05.026.
18. Buoli M, Caldiroli A, Caletti E, Zugno E, Altamura AC. The impact of mood episodes and duration of illness on cognition in bipolar disorder. Compr Psychiatry. 2014;55(7):1561–6. https://doi.org/10.1016/j.comppsych.2014.06.001.
19. Kessing LV, Andersen PK. Does the risk of developing dementia increase with the number of episodes in patients with depressive disorder and in patients with bipolar disorder? J Neurol Neurosurg Psychiatry. 2004;75(12):1662–6. https://doi.org/10.1136/jnnp.2003.031773.
20. Martínez-Arán A, Vieta E, Reinares M, et al. Cognitive function across manic or hypomanic, depressed, and euthymic states in bipolar disorder. Am J Psychiatry. 2004;161(2):262–70. https://doi.org/10.1176/appi.ajp.161.2.262.
21. Moorhead TWJ, McKirdy J, Sussmann JED, et al. Progressive gray matter loss in patients with bipolar disorder. Biol Psychiatry. 2007;62(8):894–900. https://doi.org/10.1016/j.biopsych.2007.03.005.
22. Depp CA, Savla GN, Moore DJ, et al. Short-term course of neuropsychological abilities in middle-aged and older adults with bipolar disorder. Bipolar Disord. 2008;10(6):684–90. https://doi.org/10.1111/j.1399-5618.2008.00601.x.
23. Gildengers AG, Mulsant BH, Begley A, et al. The longitudinal course of cognition in older adults with bipolar disorder. Bipolar Disord. 2009;11(7):744–52. https://doi.org/10.1111/j.1399-5618.2009.00739.x.
24. Rosa AR, Reinares M, Michalak EE, et al. Functional impairment and disability across mood states in bipolar disorder. Value Health. 2010;13(8):984–8. https://doi.org/10.1111/j.1524-4733.2010.00768.x.

25. Ng B, Camacho A, Lara DR, Brunstein MG, Pinto OC, Akiskal HS. A case series on the hypothesized connection between dementia and bipolar spectrum disorders: bipolar type VI? J Affect Disord. 2008;107(1–3):307–15. https://doi.org/10.1016/j.jad.2007.08.018.
26. Association AP. Diagnostic and statistical manual of mental disorders (DSM-III). 3rd ed. Washington: APA; 1980.
27. Chen M-H, Li C-T, Tsai C-F, et al. Risk of subsequent dementia among patients with bipolar disorder or major depression: a Nationwide longitudinal study in Taiwan. J Am Med Dir Assoc. 2015;16(6):504–8. https://doi.org/10.1016/j.jamda.2015.01.084.
28. Cipriani G, Lucetti C, Danti S, Ulivi M, Nuti A. Uncommon and/or bizarre features of dementia. Acta Neurol Belg. 2015;115(1):19–25. https://doi.org/10.1007/s13760-014-0306-y.
29. Freter S, Bergman H, Gold S, Chertkow H, Clarfield AM. Prevalence of potentially reversible dementias and actual reversibility in a memory clinic cohort. CMAJ. 1998;159(6):657–62. http://www.ncbi.nlm.nih.gov/pubmed/9780965. Accessed 31 Aug 2017
30. Sabat SR. Dementia in developing countries: a tidal wave on the horizon. Lancet (London, England). 2009;374(9704):1805–6. https://doi.org/10.1016/S0140-6736(09)62037-7.
31. Sousa RM, Ferri CP, Acosta D, et al. Contribution of chronic diseases to disability in elderly people in countries with low and middle incomes: a 10/66 dementia research group population-based survey. Lancet. 2009;374(9704):1821–30. https://doi.org/10.1016/S0140-6736(09)61829-8.
32. Prince M, Bryce R, Albanese E, Wimo A, Ribeiro W, Ferri CP. The global prevalence of dementia: a systematic review and metaanalysis. Alzheimers Dement. 2013;9(1):63–75.e2. https://doi.org/10.1016/j.jalz.2012.11.007.
33. Djukic M, Wedekind D, Franz A, Gremke M, Nau R. Frequency of dementia syndromes with a potentially treatable cause in geriatric in-patients: analysis of a 1-year interval. Eur Arch Psychiatry Clin Neurosci. 2015;265(5):429–38. https://doi.org/10.1007/s00406-015-0583-3.
34. Martin Prince A, Wimo A, Guerchet M, World Alzheimer Report, et al. The global impact of dementia an analysis of prevalence, incidence, cost and trends. London: Alzheimer's Disease International; 2015. https://www.alz.co.uk/research/WorldAlzheimerReport2015.pdf. Accessed 31 Aug 2017
35. Altamura AC. Fondamenti Di Psichiatria E Psicofarmacologia Clinica. Rome: Il Pensiero Scientifico ed; 2009.
36. Saczynski JS, Beiser A, Seshadri S, Auerbach S, Wolf PA, Au R. Depressive symptoms and risk of dementia: the Framingham heart study. Neurology. 2010;75(1):35–41. https://doi.org/10.1212/WNL.0b013e3181e62138.
37. Wu K-Y, Chang C-M, Liang H-Y, et al. Increased risk of developing dementia in patients with bipolar disorder: a nested matched case-control study. Bipolar Disord. 2013;15(7):787–94. https://doi.org/10.1111/bdi.12116.
38. Galimberti D, Dell'Osso B, Altamura AC, Scarpini E. Psychiatric symptoms in Frontotemporal dementia: epidemiology, phenotypes, and differential diagnosis. Biol Psychiatry. 2015;78(10): 684–92. https://doi.org/10.1016/j.biopsych.2015.03.028.
39. Papazacharias A, Lozupone M, Barulli MR, et al. Bipolar disorder and Frontotemporal dementia: an intriguing association. J Alzheimers Dis. 2016;55(3):973–9. https://doi.org/10.3233/JAD-160860.
40. Rubino E, Vacca A, Gallone S, et al. Late onset bipolar disorder and frontotemporal dementia with mutation in progranulin gene: a case report. Amyotroph Lateral Scler Front Degener. 2017:1–3. https://doi.org/10.1080/21678421.2017.1339716.
41. Akiskal HS, Pinto OC, Lara DR. "Bipolarity" in the setting of dementia: bipolar type VI. Medscape Fam Med/PrimCare. http//www.medscape.com/viewarticle/495795. Accessed 7 Apr 2005.
42. Akiskal HS, Pinto O. The evolving bipolar spectrum. Prototypes I, II, III, and IV. Psychiatr Clin North Am 1999;22(3):517–534, .vii. http://www.ncbi.nlm.nih.gov/pubmed/10550853. Accessed 31 Aug 2017.

43. Feinberg T, Goodman B. Affective illness, dementia, and pseudodementia. J Clin Psychiatry. 1984;45(3):99–103. http://www.ncbi.nlm.nih.gov/pubmed/6365898. Accessed 31 Aug 2017
44. Caine ED. Pseudodementia. Current concepts and future directions. Arch Gen Psychiatry. 1981;38(12):1359–64. http://www.ncbi.nlm.nih.gov/pubmed/7316680. Accessed 31 Aug 2017
45. Jorm AF. History of depression as a risk factor for dementia: an updated review. Aust NZ J Psychiatry. 2001;35(6):776–81. https://doi.org/10.1046/j.1440-1614.2001.00967.x.
46. Vinet-Couchevellou M, Sauvagnat F. Pseudodémence, de quoi parle-t-on? Partie I : à la recherche de la pseudodémence de Wernicke. Encéphale. 2015;41(2):130–6. https://doi.org/10.1016/j.encep.2013.10.005.
47. Vinet-Couchevellou M, Sauvagnat F. Pseudodémence, de quoi parle-t-on? Partie II : de Stertz à Alzheimer : une maladie psychogène après traumatisme. Encéphale. 2015;41:S37–43. https://doi.org/10.1016/j.encep.2014.08.015.
48. Patterson C. The diagnosis and differential diagnosis of dementia and pseudo-dementia in the elderly. Can Fam Physician. 1986;32:2607–10. http://www.ncbi.nlm.nih.gov/pubmed/20469450. Accessed 31 Aug 2017
49. Alexopoulos GS, Meyers BS, Young RC, Mattis S, Kakuma T. The course of geriatric depression with "reversible dementia": a controlled study. Am J Psychiatry. 1993;150(11):1693–9. https://doi.org/10.1176/ajp.150.11.1693.
50. Kittel-Schneider S, Weigl J, Volkert J, et al. Further evidence for plasma progranulin as a biomarker in bipolar disorder. J Affect Disord. 2014;157:87–91. https://doi.org/10.1016/j.jad.2014.01.006.
51. Preuss U, Wong J, Koller G. Treatment of behavioral and psychological symptoms of dementia: a systematic review. Psychiatr Pol. 2016;50(4):679–715. https://doi.org/10.12740/PP/64477.
52. Nussbaum PD. Pseudodementia: a slow death. Neuropsychol Rev. 1994;4(2):71–90. http://www.ncbi.nlm.nih.gov/pubmed/8061683. Accessed 31 Aug 2017
53. Muangpaisan W, Petcharat C, Srinonprasert V. Prevalence of potentially reversible conditions in dementia and mild cognitive impairment in a geriatric clinic. Geriatr Gerontol Int. 2012; 12(1):59–64. https://doi.org/10.1111/j.1447-0594.2011.00728.x.
54. Cotrena C, Damiani Branco L, Ponsoni A, Milman Shansis F, Paz Fonseca R. Neuropsychological clustering in bipolar and major depressive disorder. J Int Neuropsychol Soc. 2017; 23(7):584–93. https://doi.org/10.1017/S1355617717000418.
55. Martínez-Arán A, Penadés R, Vieta E, et al. Executive function in patients with remitted bipolar disorder and schizophrenia and its relationship with functional outcome. Psychother Psychosom. 71(1):39–46. http://www.ncbi.nlm.nih.gov/pubmed/11740167. Accessed 31 Aug 2017
56. Quraishi S, Frangou S. Neuropsychology of bipolar disorder: a review. J Affect Disord. 2002;72 (3):209–26. http://www.ncbi.nlm.nih.gov/pubmed/12450638. Accessed 31 Aug 2017
57. Caixeta L, Soares VLD, Vieira RT, et al. Executive function is selectively impaired in old age bipolar depression. Front Psychol. 2017;8:194. https://doi.org/10.3389/fpsyg.2017.00194.
58. Jorm AF. Is depression a risk factor for dementia or cognitive decline? A review. Gerontology. 46(4):219–27. https://doi.org/10.1159/000022163.
59. Kessing LV, Nilsson FM. Increased risk of developing dementia in patients with major affective disorders compared to patients with other medical illnesses. J Affect Disord. 2003;73(3):261–9. http://www.ncbi.nlm.nih.gov/pubmed/12547295. Accessed 31 Aug 2017
60. Ownby RL, Crocco E, Acevedo A, John V, Loewenstein D. Depression and risk for Alzheimer disease: systematic review, meta-analysis, and meta-regression analysis. Arch Gen Psychiatry. 2006;63(5):530–8. https://doi.org/10.1001/archpsyc.63.5.530.
61. Lamberty GJ, Bieliauskas LA. Distinguishing between depression and dementia in the elderly: a review of neuropsychological findings. Arch Clin Neuropsychol. 1993;8(2):149–70. http://www.ncbi.nlm.nih.gov/pubmed/14589672. Accessed 31 Aug 2017
62. Sáez-Fonseca JA, Lee L, Walker Z. Long-term outcome of depressive pseudodementia in the elderly. J Affect Disord. 2007;101(1–3):123–9. https://doi.org/10.1016/j.jad.2006.11.004.
63. Masouy A, Chopard G, Vandel P, et al. Bipolar disorder and dementia: where is the link? Psychogeriatrics. 2011;11(1):60–7. https://doi.org/10.1111/j.1479-8301.2010.00348.x.

64. Akiskal HS, Mallya G. Criteria for the "soft" bipolar spectrum: treatment implications. Psychopharmacol Bull. 1987;23(1):68–73. http://www.ncbi.nlm.nih.gov/pubmed/3602332. Accessed 14 Oct 2017
65. Sternberg DE, Jarvik ME. Memory functions in depression. Arch Gen Psychiatry. 1976;33(2): 219–24. http://www.ncbi.nlm.nih.gov/pubmed/1252098. Accessed 31 Aug 2017
66. Hibbard MR, Ashman TA, Spielman LA, Chun D, Charatz HJ, Melvin S. Relationship between depression and psychosocial functioning after traumatic brain injury. Arch Phys Med Rehabil. 2004;85(4 Suppl 2):S43–53. http://www.ncbi.nlm.nih.gov/pubmed/15083421. Accessed 7 Oct 2017
67. Bench CJ, Friston KJ, Brown RG, Frackowiak RS, Dolan RJ. Regional cerebral blood flow in depression measured by positron emission tomography: the relationship with clinical dimensions. Psychol Med. 1993;23(3):579–90. http://www.ncbi.nlm.nih.gov/pubmed/7901863. Accessed 31 Aug 2017
68. Dolan RJ, Bench CJ, Brown RG, Scott LC, Frackowiak RS. Neuropsychological dysfunction in depression: the relationship to regional cerebral blood flow. Psychol Med. 1994;24(4):849–57. http://www.ncbi.nlm.nih.gov/pubmed/7892353. Accessed 31 Aug 2017
69. Delvecchio G, Fossati P, Boyer P, et al. Common and distinct neural correlates of emotional processing in bipolar disorder and major depressive disorder: a voxel-based meta-analysis of functional magnetic resonance imaging studies. Eur Neuropsychopharmacol. 2012;22(2): 100–13. https://doi.org/10.1016/j.euroneuro.2011.07.003.
70. Brooks JO, Vizueta N. Diagnostic and clinical implications of functional neuroimaging in bipolar disorder. J Psychiatr Res. 2014;57:12–25. https://doi.org/10.1016/j.jpsychires.2014.05.018.
71. Birur B, Kraguljac NV, Shelton RC, Lahti AC. Brain structure, function, and neurochemistry in schizophrenia and bipolar disorder-a systematic review of the magnetic resonance neuroimaging literature. NPJ Schizophr. 2017;3(1):15. https://doi.org/10.1038/s41537-017-0013-9.
72. Brambilla P, Nicoletti MA, Harenski K, et al. Anatomical MRI study of subgenual prefrontal cortex in bipolar and unipolar subjects. Neuropsychopharmacology. 2002;27(5):792–9. https://doi.org/10.1016/S0893-133X(02)00352-4.
73. Ferro A, Bonivento C, Delvecchio G, et al. Longitudinal investigation of the parietal lobe anatomy in bipolar disorder and its association with general functioning. Psychiatry Res. 2017;267: 22–31. https://doi.org/10.1016/j.pscychresns.2017.06.010.
74. Dolek N, Saylisoy S, Ozbabalik D, Adapinar B. Comparison of hippocampal volume measured using magnetic resonance imaging in Alzheimer's disease, vascular dementia, mild cognitive impairment and pseudodementia. J Int Med Res. 2012;40(2):717–25. https://doi.org/10.1177/147323001204000236.
75. Shulman KI, Herrmann N. Bipolar disorder in old age. Can Fam Physician. 1999;45:1229–37. http://www.ncbi.nlm.nih.gov/pubmed/10349067. Accessed 31 Aug 2017
76. Kessing LV, Forman JL, Andersen PK. Does lithium protect against dementia? Bipolar Disord. 2010;12(1):87–94. https://doi.org/10.1111/j.1399-5618.2009.00788.x.
77. Kessing LV, Søndergård L, Forman JL, Andersen PK. Lithium treatment and risk of dementia. Arch Gen Psychiatry. 2008;65(11):1331–5. https://doi.org/10.1001/archpsyc.65.11.1331.

Cannabis-Induced Psychosis

7

Lucio Oldani and Benedetta Grancini

Abstract

The association between mental illness and drug abuse is complex and has been investigated by a wide scientific literature. Such comorbidity has a negative impact on both persistence and severity of illness. Nonetheless, a causal relationship still needs to be univocally defined, and the sole chronological criterion is not sufficient to determine a cause-effect relationship.

Among different illicit drugs, cannabis is the most commonly used in Europe, with almost 26.3% of adults using it in their lifetime. Although cannabis is generally regarded as a substance with low acute toxicity, its THC content and relative potency have increased over time, leading to a more frequent onset of psychotic pictures and an increased number of hospital admissions and of anxiety symptoms. As many experimental studies on healthy humans report, THC cannot only induce transient, dose-dependent psychotic symptoms but also affective, behavioral, cognitive, neurovegetative, and psychophysiological effects.

This chapter will present a case of a young adult with a history of cannabis abuse and who was admitted to the psychiatric ward for an acute psychotic symptomatology. His clinical picture was characterized by psychomotor agitation, aggressiveness, logorrhea, and a florid psychotic dimension. A toxicology screen on his urine resulted positive for cannabinoids. He underwent EKG, blood tests, magnetic resonance imaging, and positron emission tomography, all resulted in normal range.

L. Oldani (✉) · B. Grancini
Department of Neurosciences and Mental Health, Fondazione IRCCS Ca' Granda Ospedale Maggiore Policlinico, University of Milan, Milan, Italy
e-mail: lucio.oldani@policlinico.mi.it

A. C. Altamura, P. Brambilla (eds.), *Clinical Cases in Psychiatry: Integrating Translational Neuroscience Approaches*,
https://doi.org/10.1007/978-3-319-91557-9_7

In clinical cases similar to the current one, it seems crucial to distinguish between a primary psychiatric syndrome and a substance-induced disorder in order to establish an adequate acute treatment and follow-up.

Keywords
Cannabis-induced psychosis · Substance-induced psychosis · Substance abuse

7.1 Introduction

7.1.1 Epidemiology

According to the European Monitoring Centre for Drugs and Drug Addiction (EMCDDA), over 90 million adults (a quarter of the European population) have tried illicit drugs at least once in their lifetime, particularly during adolescence. Cannabis is the most commonly used illicit drug in Europe with almost 87.7 million Europeans (26.3% of adults aged 15–64 years) who have used it in their lifetime and 17.1 million (13.9% of adults) in the last year. Considering young subjects who used cannabis in the last year, the sex ratio (M/F) is 2:1 (Table 7.1). The cannabis product market has an estimated value of EUR 9.3 billion in Europe and represents the largest share of the illicit drug retail market (38%) [1].

Although cannabis is generally regarded as having low acute toxicity, the potency of both herbal and resin has increased since 2006, as well as the tetrahydrocannabinol (THC) content, with some evidence of increased availability and higher potency [2].

Specific cannabinoid compounds have distinct effects on mental health and brain function [3]. The psychoactive and addictive properties of cannabis are primarily due to THC [4], while cannabidiol is responsible for the anxiolytic effect [3]. Increased availability of cannabis varieties that are high in THC (e.g., "skunk") has consistently been linked to onset of psychosis [5], increased hospital admissions, and anxiety symptoms [6, 7]. Moreover, preclinical studies showed that THC is neurotoxic to brain areas rich in cannabinoid type 1 receptors, including the hippocampus [8], amygdala [9], striatum [10], and prefrontal cortex (PFC) [11, 12].

Table 7.1 Some numbers concerning European population using Cannabis

		Millions of users	Population percent (%)
Adults (15–64 years)	Last year	23.5	7
	Lifetime	87.7	26.3
Young adults (15–34 years)	Last year	17.1	13.9
Adolescents and early twenties (15–24 years)	Last year	10.0	17.7

Some numbers concerning European population using cannabis (adapted from European Drug Report 2017 [1])

Table 7.2 Facts

Around 1% of European adults have used cannabis on 20 days or more in the last month
Around 30% of such users are aged 35 to 64 years
Over three quarters are male
The number of subjects entering a treatment for cannabis-related issues has almost doubledin about 10 years (from 43,000 in 2006 to 76,000 in 2015, corresponding to an increase of (qualifying them as daily cannabis users) about 77%)

Adapted from European Drug Report 2017 [1]

7.1.2 Cannabis Abuse and Psychiatric Emergencies

The European Drug Emergencies Network (Euro-DEN) aims to collect data on acute harm related to recreational drugs through a network of 15 centers located in 11 European Union countries [13].

More in detail, 16.2% of the total presentations reported by Euro-DEN centers involved cannabis either alone or together with other drugs/alcohol [14]. Cannabis is the most common illicit drug across all ages; usually, it is smoked together with tobacco while rarely taken orally [1, 14].

Considering Euro-DEN records of presentation to the emergency department, the majority of users were male (76%), and the median age was 26 years. On mental status examination, the most common symptoms were agitation/aggression (22.9%), psychosis (20.0%), anxiety (20.0%), and vomiting (17.1%). The majority of them received no pharmacological treatment (71.4%) and were discharged/self-discharged from the ED (85.7%) [14] (Table 7.2).

7.1.3 Drug Abuse and Psychiatric Disorders

Wide scientific literature has investigated the strong relationship between mental disorders and substance abuse. The latter comorbidity has a negative impact on both persistence and severity of illnesses. Many literature studies have been focused on identifying possible causal effects between substance abuse and different mental disorders, and future discoveries might have an impact on refining substance use prevention and treatment strategies [15].

The association between mental illnesses and drug abuse is complex, and the chronological criterion itself is not sufficient to determine a cause-effect relationship. The coexistence of a psychiatric disorder and drug abuse might be explained by different reciprocal relationships, and it might be hard to discriminate between which one comes first, if so. More in detail, the development of a mental disorder and drug abuse may share common risk factors, including shared genetic vulnerability and environmental predisposition, as in the case of comorbidity. Alternatively, substance abuse can determine an acute psychotic state without a preexisting psychiatric condition, as in the case of substance-induced disorder (psychotic or not psychotic). Finally, drug abuse can occur within the course of an already existing psychiatric disorder, as in "self-medication hypothesis," and therefore precipitate the psychopathological picture and change its clinical features [16]. In those with an established psychotic disorder, ongoing cannabis use is associated with more severe psychotic symptoms, worse medication adherence, more frequent relapse and hospital admission, worse general functioning, and greater risk of serious aggression [17].

Therefore, with classic mental disorders being induced by both intoxication and withdrawal and chronic drug use, it is important to distinguish between a primary psychiatric syndrome and a substance-induced disorder in order to establish an adequate treatment and follow-up [18].

7.2 Case Presentation

7.2.1 Clinical History

Mr. B is a 22-year-old Caucasian man with no previous psychiatric contacts, who came to our attention for persecutory delusions and verbal and physical aggressiveness.

Mr. B was born by C section delivery because of fever during labor; he was breastfed and had a regular early life development. He had no significant illnesses in his medical history, excluding a past wrist fracture because of a car accident and an ankle fracture during sport activity. No allergy was reported, neither previous hospitalization nor surgery. Mr. B had a positive family history for psychiatric disorders, particularly his grandfather and his father's cousin affected by major depressive disorder, and his father's cousin was a drug abuser.

His parents outlined a good premorbid functioning and described Mr. B as a lively boy, with several friends, with a fair academic performance despite bad school conduct. He is described as a "negative leader," with good leadership skills but also intolerance toward society rules. At school, he was often nervous and restless, showing moderate disrespect toward authority and playing tricks on his teachers. His drug abuse started at the age of 14 and consisted mainly in cannabis use; he also occasionally took crack, cocaine, and hallucinogenic mushrooms during parties with friends.

At 14 years of age, he started a daily cannabis use, both alone and with friends, together with daily alcohol use. He also started growing marihuana plants on his balcony for self-consumption.

At 17 years of age, he underwent psychotherapy after being investigated for selling illicit drugs, but he quit after a few months due to lack of insight.

At 18 years of age, he returned home earlier from a vacation with friends, showing withdrawal from social life, apathy, indifference, and affective flattening, and he started to spend his entire day on the internet. In a few weeks, Mr. B returned to his usual life, finishing high school and starting to attend university. He did two different university courses, taking just a few exams and quitting for lack of interest. After these attempts, he started to work as a barman.

From 19 till 21 years, Mr. B quit the use of cannabis, starting again at the age of 21, after a holiday in Barcelona. On that occasion, he first experienced psychotic symptoms such as grandiose delusions ("I am God") and psychomotor agitation. The latter symptoms reoccurred on different occasions after taking illicit drugs, with spontaneous remission and without taking any psychotropic medication. He became more nervous and irritable over time; a few weeks before hospitalization, he broke several objects such as paintings at home, and some persecutory thoughts began to surface in his mind. During the latter crisis, he also experienced disorganized thoughts and ideas of reference.

The first pharmacological treatment was administered at the age of 22, when Mr. B came back from a holiday trip in London experiencing persecutory thoughts and impulsiveness. He was treated with low doses of haloperidol and alprazolam. His psychiatrist also proposed hospitalization, but he refused.

7.2.2 Acute Presentation and Hospitalization

After a few days, Mr. B was finally admitted to our ward as an involuntary commitment, conducted by the police after agitation and aggressiveness toward his parents at home. On mental status examination, he appeared restless and moderately agitated, carrying a suspicious glance. He showed challenging behavior toward medical staff, with a low level of insight and judgment. His speech was pressured with elevated volume. No affective symptoms were reported, but a moderate/severe level of anxiety and irritability could be elicited in the clinical interview. He presented persecutory delusions but, apparently, was not experiencing hallucinations. Racing thoughts and flights of ideas were present. Mr. B described his last week as characterized by derealization ("it felt like being in a movie"). The patient denied any suicidal and/or homicidal ideation, intent, or plan.

He was initially treated with haloperidol 6 mg per day, switched after a few days to olanzapine 20 mg per day in order to reduce irritability and agitation.

During hospitalization, he had his first complete routine screening, consisting of EKG (in normal range), blood tests (i.e., hepatic function panel, renal function panel, blood count; in normal range), and urine toxicology screen (positive for cannabinoids; negative for other illicit drugs or legal medications, such as amphetamines, barbiturates, benzodiazepines, cocaine, methadone, opioids). Furthermore, the patient underwent magnetic resonance imaging (MRI) and positron emission tomography (PET). None of these exams showed a significant alteration (Fig. 7.1).

His attitude toward the medical staff and other patients was hostile and suspicious. He spent most of the day controlling his personal belongings and hiding from other patients, avoiding most of the social interaction. On the third day of hospitalization, Mr. B had a state of agitation, with persecutory thoughts, high levels of anxiety, and instinct of flight. He kicked the entrance door with the intention to escape, but soon the nurses and the medical staff persuaded him to stop. Therefore,

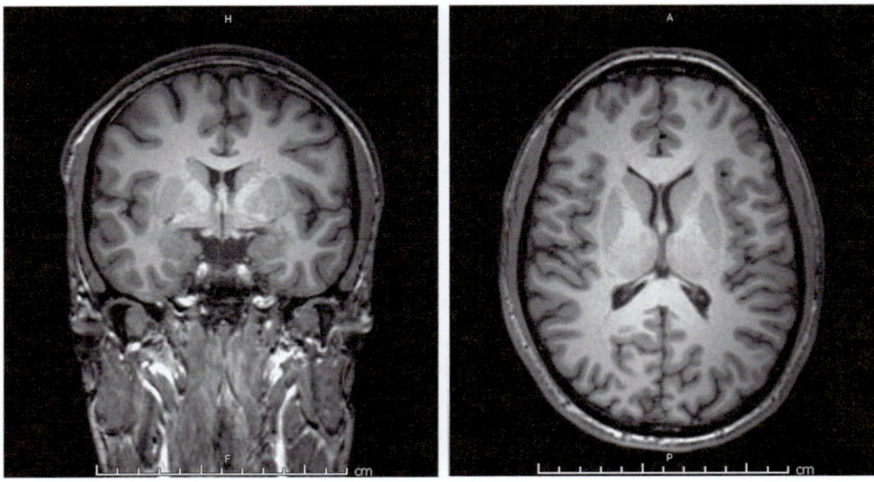

Fig. 7.1 Magnetic resonance imaging (MRI), T1-weighted images

the pharmacological treatment was changed from haloperidol to olanzapine, due to the more sedative profile of the latter compound (Fig. 7.2).

The following clinical psychometric tests were performed, at baseline (T0), after 7 (T1), and after 14 days (T2) of admission to the inpatient unit: Brief Psychiatric

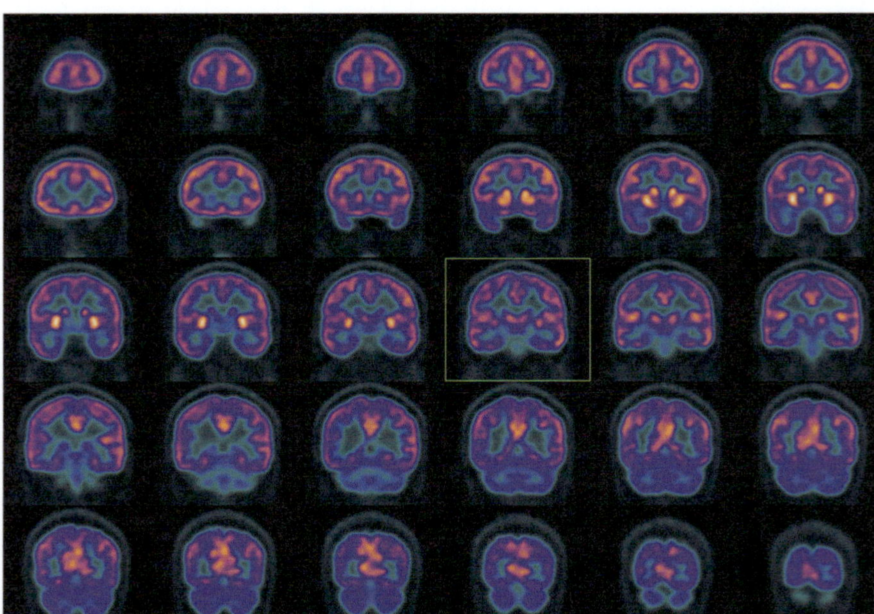

Fig. 7.2 Positron emission tomography (PET)

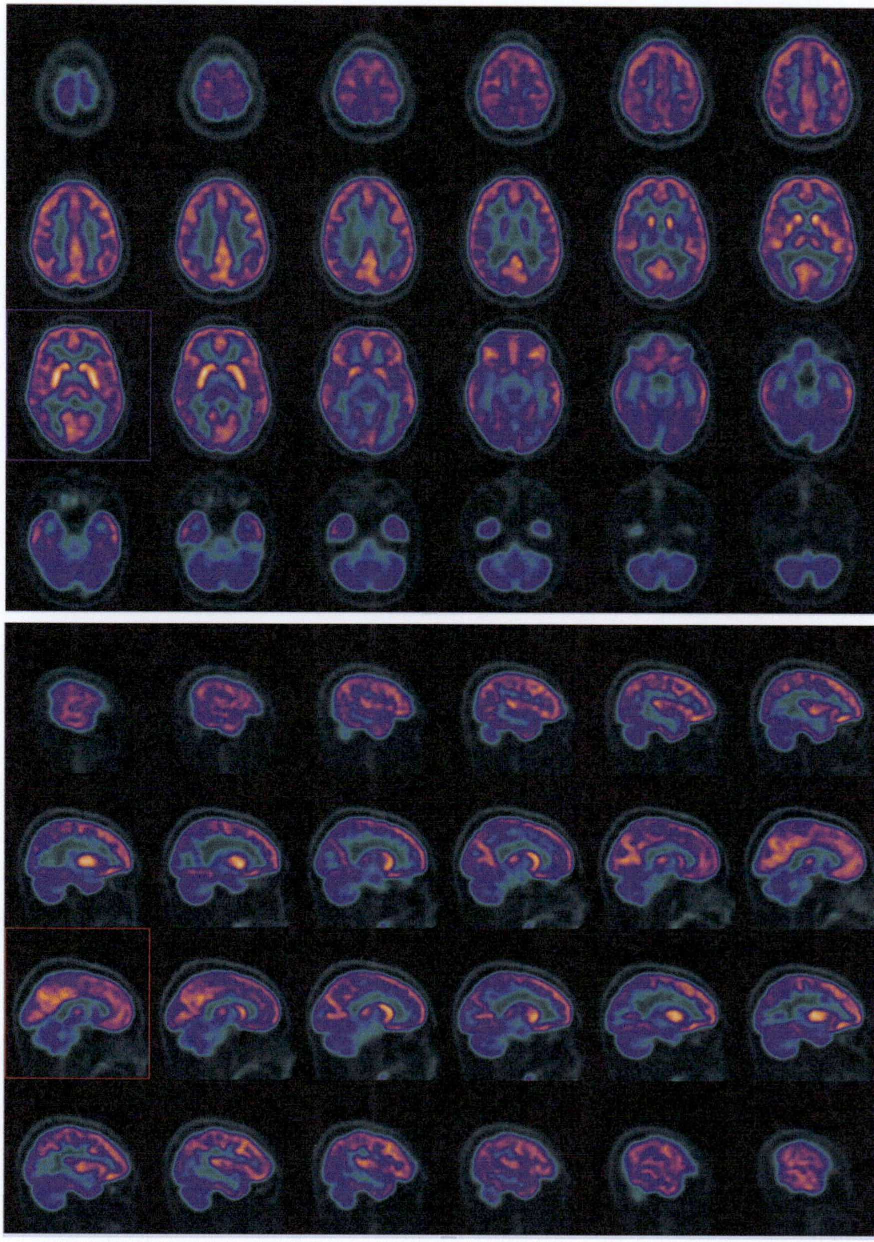

Fig. 7.2 (continued)

Table 7.3 Psychometric test scores at baseline (T0), after 7 days (T1), and after 14 days (T2) from hospitalization

	BPRS	YMRS	HDRS
T0	43	13	13
T1	36	9	6
T2	27	4	7

BPRS Brief Psychiatric Rating Scale, *YMRS* Young Mania Rating Scale, *HDRS* Hamilton Depression Rating Scale

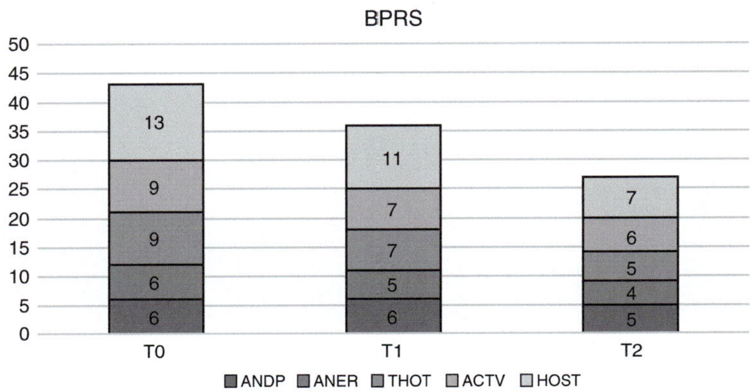

Fig. 7.3 Brief Psychiatric Rating Scale (BPRS) total and domain scores at baseline (T0), after 7 days (T1), and after 14 days (T2) from hospitalization

Rating Scale (BPRS) [19], Hamilton Depression Rating Scale (HDRS) [20], and Young Mania Rating Scale (YMRS) [21]. The BPRS domains were calculated, and the following ones resulted as predominant: hostile-suspiciousness (HOST), thought disturbance (THOT), and activation (ACTV) domains compared to anxiety-depression (ANDP) and anergia (ANER). The values are reported in Table 7.3 and graphically represented in Fig. 7.3.

Moreover, as soon as he entered the psychiatric ward, medical and nursing staff completed the Brøset Violence Checklist (BVC) with a total score of 3/6, reduced to 1/6 at T1 and 0/6 at T2.

Mr. B underwent cognitive assessment with BACS battery (Brief Assessment of Cognition in Schizophrenia), and the cognitive performance resulted globally impaired. More in detail, verbal memory (30.25; NV > 33.01), token motor task (47.25; NV > 68.77), symbol-coding task (21.50; NV > 40.49), and working memory (12.50; NV > 14.39) resulted clearly compromised, while verbal fluency (37.25; NV > 31.68) and Tower of London (13.75; NV > 12.37) were just sufficient [22]. In addition, the patient was tested with Beck Cognitive Insight Scale that showed lack of insight [23]. The latter tests were performed during hospitalization, and they refer to acute psychotic phase; retest after 6 months showed normalization of the results.

Clinical resolution of psychotic symptoms was relatively fast, happening in approximately 10 days of intramuscular antipsychotic treatment. After 14 days Mr. B was calm and cooperative with the staff. He showed an appropriate behavior within the psychiatric ward and with his relatives. He did not verbalize any delusional idea and developed a good degree of insight toward the symptomatology that led him to hospitalization, yet poor into the nature of his substance abuse. He remained slightly reluctant to accept the idea of taking a pharmacological treatment, although he was aware he needed to be followed by the psychiatric services. This state allowed us to discharge Mr. B from the ward after 2 weeks of hospitalization.

Resolution of the psychotic picture was complete and not followed by residual symptoms; recovery to usual life activities happened in a couple of weeks after discharge, and the psychotic breakdown did not result in any impairment in social, familial, or working functioning.

7.3 Discussion

7.3.1 Cannabis Use and Access to Emergency Services

Mr. B's clinical presentation is very often observed in late adolescents and young adults in acute condition accessing the hospital's ER. This clinical picture, characterized by psychomotor agitation, aggressiveness, logorrhea, and a florid psychotic dimension, is polymorph and often misleading. It poses a certain number of issues both in terms of differential diagnosis, acute and long-term treatment, and final outcome. In fact, the acute clinical presentation often does not allow formulating an accurate diagnosis. The latter is normally achieved at a later time.

The main target, in cases of acute psychomotor agitation in a psychotic patient, should be the prevention of the patient's potentially harmful behavior toward himself and/or third parties and the exclusion of organic comorbidity [24]. In fact, both the internal doctor and the psychiatrist are called to provide assistance. Teamwork is crucial to identify whether there is an underlying organic condition determining or contributing to the acute clinical picture or whether a psychiatric condition is more likely to be responsible.

As always in medical sciences, the collection of history data is crucial. As in Mr. B's case, the awareness of substance abuse immediately points out a direction to follow in terms of diagnostic exams and therapeutic options. Nonetheless, substance abuse can be a precipitating factor of any other medical or psychiatric condition, rather than the only responsible factor determining an acute psychotic picture, with (or without) psychomotor agitation. Therefore, a preliminary diagnostic panel is routinely carried out in such patients, e.g., blood tests (including renal and liver function, alcohol concentrations, ammonia, C-reactive protein/erythrocyte sedimentation rate, complete blood count, thyroid hormones, etc.), urine toxicology test, EKG, brain CT, and EEG.

When the overall physical health of the patient has been assessed, and an acute medical condition has been excluded, hospitalization in a psychiatric ward might be the next step. A protected and safe environment is, in fact, necessary to allow further investigations in order to assess whether the substance abuse is an epiphenomenon of an underlying psychiatric condition or is primarily responsible for the acute psychotic presentation.

7.3.2 Admission to Hospital

7.3.2.1 Medical Case History

Within a psychiatric ward, a more accurate diagnostic process can be outlined. In the first place, admission to hospital allows a detailed reconstruction of the patient's clinical history. The importance of an accurate collection of data regarding the psychopathological onset and course of illness is not the purpose of this work. Herein, we consider it of particular importance to investigate two fields of information: abuse habits and family.

A definite and strict relationship between the intensity of cannabis use and the severity of a clinical syndrome cannot be established. Nonetheless it appears common good sense to ascertain if such use is actually sporadic and circumstantial or rather constant and pervasive. Specific questionnaires can detect the quantity, quality, and frequency of abuse [25]. In addition, the nature of the cannabis itself plays a relevant role. In fact, many variations of this substance are currently available on the market, and the concentration of THC, responsible for the psychotropic effect, may vary significantly. So-called skunk, for example, is known to have a high potency and has been proven to determine an increased risk of psychosis compared with traditional low-potency cannabis (hash), especially in the case of daily use [26].

Family history of psychiatric disorders or abuse habits is another important element to collect. Often, such data might not be readily available. Patients might not know whether their relatives ever experienced a psychiatric disorder in their lives before, as such information might be unlikely to be shared within certain families, often due to the discomfort and stigma still connected to psychiatric illness. In addition, some patients, when interviewed, might remain reticent and contradictory toward such data, in the attempt to protect their relatives' privacy [27, 28]. Indeed, the genetic transmission of abuse habits has been discussed, and evidence for strong family aggregation and further risk factors in substance use—without or without comorbid psychiatric conditions—has been pinpointed [29].

7.3.2.2 Cannabis-Induced Psychosis: Clinical Features

The relationship between cannabis consumption and the development of psychotic symptoms has been documented [30, 31]. A causal relationship has been proposed, but the pathophysiological mechanism remains to be explored. The main theory suggests that cannabis may have an effect on neurodevelopmental processes such as synaptic plasticity. Adolescence in particular is a crucial phase for brain development, consisting of neuronal maturation and rearrangement processes (i.e., myelination, synaptic pruning, and dendritic plasticity). Cannabis consumption occurring on a regular basis during adolescence can therefore interfere with such mechanisms [32].

As many experimental studies on healthy humans report, THC (cannabis active ingredient) can induce not only transient, dose-dependent psychotic symptoms but also behavioral, cognitive, and psychophysiological effects [33]. The latter reflects a change in regular functioning and may include positive symptoms, such as delusions, hallucinations, feelings of paranoia and suspiciousness, disorganized thinking, and disorganized speaking, and negative symptoms, such as loss of or decreased motivation, loss of or decrease in ability to take initiative or come up with new ideas, loss of or decreased speech, difficulties expressing emotion, and difficulties thinking and/concentrating. Further psychopathological manifestations may consist in changes in the affective area, such as anxiety, panic, and depressed and/or elevated and/or dysphoric mood, and neurovegetative alterations, such as in sleep pattern, appetite, motor activity, and endocrine regulation [34].

In addition, heavy cannabis use has been shown to determine perceptual, psychomotor, attentional, and mnemonic deficits as well as affect learning performance both in healthy individuals and in patients suffering from psychosis [26].

Sustained cannabis consumption during adolescence may lead to permanent affective and cognitive impairments in adulthood, such as avolition, alexithymia, impaired informational processing, sustained and distributed attention, spatial working memory, verbal fluency, decision making, and executive functions [35].

Finally, cannabis may determine a variety of behavioral changes, spanning from mild transient relaxation to impaired psychomotor performances and severe agitation, also depending on the quality of the compound, the severity of abuse, and any comorbid consumption of other illicit drugs or alcohol (the latter being the rule rather than the exception) [36].

7.3.2.3 Clinical Assessment Tools

A number of clinical questionnaires may be administered in order to assess the severity of the clinical picture.

Worldwide-known psychometric questionnaires can, in fact, measure the overall psychopathology severity (Brief Psychiatric Rating Scale [19]) and, more in detail, the concomitant presence of depressive symptoms or elevated/irritable mood (Hamilton-Depression Rating Scale, Montgomery-Asberg Depression Rating Scale, Young Mania Rating Scale [20, 21, 37]). When aggressiveness and/or agitation are dominant features, the Brøset Violence Checklist [38] or the Overt Aggression Scale [39] might become useful.

A close link between acute cannabis use and risk of suicidality has not been proven.

Recent evidence suggests that chronic cannabis use is correlated with suicidality, but the lack of homogeneity in measuring cannabis consumption and of systematic control for known risk factors has to be taken into account [40].

Therefore, the assessment of suicidal risk in patients with an acute cannabis-induced psychosis seems to be a sensible action in clinical practice. Specific rating scales are used worldwide for this purpose [41, 42].

The SCID 1 and SCID 2 [43] are crucial to determining whether a concomitant or past primary psychiatric disorder can be diagnosed for that patient. In fact, as previously highlighted, substance abuse might be a condition that facilitates the development of an acute psychosis, or, alternatively, the habit itself of consuming cannabis might be used to mitigate a preexisting psychiatric condition. Much as what happens with alcohol, cannabis might decrease the level of anxiety and/or be employed as a social facilitator and therefore mask a former generalized anxiety disorder or social phobia [44].

Neuropsychological tests can be administered in order to assess any cognitive deficits. Indeed, the administration of such investigations during an acute psychotic episode is not suggested, but repetitive measures at regular intervals from the index episode can be useful in monitoring the cognitive functions of cannabis users—a population at risk of developing cognitive impairment, especially in adulthood. The Brief Assessment of Cognition in Schizophrenia (BACS) is routinely used in schizophrenic patients but can be administered in non-affective psychotic syndromes. It is an effective tool with the following characteristics: capable of testing the main cognitive domains, including executive functions, and reasonably short and therefore suitable for such patients, who often feel reluctant to comply with long and demanding clinical assessments; it is corrected for age and education [22]. In addition, social cognition should be investigated. Within social cognition, the theory of mind (ToM) refers to the ability to infer mental states (e.g., beliefs, desires, intentions, imagination, emotions) that cause actions. An individual with a valid ToM shows the ability to reflect on the content of his/her own and others' minds [45]. Such skills have been proven faulty in chronic cannabis abusers. Therefore, tools that investigate this cognitive function, such as the Reading the Mind in the Eyes and the Faux Pas Recognition tests, might be employed in clinical practice [46, 47].

7.3.2.4 Brain Imaging

When Mr. B's conditions allowed, he underwent a magnetic resonance (MR) and a positron emission tomography (PET), in order to study the morphology and the metabolism of his brain. Such exams, although not indispensable to formulating a final diagnosis, can help the clinician to exclude underlying organic conditions, such as altered cortical tropism or vascular alterations, as well as showing suggestive patterns of cerebral metabolism.

Especially when used in such a critical time as adolescence, illicit drugs seem to affect the normal process of neurodevelopment, both in terms of the risk of developing a psychosis and of functional and structural modifications. The cognitive impairment observed in younger abusers tends to become stable over time, even a long time after withdrawal [48].

As previously reported in literature, chronic exposure to THC leads to neuroanatomic alterations, both in experimental animals and humans [11, 12]. However, methodological differences may occur across the studies, making results hard to interpret: so far, no final results have been obtained to explain the interplay between chronic use of illicit drugs and brain aging. Nonetheless, prior neuroimaging studies have shown that cannabis-induced psychosis may lead to a distinct pattern of structural and functional brain alterations, with volume reductions in the dorsolateral prefrontal cortex and posterior cingulate and concomitant [49, 50] increase of striatal metabolism [51, 52].

A chronic use of cannabinoids during adolescence may foster premature aging, ultimately worsening cognition and social functioning in the elderly. Regular exposure to cannabis is associated with neuroanatomic alterations in several cortical and subcortical brain regions, compared to healthy controls, including the superior and middle frontal gyri (Brodmann areas—BA6 and BA11), inferior parietal gyrus (BA39), cuneus (BA23), subgenual anterior cingulate cortex (BA25), thalamus, and hippocampus. Such regions are known to be rich in cannabinoid receptors, to which THC binds to exert its psychoactive effects, but are also accountable for several cognitive functions, which decline during normal aging.

Global cortical effects of cannabis have been documented, including global cerebral glucose metabolism after infusion of THC, increased cerebral blood flow in frontal regions with greater increases detected in the right hemisphere among individuals reporting a history of use, lower volumes of ventricular cerebrospinal fluid (CSF) among users when compared to controls, denser gray matter in the parahippocampal gyrus, and denser white matter in the left parietal lobe among heavy users compared to nonusers [53–57].

Finally, cannabis affects the mesolimbic dopamine system, including the putamen and caudate, brain areas central in reward processing and drug addiction. Indeed, previous studies have underlined that a dysfunction in the reward system is connected to increased risk of substance abuse [52].

7.3.3 Differential Diagnosis

The differential diagnosis between a primary psychiatric syndrome and a substance-induced disorder can be very challenging and, at the same time, essential to planning an adequate treatment and follow-up. Clinicians' awareness of the importance of defining the role of drug abuse within mental disorders is constantly rising, together with literature studies.

Focusing on the distinction between cannabis-induced psychosis and schizophrenia-related disorders, Tennant and colleagues have observed that comparing psychotic patients without substance abuse and cannabis-induced psychotic subjects, the latter showed more symptoms of agitation and aggressiveness and less auditory hallucinations, flattening of affect, and incoherent speech. They also showed rapid improvements after pharmacological treatment (within 1 week) [58]. Rubio and colleagues focused on premorbid personality disorders and psychopathologic symptoms, reporting that patients with a cannabis-induced disorder had a higher prevalence of social phobia and showed more often a neurotic (somatizations, obsessive-compulsive, interpersonal sensitivity, depression, anxiety, and phobic anxiety) rather than a psychotic profile, in comparison to those affected by a primary mental disorder without substance abuse. The authors suggest that the clinical picture of a cannabis-induced disorder might be the result of the interaction between personality traits (e.g., interpersonal sensitivity and phobic anxiety) and drug consumption [44].

Fewer literature studies have investigated the relationship between manic symptoms and cannabis use. Gibbs and colleagues reported that cannabis use may worsen the occurrence of manic symptoms in bipolar patients but also represent a causal risk factor for developing manic symptoms per se. More in detail, the manic phase of bipolar disorder and a cannabis-induced disorder may share several psychopathological characteristics, such as euphoria, agitation, grandiosity, and aggressiveness [59]. Again, an accurate collection of clinical history seems crucial to distinguish between a manic episode in a bipolar disorder and comorbid drug abuse and a cannabis-induced psychotic picture with activation symptoms (i.e., previous substance-free manic/hypomanic episodes, family history, etc.).

Altogether, longitudinal observation of the patient and the course of the disease appear to be the most important factors in identifying the most correct diagnosis. In fact, the cross-sectional picture is often confused, and several symptoms may contribute to delineating a polymorph clinical syndrome that hardly meets the usual diagnostic criteria. It is not uncommon for a final diagnosis to be postponed, as only time and full clinical expression can better discriminate between a primary progressive and often regressive pathology and a time-limited induced psychotic picture.

Figure 7.4 aims to summarize common and peculiar clinical characteristics of cannabis-induced psychosis and primary psychosis, such as schizophrenia and bipolar disorder.

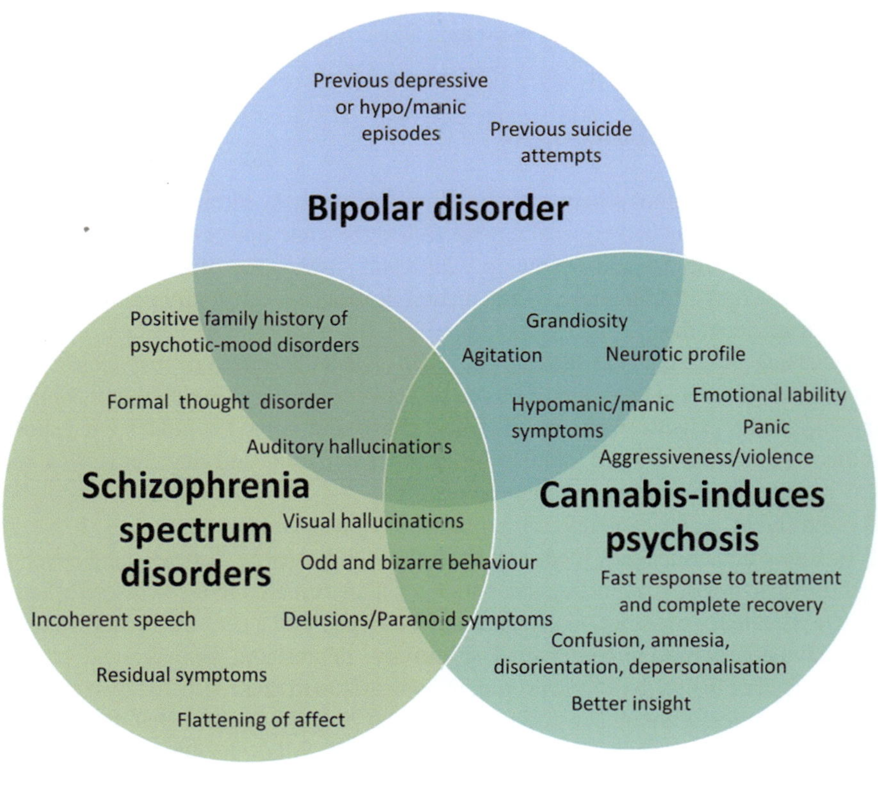

Fig. 7.4 (1) Cannabis-induced, (2) behavior and (3) depersonalization

7.3.4 Options of Treatment

7.3.4.1 Pharmacological Options

As mentioned previously, subjects with a cannabis-induced psychosis are usually young patients, who often access a psychiatric emergency room or are admitted to a ward for the first time in their life. It seems therefore sensible to think that atypical antipsychotics should be preferred in the treatment of such subjects, in consideration of the more favorable side effect profile of such compounds. Indeed, atypical antipsychotics have been described either as effective as or more effective than typical antipsychotics (and haloperidol in particular) in open-label and double-blind randomized controlled studies and, however, preferable due to their lower rate of extrapyramidal side effects [60]. Nonetheless, a recent review conducted by Wilson and colleagues failed to find clear differences between antipsychotics belonging to different classes (i.e., risperidone, olanzapine, haloperidol, zuclopenthixol, plus clozapine as an add-on) in treating psychotic symptoms. Clozapine has instead shown some superiority, in comparison to other antipsychotics, in reducing craving for cannabis [17].

When it comes to the treatments that might help with quitting the habit of consuming cannabis, regardless of the presence of a psychotic picture, results are even more discordant, and there is incomplete evidence for all the molecules so far investigated (i.e., THC, selective serotonin reuptake and mixed-action antidepressants, anticonvulsants and mood stabilizers, atypical antidepressants, anxiolytic, norepinephrine reuptake inhibitors, and glutamatergic modulators) [61].

7.3.4.2 Non-pharmacological Options

Cannabis use increases the risk of nonadherence; therefore quitting cannabis use may help with compliance to antipsychotics. Thus, cannabis use may represent a potential target for intervention to improve medication adherence in those with psychosis [62]. As a matter of fact, patients and family psychoeducation as well as cognitive-behavioral treatments have been proposed.

More in detail, psychoeducational goals consist of supporting patients and their families in understanding the disease and the treatment, cooperation with caregivers, living a healthier life, and maintaining or improving their quality of life. Psychoeducation aims to influence several functional domains, such as service engagement (active participation in defining treatment plans, self-seeking for medical help, etc.), resilience, and self-stigma [63].

Additionally, cognitive-behavioral therapy has been used in cannabis users and has been experimented in patients with first-episode psychosis in particular, showing encouraging results in terms of overall outcome and, more in detail, frequency of cannabis use per week, addiction severity, number of disability days, and overall level of psychopathology, over a 6-month follow-up period [64–66].

Table 7.4 consists of a point-by-point list of steps to keep in mind when assessing a patient with substance-induced psychosis.

Table 7.4 What to do with a patient experiencing a substance-induced psychosis? An assessment chart

Treating behaviors that might lead to harming self or others
Excluding acute nonpsychiatric conditions by means of preliminary routine tests (e.g., blood tests, urine toxicology screen, CT)
Proceeding with the admission to an inpatient unit in order to provide a safe environment to perform the next steps
Collecting a detailed clinical history, including quantity and quality of the abuse
Performing further diagnostic procedures, when necessary
Starting an acute antipsychotic therapy
Assessing the risk of relapsing into abuse habits
Discharge and activation of specific treatments to prevent relapse into substance abuse

7.3.5 Outcome

If the resolution of a cannabis-induced psychotic picture may be complete and the patient's level of functioning returns to satisfying levels in a reasonably short amount of time, the long-term outcome of a cannabis abuser might be uncertain.

It has been estimated that about 10% of those who have used cannabis at least once will develop cannabis dependence [67]. Based on a large epidemiological survey in the USA, it has been estimated that, among those exposed once to cannabis, 7.0% of males and 5.3% of females will develop cannabis dependence at some point in their life, while 47.4% of males and 32.5% of females will develop cannabis use disorders (abuse or dependence) at some point in their life [68]. As with other drugs of dependence, the risk of developing dependency is influenced by multiple factors. However, intensive use of cannabis, that is, daily or near daily use, is likely to increase the risk of cannabis dependence. In addition, polydrug abuse seems to be a common condition, and often cannabis use is complicated with alcohol, cocaine, and other regular intake substances [1].

Once a substance-induced disorder is outlined, it seems imperative to set up pharmacological and non-pharmacological strategies, with the aim to reduce and possibly suspend abuse habits. The risk of developing dependence and a chronic illness is high, with a consequence over quality of life and social, work, and family functioning. Substance abuse is a very topical area of interest and should be thoroughly assessed in routine clinical practice, especially in contemporary times, when the availability of illicit substances is high, in rapid development, and easily accessible by the younger population.

Key Points

1. Wide scientific literature has investigated the strong link between mental disorders and substance abuse; such relationship has a negative impact on both persistence and severity of illness. Potential causal effects between substance abuse and different mental disorders still need to be clearly defined, and future

discoveries might have an impact on defining substance use prevention and treatment strategies.
2. Cannabis is the most commonly used illicit drug in Europe with almost 26.3% of adults using it in their lifetime. The psychoactive and addictive properties of cannabis are primarily due to THC, and the increased availability of cannabis varieties with high THC content have consistently been linked to a more frequent onset of psychotic pictures and an increased number of hospital admissions and of anxiety symptoms. THC cannot only induce transient, dose-dependent psychotic symptoms but also affective, behavioral, cognitive, neurovegetative, and psychophysiological effects.
3. An acute substance-related clinical picture can be characterized by psychomotor agitation, aggressiveness, logorrhea, and a florid psychotic dimension. Such presentation often does not allow formulating an accurate differential diagnosis between primary psychiatric syndrome and a substance-induced disorder. A final diagnosis is usually achieved in a later time.
4. A preliminary diagnostic panel is routinely carried out in such patients, including blood tests, urine toxicology test, EKG, brain CT, and EEG. A more accurate diagnostic process might include a detailed reconstruction of patient's clinical history, with particular attention to their psychopathological onset, course of illness, abuse habits, and familiarity, structured clinical interview for psychiatric disorders (i.e., SCID-5), MRI scan, PET scan, and neuropsychological tests.
5. Once a substance-induced disorder is outlined, it seems imperative to set up pharmacological and non-pharmacological strategies, with the aim to reduce and possibly suspend abuse habits and prevent from developing induced psychiatric symptoms.

Self-Assessment Questionnaire

1. What percentage of the total adult European population has tried cannabis at least once in their life?
   (A) 15%
   (B) 26%
   (C) 13%
2. Why might neuroimaging be so important in cannabis psychosis?
   (A) To exclude medical conditions
   (B) To help a differential diagnosis
   (C) To choose the most suitable pharmacological treatment
3. Tetrahydrocannabinol in cannabis represents:
   (A) The calming component
   (B) The psychotropic component
   (C) An inactive component
4. Collecting medical history, in a psychotic patient and substance abuse, is:
   (A) Not strictly necessary, as further diagnostic tools are more relevant to formulate a final diagnosis

(B) Very valuable in discriminating between a primary psychiatric disorder and a substance-induced clinical picture

(C) A process that has to be performed accurately, as only poor tools exist for detecting data such as intensity of abuse, clinical symptomatology, etc.

References

1. European Monitoring Centre for Drugs and Drug Addiction. European Drug Report. Trends and developments. Luxembourg: Publications Office of the European Union; 2017. p. 2017.
2. Dines AM, Wood DM, Yates C, et al. Acute recreational drug and new psychoactive substance toxicity in Europe: 12 months data collection from the European drug emergencies network (euro-DEN). Clin Toxicol. 2015;53(9):893–900.
3. Bhattacharyya S, Atakan Z, Martin-Santos R, Crippa JA, McGuire PK. Neural mechanisms for the cannabinoid modulation of cognition and affect in man: a critical review of neuroimaging studies. Curr Pharm Des. 2012;18(32):5045–54.
4. Lupica CR, Riegel AC, Hoffman AF. Marijuana and cannabinoid regulation of brain reward circuits. Br J Pharmacol. 2004;143(2):227–34.
5. Di Forti M, Sallis H, Allegri F, et al. Daily use, especially of high-potency Cannabis, drives the earlier onset of psychosis in Cannabis users. Schizophr Bull. 2014;40(6):1509–17.
6. Bhattacharyya S, Morrison PD, Fusar-Poli P, et al. Opposite effects of delta-9-tetrahydrocannabinol and cannabidiol on human brain function and psychopathology. Neuropsychopharmacology. 2010;35(3):764–74.
7. Martin-Santos R, Crippa JA, Batalla A, et al. Acute effects of a single, oral dose of d9-tetrahydrocannabinol (THC) and cannabidiol (CBD) administration in healthy volunteers. Curr Pharm Des. 2012;18(32):4966–79.
8. Schacht JP, Hutchison KE, Filbey FM. Associations between cannabinoid Receptor-1 (CNR1) variation and hippocampus and amygdala volumes in heavy cannabis users. Neuropsychopharmacology. 2012;37(11):2368–76.
9. Gilman JM, Kuster JK, Lee S, et al. Cannabis use is quantitatively associated with nucleus Accumbens and amygdala abnormalities in young adult recreational users. J Neurosci. 2014;34 (16):5529–38.
10. Tomasi D, Wang G-J, Volkow ND. Balanced modulation of striatal activation from D_2/D_3 receptors in caudate and ventral striatum: disruption in cannabis abusers. Hum Brain Mapp. 2015;36(8):3154–66.
11. Lorenzetti V, Solowij N, Yücel M. The role of cannabinoids in Neuroanatomic alterations in Cannabis users. Biol Psychiatry. 2016;79(7):e17–e31.
12. Renard J, Vitalis T, Rame M, et al. Chronic cannabinoid exposure during adolescence leads to long-term structural and functional changes in the prefrontal cortex. Eur Neuropsychopharmacol. 2016;26(1):55–64.
13. Wood DM, Heyerdahl F, Yates CB, et al. The European drug emergencies network (euro-DEN). Clin Toxicol. 2014;52(4):239–41.
14. Dines AM, Wood DM, Galicia M, et al. Presentations to the emergency department following Cannabis use—a multi-Centre case series from ten European countries. J Med Toxicol. 2015;11 (4):415–21.
15. Swendsen J, Conway KP, Degenhardt L, et al. Mental disorders as risk factors for substance use, abuse and dependence: results from the 10-year follow-up of the National Comorbidity Survey. Addiction. 2010;105(6):1117–28.
16. Edwards G, Arif A, Hadgson R. Nomenclature and classification of drug- and alcohol-related problems: a WHO memorandum. Bull World Health Organ. 1981;59(2):225–42.
17. Wilson RP, Bhattacharyya S. Antipsychotic efficacy in psychosis with co-morbid cannabis misuse: a systematic review. J Psychopharmacol. 2016;30(2):99–111.

18. Santucci K. Psychiatric disease and drug abuse. Curr Opin Pediatr. 2012;24(2):233–7.
19. Overall J, Gorham D. The brief psychiatric rating scale. Psychol Rep. 1962;10:799–812.
20. Hamilton M. A rating scale for depression. J Neurol Neurosurg Psychiatry. 1960;23:56–62.
21. Young RC, Biggs JT, Ziegler VE, Meyer DA. A rating scale for mania: reliability, validity and sensitivity. Br J Psychiatry. 1978;133:429–35.
22. Keefe R. The brief assessment of cognition in schizophrenia: reliability, sensitivity, and comparison with a standard neurocognitive battery. Schizophr Res. 2004;68(2–3):283–97.
23. Beck A, Baruch E, Balter JM, Steer RA, Warman DM. A new instrument for measuring insight: the Beck cognitive insight scale. Schizophr Res. 2004;68(2–3):319–29.
24. Trezza GR, Popp SM. The substance user at risk of harm to self or others: assessment and treatment issues. J Clin Psychol. 2000;56(9):1193–205.
25. Cuttler C, Spradlin A. Measuring cannabis consumption: psychometric properties of the daily sessions, frequency, age of onset, and quantity of Cannabis use inventory (DFAQ-CU). PLoS One. 2017;12(5):e0178194.
26. Di Forti M, Marconi A, Carra E, et al. Proportion of patients in South London with first-episode psychosis attributable to use of high potency cannabis: a case-control study. The Lancet Psychiatry. 2015;2(3):233–8.
27. Kendler KS. Family history information in biomedical research. J Contin Educ Heal Prof. 2001;21(4):215–23.
28. Gray AJ. Stigma in psychiatry. J R Soc Med. 2002;95(2):72–6.
29. Steinhausen H-C, Jakobsen H, Munk-Jørgensen P. Family aggregation and risk factors in substance use disorders over three generations in a nation-wide study. PLoS One. 2017;12(5): e0177700.
30. Mackie K. Understanding cannabinoid Psychoactivity with mouse genetic models. PLoS Biol. 2007;5(10):e280.
31. Parrott AC, Hayley AC, Downey LA. Recreational stimulants, herbal, and spice cannabis: the core psychobiological processes that underlie their damaging effects. Hum Psychopharmacol Clin Exp. 2017;32(3):e2594.
32. Shrivastava A, Johnston M, Terpstra K, Bureau Y. Cannabis and psychosis: neurobiology. Indian J Psychiatry. 2014;56(1):8–16.
33. Murray RM, Englund A, Abi-Dargham A, et al. Cannabis-associated psychosis: neural substrate and clinical impact. Neuropharmacology. 2017;124:89–104.
34. Wilson N, Cadet JL. Comorbid mood, psychosis, and marijuana abuse disorders: a theoretical review. J Addict Dis. 2009;28(4):309–19.
35. Bortolato M, Bini V, Tambaro S. Vulnerability factors for the psychiatric and Behavioral effects of Cannabis. Pharmaceuticals. 2010;3(9):2799–820.
36. Barthelemy OJ, Richardson MA, Cabral HJ, Frank DA. Prenatal, perinatal, and adolescent exposure to marijuana: relationships with aggressive behavior. Neurotoxicol Teratol. 2016;58:60–77.
37. Montgomery SA, Asberg M. A new depression scale designed to be sensitive to change. Br J Psychiatry. 1979;134:382–9.
38. Woods P, Almvik R. The Brøset violence checklist (BVC). Acta Psychiatr Scand Suppl. 2002;412:103–5.
39. Yudofsky SC, Silver JM, Jackson W, Endicott J, Williams D. The overt aggression scale for the objective rating of verbal and physical aggression. Am J Psychiatry. 1986;143(1):35–9.
40. Borges G, Bagge CL, Orozco R. A literature review and meta-analyses of cannabis use and suicidality. J Affect Disord. 2016;195:63–74.
41. Ghasemi P, Shaghaghi A, Allahverdipour H. Measurement scales of suicidal ideation and attitudes: a systematic review article. Heal Promot Perspect. 2015;5(3):156–68.
42. Patterson WM, Dohn HH, Bird J, Patterson GA. Evaluation of suicidal patients: the SAD PERSONS scale. Psychosomatics. 1983;24(4):343–9.
43. First MB, Williams JB, Karg RS, Spitzer R. SCID-5-CV : structured clinical interview for DSM-5 disorders, clinician version. Washington: American Psychiatric Association.

44. Rubio G, Marín-Lozano J, Ferre F, et al. Psychopathologic differences between cannabis-induced psychoses and recent-onset primary psychoses with abuse of cannabis. Compr Psychiatry. 2012;53(8):1063–70.
45. Poulsen D, Kintsch E, Kintsch W, Premack D. Children's comprehension and memory for stories. J Exp Child Psychol. 1979;28(3):379–403.
46. Stone VE, Baron-Cohen S, Knight RT. Frontal lobe contributions to theory of mind. J Cogn Neurosci. 1998;10(5):640–56.
47. Baron-Cohen S, Wheelwright S, Hill J, Raste Y, Plumb I. The "reading the mind in the eyes" test revised version: a study with normal adults, and adults with Asperger syndrome or high-functioning autism. J Child Psychol Psychiatry. 2001;42(2):241–51.
48. Jacobus J, Squeglia LM, Meruelo AD, et al. Cortical thickness in adolescent marijuana and alcohol users: a three-year prospective study from adolescence to young adulthood. *Dev. Cogn.* Neurosci. 2015;16:101–9.
49. Parkar S, Ramanathan S, Nair N, et al. Are the effects of cannabis dependence on glucose metabolism similar to schizophrenia? An FDG PET understanding. *Indian.* J Psychiatry. 2011;53(1):13.
50. Schweinsburg AD, Brown SA, Tapert SF. The influence of marijuana use on neurocognitive functioning in adolescents. Curr Drug Abuse Rev. 2008;1(1):99–111.
51. Dragogna F, Mauri MC, Marotta G, Armao FT, Brambilla P, Altamura AC. Brain metabolism in substance-induced psychosis and schizophrenia: a preliminary PET study. Neuropsychobiology. 2014;70(4):195–202.
52. Altamura AC, Delvecchio G, Marotta G, et al. Structural and metabolic differentiation between bipolar disorder with psychosis and substance-induced psychosis: an integrated MRI/PET study. Eur Psychiatry J Assoc Eur Psychiatr. 2017;41:85–94.
53. Breivogel CS, Childers SR. The functional neuroanatomy of brain cannabinoid receptors. Neurobiol Dis. 1998;5(6 Pt B):417–31.
54. Volkow ND, Gillespie H, Mullani N, et al. Cerebellar metabolic activation by delta-9-tetrahydro-cannabinol in human brain: a study with positron emission tomography and 18F-2-fluoro-2-deoxyglucose. Psychiatry Res. 1991;40(1):69–78.
55. Mathew RJ, Wilson WH, Humphreys DF, Lowe JV, Wiethe KE. Regional cerebral blood flow after marijuana smoking. J Cereb Blood Flow Metab. 1992;12(5):750–8.
56. Block RI, O'Leary DS, Ehrhardt JC, et al. Effects of frequent marijuana use on brain tissue volume and composition. Neuroreport. 2000;11(3):491–6.
57. Matochik JA, Eldreth DA, Cadet J-L, Bolla KI. Altered brain tissue composition in heavy marijuana users. Drug Alcohol Depend. 2005;77(1):23–30.
58. Tennant C. Marijuana and madness. Bull World Health Organ. 2005;83(6):479.
59. Gibbs M, Winsper C, Marwaha S, Gilbert E, Broome M, Singh SP. Cannabis use and mania symptoms: a systematic review and meta-analysis. J Affect Disord. 2015;171:39–47.
60. Isaac M, Isaac M, Holloway F. Is cannabis an anti-antipsychotic? The experience in psychiatric intensive care. Hum Psychopharmacol. 2005;20(3):207–10.
61. Marshall K, Gowing L, Ali R, Le Foll B. Pharmacotherapies for cannabis dependence. Cochrane Database Syst Rev. 2014:CD008940.
62. Foglia E, Schoeler T, Klamerus E, Morgan K, Bhattacharyya S. Cannabis use and adherence to antipsychotic medication: a systematic review and meta-analysis. Psychol Med. 2017;47(10):1691–705.
63. Morin L, Franck N. Rehabilitation interventions to promote recovery from schizophrenia: a systematic review. Front Psych. 2017;8:100.
64. Cooper K, Chatters R, Kaltenthaler E, Wong R. Psychological and psychosocial interventions for cannabis cessation in adults: a systematic review short report. Health Technol Assess (Rockv). 2015;19(56):1–130.
65. González-Ortega I, Echeburúa E, García-Alocén A, Vega P, González-Pinto A. Cognitive behavioral therapy program for cannabis use cessation in first-episode psychosis patients: study protocol for a randomized controlled trial. Trials. 2016;17(1):372.

66. Hoch E, Noack R, Henker J, et al. Efficacy of a targeted cognitive–behavioral treatment program for cannabis use disorders (CANDIS*). Eur Neuropsychopharmacol. 2012;22 (4):267–80.
67. Wagner FA, Anthony JC. From first drug use to drug dependence; developmental periods of risk for dependence upon marijuana, cocaine, and alcohol. Neuropsychopharmacology. 2002;26(4):479–88.
68. Lev-Ran S, Le Foll B, McKenzie K, George TP, Rehm J. Cannabis use and cannabis use disorders among individuals with mental illness. Compr Psychiatry. 2013;54(6):589–98.

8 Psychotic Disorders Due to Traumatic Brain Injury (PD-TBI)

Matteo Lazzaretti, Gian Mario Mandolini, and Silvia Grassi

Abstract

In this chapter we debate the case report of a patient diagnosed with psychotic disorder due to traumatic brain injury. Firstly, we describe the classification of psychotic disorders in the DSM-5 with particular regard to its innovations in the diagnostic process. In the second part of the chapter, we focus on the DSM-5 criteria for "psychotic disorder due to another medical condition". Specifically, in some cases, a traumatic brain injury could explain the onset of psychotic symptoms. A 34-year-old patient, previously diagnosed with "paranoia schizophrenia," was admitted to our inpatient ward for diagnostic and therapeutic reassessment. An accurate diagnostic strategy including medical history, neurological examination, MRI and PET scans allowed clinicians to revise the diagnosis of schizophrenia and to identify previous traumatic brain injuries and consequent focal cerebral frontotemporal lesions, leading to a diagnosis of psychotic disorder due to traumatic brain injury (PD-TBI). The identification of previous traumatic brain injuries is essential in the clinical evaluation of a psychiatric disorder in order to perform a correct differential diagnosis with psychosis. Indeed, this has relevant implications in both treatment and prognosis.

Keywords

Traumatic brain injury (TBI) · Psychotic disorder due to traumatic brain injury (PD-TBI) · Paranoid schizophrenia · Frontal lobe · Temporal region · Psychotic spectrum · Magnetic resonance imaging (MRI) · Positron emission tomography (PET) · Integration PET - MRI

M. Lazzaretti (✉) · G. M. Mandolini · S. Grassi
Department of Neurosciences and Mental Health, Fondazione IRCCS Ca' Granda Ospedale Maggiore Policlinico, University of Milan, Milan, Italy
e-mail: matteolazzaretti@hotmail.com

A. C. Altamura, P. Brambilla (eds.), *Clinical Cases in Psychiatry: Integrating Translational Neuroscience Approaches*,
https://doi.org/10.1007/978-3-319-91557-9_8

8.1 Introduction

The presentation of the following case report arises from the need to better elucidate some unique perspectives of psychotic disorders. Specifically, nowadays, the phases of the diagnostic process have acquired particular relevance in order to exclude organic etiology of psychosis with potential implications concerning treatment and prognosis.

The fifth edition of the *Diagnostic and Statistical Manual of Mental Disorders* (DSM-5) preserves a categorical classification system of psychiatric disorders such as psychosis, since there is still a lack of research for the establishment of a new psychiatric nosology [1]. However, DSM-5 aims to give more value to the dimensional features of psychosis, thus referring to a greater flexibility in the psychopathological symptom description and in the subsequent diagnostic process [2]. Indeed, regarding the chapter "Schizophrenia Spectrum and Other Psychotic Disorders," a new dimensional perspective of psychosis has been partially proposed [3]. The five main psychopathological domains that were historically used in DSM-4 for the categorical assessment of schizophrenia (SKZ) are now presented, in DSM-5, at the beginning of the chapter, thus giving a smaller emphasis to SKZ [6]. These domains, which include hallucinations, delusions, disorganized thought (speech), disorganized or abnormal motor behavior (including catatonia), and negative symptoms, now seem to hold an independent role in the mental state examination, clinical presentation, and life history of a psychotic patient [2]. However, despite the attempt at a paradigm shift, DSM-5 still strictly maintains a categorical approach, describing the severity of psychosis through level, number, and duration of psychotic symptoms and signs [2]. Heckers and colleagues suggested that the assessment of these psychopathological features in a psychotic patient should pass through eight specific dimensions, consisting of the above mentioned five domains and including also cognition impairment, depression, and manic symptoms [2, 4] (Fig. 8.1). This integration of dimensions with clinical practice could help psychiatrists to plan treatments and better predict course and outcome of the psychotic disorder [5].

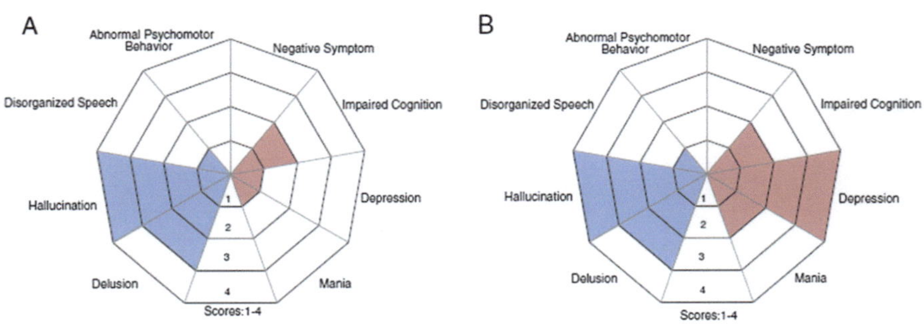

Fig. 8.1 Example of assessment of psychotic disorder in two different patients (**a**, **b**). A five-point scale ranging from 0 to 4 for each domain. Dimensional assessment of psychosis from Heckers et al. [2]

DSM-5 identified the SKZ spectrum psychotic disorders including delusional disorder, brief psychotic disorder, schizophreniform disorder, SKZ, schizoaffective disorder, substance–/medication-induced psychotic disorder, and psychotic disorder due to another medical condition [6]. It also includes the schizotypal personality disorder, unless it is better described in the personality disorders chapter [6]. Regarding psychotic disorder due to another medical condition, the diagnostic criteria are shown in Box 8.1.

Box 8.1 DSM-5 Diagnostic Criteria for Psychotic Disorder Due to Another Medical Condition [6]

A. Prominent hallucinations or delusions.
B. There is evidence from the history, physical examination, or laboratory findings that the disturbance is the direct pathophysiological consequence of another medical condition.
C. The disturbance is not better explained by another mental disorder.
D. The disturbance does not occur exclusively during the course of a delirium.
E. The disturbance causes clinically significant distress or impairment in social, occupational, or other important areas of functioning.

While the lifetime prevalence of psychotic disorders in the general population has been estimated around 3%, the prevalence of psychotic disorders due to another medical condition is approximately 0.21% [7]. There are several medical conditions whose first clinical presentation may be psychosis, such as endocrine (ACTH-producing lung carcinoma, Cushing's disease, diabetes mellitus type 1 or 2, parathyroid disease) and genetic disorders (Huntington's disease, Lewy body disease, Parkinson's disease, Wilson's disease) but also infective (encephalitis, HIV, neurosyphilis) and neurological diseases (brain tumors, dementia, Alzheimer's disease, epilepsy) or nutritional deficiencies (niacin, thiamine and vitamin B12) [8]. As regards neurological diseases, a problem of major concern to public health is represented by traumatic brain injury. TBI has been defined as "an alteration in brain function, or other evidence of brain pathology, caused by an external force" [9], and, according to the Glasgow Coma Scale, a rapid assessment permits differentiating mild and moderate TBIs from severe ones [10]. Interestingly, the onset of behavioral alterations after TBI has been known since 1848, from the famous description of the incredible case of Phineas Gage, a 25-year-old railroad laborer who was hit by an iron rod which penetrated through the frontal part of his head without killing him [11] (Fig. 8.2). John Martyn Harlow was the doctor who first assisted him by blocking the hemorrhage, eliminating skull fragments and treating the consequent infection that occurred in the weeks after the accident [11]. Twenty years later, Dr. Harlow [12] also documented the behavioral changes undergone by Mr. Gage after the accident: "*His contractors, who regarded him as the most efficient and capable foreman in their employ previous to his injury,*

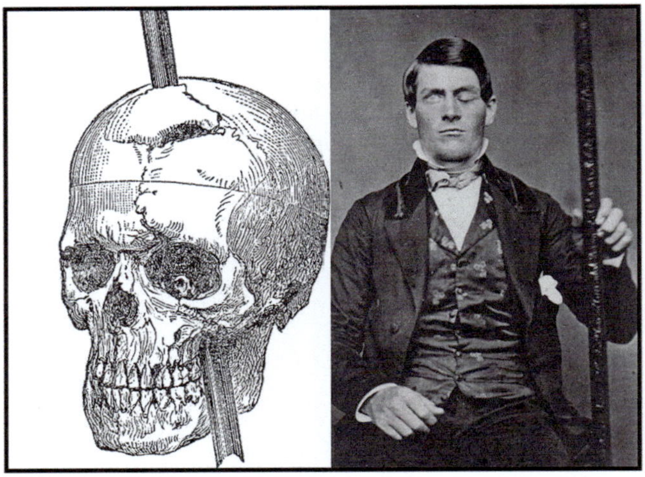

Fig. 8.2 Mr. Phineas Gage (on the right) and a representation of the skull trauma after the accident

considered the change in his mind so marked that they could not give him his place again. The balance between his intellectual faculties and animal propensities seems to have been destroyed. He is fitful, irreverent, indulging at times in the grossest profanity (which was not previously his custom), manifesting but little deference for his fellows, impatient of restraint or advice when it conflicts with his desires, at times pertinaciously obstinate, yet capricious and vacillating, devising many plans of future operation, which are no sooner arranged than they are abandoned in turn for others appearing more feasible".

It is therefore well documented that prior to the injury Mr. Gage was a trustworthy, methodical, and diligent person and that the personality changes he experienced after the accident seem to have involved the executive brain functions, resulting in a loss of social inhibition, emotion processing, and decision-making [13]. Dr. Harlow attributed these behavioral alterations directly to the damage of the left frontal lobe, while more recent researches have demonstrated the damage of specific white matter networks, such as "*the uncinate fasciculus linking the orbitofrontal to the anterior temporal lobe; the frontal intralobar networks (connections between frontal regions) and the fronto-striatal-thalamo-frontal network*" [14].

The correlation between cerebral alterations and the onset of psychiatric symptoms has also been suggested for psychotic disorders, even if it is still not certain if these changes represent an early biomarker of psychosis or the consequence of the disease itself [15]. Interestingly, Dukart et al. [15] identified, through magnetic resonance imaging (MRI), similar patterns of brain alterations in both at-risk mental state (ARMS) subjects and first-episode psychosis (FEP) patients, thus suggesting that structural brain abnormalities may represent an early endophenotype of psychosis.

Nowadays, TBI includes several causes depending on age group: the most frequent cause is trauma after fall accidents which occur mostly among young subjects and older people, while adolescents have higher risks of experiencing traumas related to traffic accidents or violent episodes [16].

The incidence of psychotic disorders due to traumatic brain injury (PD-TBI) has been estimated from 0.1 to 9.8% [16]. However, in many cases, it is often unclear whether the psychotic disorder is strictly ascribable to the TBI or if it develops independently of it [17]. Indeed, several analyses of case studies [18–20] reported that more than half of the patients experienced PD-TBI during the first year after the injury. Contrariwise, other studies detected a temporal relationship between TBI and psychotic symptoms of about 5 [21] or 10 years [22]. However, the mean latency seems to be around 4–5 years [23]. Even if there aren't clear guidelines to prove an etiologic relationship between the onset, worsening or the remission of psychotic symptoms, and the supposed medical condition, the identification of a temporal association between them is probably the best marker to make a correct diagnosis [6]. Moreover, the presence of atypical psychotic features, such as visual/olfactory hallucinations, uncharacteristic ages of onset (younger age), lack of personal or family history of SZK, and the preexistence of cognitive deficits or visual/hearing impairments, could help the clinician in the diagnostic process [6]. Some authors [24] pointed out that the most frequent schizophrenia-like symptoms in patients who suffered TBI are predominantly persecutory delusions (55%), followed by referential (22%), control (22%), grandiose (20%), and religious ones (15%). Other symptoms consisted in hallucinations, mainly auditory (84%), visual (20%), and tactile (4%), and aggressive and agitated behavior (40%), while there was less incidence of negative symptoms, catatonia, or disorganization [24]. Moreover, on the basis of neuroimaging data, the main damaged cerebral areas were unilateral and included temporal, parietal, and frontal lobes [24], as demonstrated by other studies [18, 19, 22, 25]. The relevance of these findings is strengthened by the evidence of frontotemporal alterations in SKZ [26]. Therefore, it seems that alterations in these cerebral areas are not a specific feature of SKZ, but they could represent a pathophysiological agent triggering psychosis in several conditions, such as affective psychosis [27], cerebral vascular diseases [28], Alzheimer's disease [29], brain cancers, and epilepsy [30]. Additionally, regarding neuropsychological functioning, both patients with PD-TBI and SKZ demonstrated deficits, with the latter having a more global deficiency [31]. When compared to patients with TBI without psychotic symptoms, patients with PD-TBI showed cognitive impairments in verbal memory, general intelligence, executive functioning, and vocabulary [31], similarly to schizophrenic patients [32]. As discussed above, since the PD may occur several years after the TBI, the presence of some common prodromal symptoms has also been suggested, such as deterioration in social and work functioning, depression, and antisocial and bizarre behavior [33]. Moreover, the analysis of case reports conducted by Fujii and Ahmed [18, 19] showed that 70% of patients with TBI had electroencephalographic alterations (especially in temporal lobes), while 30% experienced seizures.

Given all these peculiar features of PD-TBI, a suitable diagnostic strategy is therefore required in the assessment of psychotic disorder (Box 8.2). First, the investigation should include questions about the use of alcohol/recreational drugs, recent traumatic brain injury, and other significant neurological/medical diseases [34], as well as information about family, culture, sociality, and previous travels. It is therefore important to start a differential diagnostic process through symptoms and signs recognition in order to discriminate between primary and secondary psychosis. If there is the suspicion of psychotic disorder due to a medical condition, laboratory testing and imaging screening are required in order to detect blood/urine alterations and focal cerebral lesions [35].

Box 8.2 Diagnostic Strategy

- Obtaining history and information about family, sociality, travels, previous head traumas, use of alcohol/recreational drugs
- Clinical examination (both physical and mental state examination): differential diagnosis through symptoms and signs recognition
- Laboratory test
- Imaging screening (CT or MRI)

Once the diagnostic procedure is completed and a diagnosis of PD-TBI has been made, it is essential to optimize treatments. However, there are no specific guidelines for the treatment of PD-TBI, since most of the researches come from clinical case reports which recommend the importance of antipsychotic therapy [16]. Atypical antipsychotics (AP) seem to be more efficient, while typical APs seem to worsen the neurological deficits occurring in TBI because of their anticholinergic and therefore more hypotensive and sedative action and also their stronger dopaminergic antagonism, which may provoke delay in neuronal recovery [16].

8.2 Case Presentation

M.P. is a 34-year-old Caucasian man who was admitted to our psychiatric ward in May 2017 in order to reassess diagnosis and pharmacological therapy. He was previously diagnosed with “paranoid schizophrenia” by the psychiatrists from territorial services which have followed his case since 2007.

Family history documented the presence of a maternal uncle diagnosed with “mental retardation,” but no other information could be retrieved. The patient was born prematurely during the 35th week; therefore, he needed to stay in the incubator for an unspecified period of time. In remote pathological history, M.P. had hypertension and previous skull traumas between the ages of 15 and 25, caused by beating, horseback riding, and alcohol poisoning. Moreover, alcohol abuse was reported, but not psychoactive drug consumption.

The onset of psychopathological symptoms was around the age of 22, when he started to manifest social withdrawal, psychomotor inhibition, avolition, and other behavioral alterations. At the age of 28, he undertook rehabilitation programs in several therapeutic communities. In 2009, 2012, and 2014, he was admitted to the psychiatric ward for psychomotor agitation or behavioral anomalies.

The family provided us with psychopathological descriptions of M.P. from previous clinical records:

- *The family insisted that he enrolled in the University, but he is not a good student [. . .] he is isolated, he has no friends, he never goes out, he does not eat much [. . .] he has anger attacks and he does not listen anymore [. . .]. Sometimes he breaks objects [. . .]*
- *After summer 2005 I found M. very much changed [. . .]. His behavior was very strange [. . .] he was just in the middle of the lawn [. . .] he did not come near us, did not speak, avoided us [. . .] he could not find the words*

Since the onset of the abovementioned symptoms, the relatives of the patient described a gradual worsening of the psychopathological conditions; in addition, they described a continuous withdrawal and failure of social functioning, without experiencing any periods of psychic well-being. The patient was treated with different antipsychotics, such as olanzapine, risperidone, aripiprazole, and other medications (i.e., valproate, mirtazapine, benzodiazepines). However, there was only a partial remission of the symptoms.

On admission to our ward, the patient appeared neglectful (especially in his personal hygiene), unable to take care of himself, and fatuous; in addition, his speech, which had to be stimulated, was perseverant, stereotyped, and partly informative. The thought was correct in form, with labile associative links, and the content was poor and concrete; in addition, his mood appeared indifferent. It is worth noting that, despite denying the presence of hallucinations, he showed a listening attitude. Lastly, the patient showed poor insight.

During his stay in our inpatient unit, both blood exams and urine drug screen were performed. The results of the blood tests were as follows: (a) serum sodium is 119 mEq/l, (b) serum chloride is 78 mEq/l, and (c) the remaining findings were normal in their values. The results of the urine drug screening were negative with respect to the common drugs of abuse. In addition, we performed electroencephalography (EEG), which resulted normal, and both magnetic resonance imaging (MRI) and fluorodeoxyglucose positron emission tomography (FDG-PET). It is worth noting that the MRI (see Fig. 8.3) revealed encephalomalacia in the front skull, near the right bilateral lap, with associated minimum T2 high-intensity signal to the adjacent white substance; analogous evidence was also observed in the right temporo-polar location. The abovementioned findings appeared to be attributable to posttraumatic contusion findings and need correlation with clinical-anamnestic data.

The PET images (Fig. 8.4) show, furthermore, a non-fixing area of the tracer at the right fronto-basal region at the co-recorded CT images of hypodensis; an analogous defect of fixation was observed on the homolateral temporal pole, both caused by a supposed expansion of liquor space.

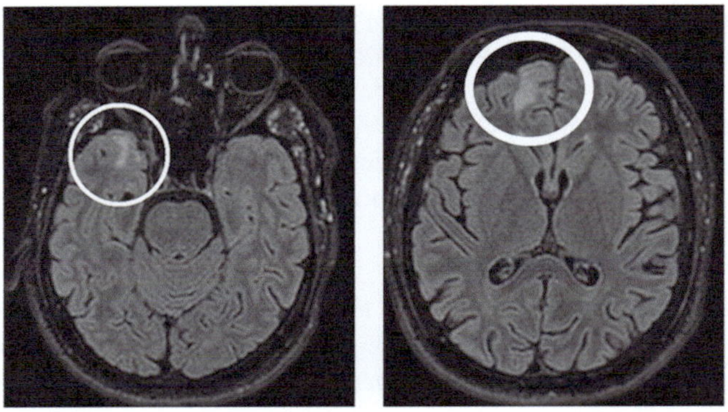

Fig. 8.3 Right panel: coronal section shows encephalomalacia with associated T2 high-intensity signal in the frontal lobe. Left panel: cerebral softening in the right temporo-polar location

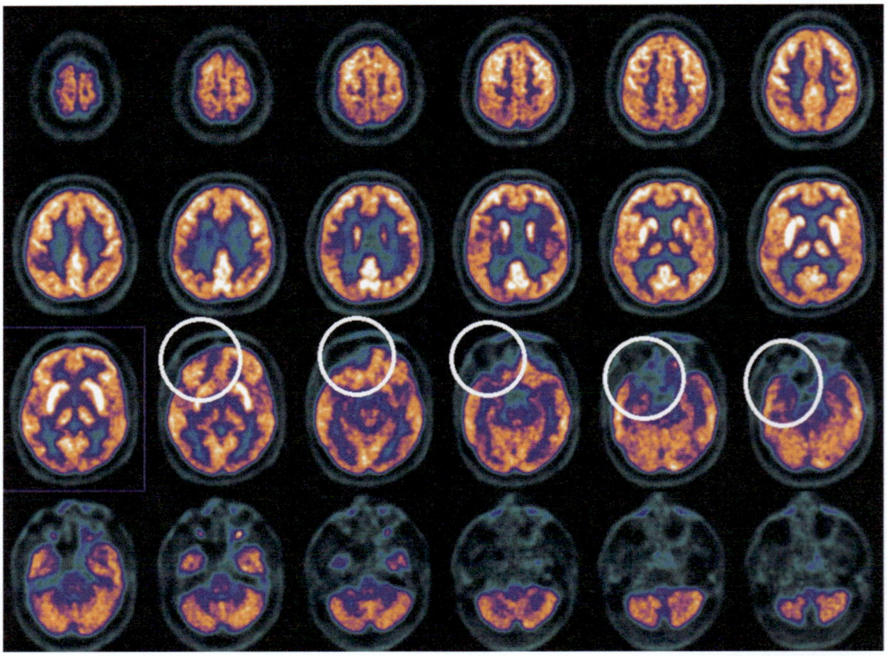

Fig. 8.4 Traversal images show non-fixing areas highlighted by empty circles in fronto-basal and temporal regions

Subsequently, images were obtained by integrating the MRI acquisition with the PET at the frontal (Fig. 8.5) and temporal (Fig. 8.6) scans: in these images it could be noticed that structural abnormalities, highlighted by magnetic resonance, overlap perfectly with the areas of hypocaptation and hence reduced functionality, obtained by scintigraphy.

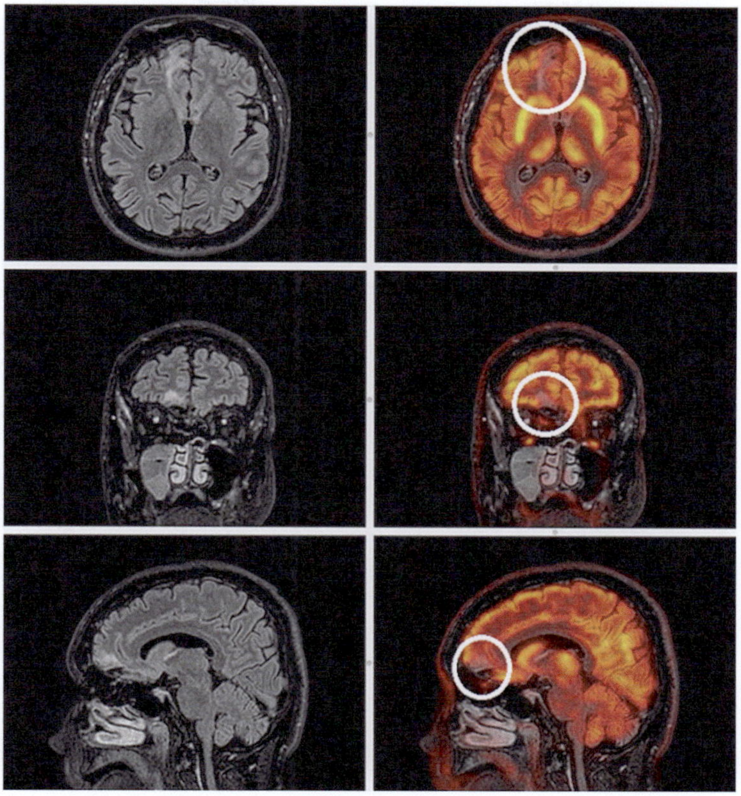

Fig. 8.5 Integration of PET and MRI images—alteration in frontal lobe

The patient subsequently underwent Brief Assessment of Cognition in Schizophrenia (BACS), which showed a deficit in all domains, in particular the frontal one (Table 8.1).

The neurologist who examined the patient suggested that the lesions might be related either to childbirth trauma (e.g., the temporal one) or to an acquired trauma (e.g., the frontal one). Hence, it seemed likely that the damaged structures might be, even partly, the cause of the patient's behavioral disorders.

We prescribed paliperidone up to 12 mg, to control the psychotic symptoms, and quetiapine up to 800 mg and valproate up to 1000 mg, to control both the mood and the behavioral alterations. The patient showed a slow and continuous improvement in psychopathology; indeed, at discharge the patient showed greater adequacy in relational behavior. The patient stayed in our department for 3 weeks; subsequently he was discharged and sent to a neuropsychiatric rehabilitation clinic. In conclusion, we formulated the diagnosis of "psychotic disorder due to another medical condition," on the basis of (a) the brain injury history, (b) the whole clinical presentation, and (c) the reports of the imaging exams.

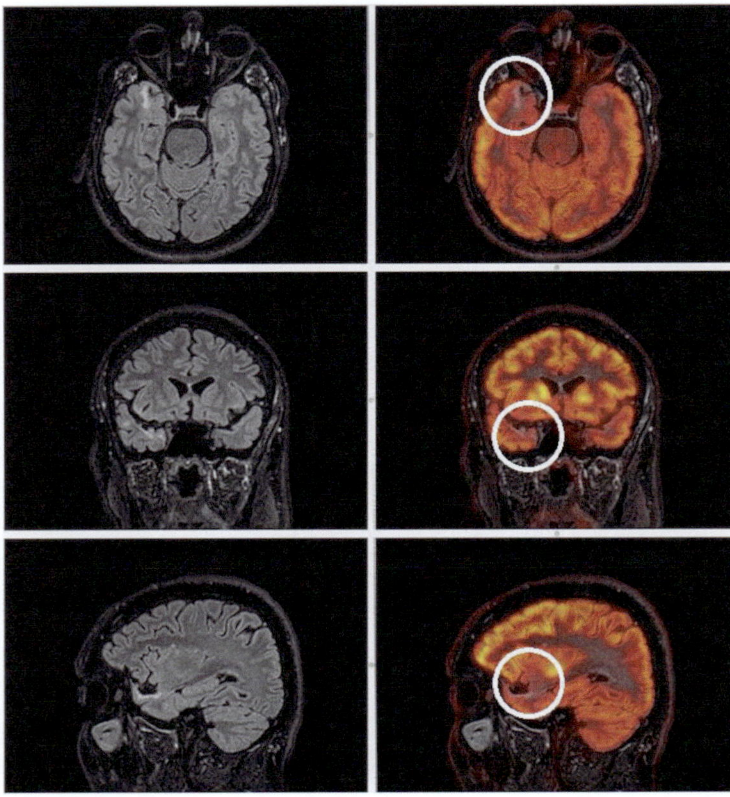

Fig. 8.6 Integration of PET and MRI images—alteration in the temporal lobe

Table 8.1 Neurocognitive evaluation

Test	Cutoff	Score	P.E.	Evaluation
Language				
Verbal fluency	v.n. $\geq$ 31.68	28.25	0	Deficit
Memory				
Verbal memory	v.n. $\geq$ 33.01	17.75	0	Deficit
Motor proficiency				
Token task	v.n. $\geq$ 68.77	30	0	Deficit
Symbol-coding task	v.n. $\geq$ 40.49	10.75	0	Deficit
Frontal proficiency				
Working memory	v.n. $\geq$ 14.93	9	0	Deficit
Tower of London	v.n. $\geq$ 12.37	8	0	Deficit

8.3 Discussion

This case report represents an opportunity to underline the importance of differential diagnosis in psychosis. Indeed, since there are several categories of psychotic syndromes, a valid diagnostic procedure and evaluation are required in order to make the correct diagnosis. Moreover, some clinicians may underestimate the prevalence of specific psychotic disorders, such as PD-TBI, and a misdiagnosis may negatively influence both treatment and prognosis.

The categorical classification in DSM-5 allows clinicians to group patients starting from their psychopathological symptomatology, thus helping the identification of a specific cluster of symptoms rather than a specific disease [36]. It is therefore useful to find those elements which may help clinicians to discriminate between the various psychotic disorders such as schizophrenia, schizoaffective disorder, schizophreniform disorder, delusional disorder, psychotic disorder NOS, brief psychotic disorder, psychotic depression, psychotic bipolar disorder, substance-induced psychotic disorder, and psychotic disorder due to a medical condition.

Historically, the concept of psychosis has been strongly associated with the concept of schizophrenia, although schizophrenia represents only a small part of the psychotic spectrum disorders (1% of the whole population) (Fig. 8.7) [36]. However, even within a specific diagnosis of psychosis, there is a large heterogeneity and variability of symptoms presentation, as well as differences in outcome and treatment response, which cannot be gathered together in the "SZK" diagnosis [36].

Interestingly, as regards PD-TBI, Fujii and Fujii [23] clearly analyzed case reports published after 2000 concerning subjects who experienced psychotic disorder after a head trauma. The research included 30 articles and 62 case reports, and it is useful in the identification of a PD-TBI. Indeed, the authors obtained information about demographics, clinical presentation, course, and laboratory findings which are shown in Box 8.3.

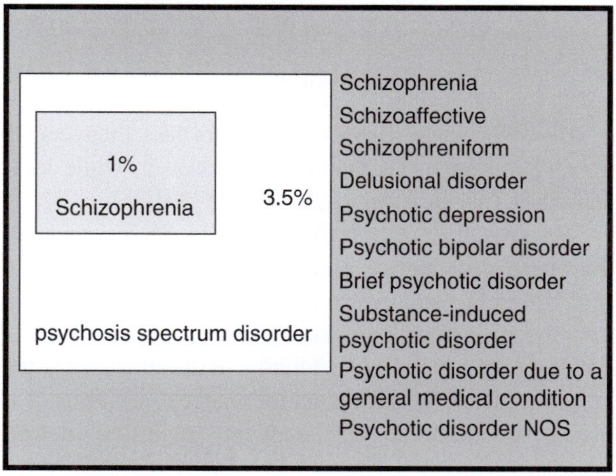

Fig. 8.7 Lifetime prevalence of psychotic disorders [36]

Box 8.3 Characteristics of PD-TBI [31]

1. *PD-TBI results from both mild and moderate head injuries.*
2. *There is a bimodal distribution of time between TBI and onset of psychosis.*
3. *Seizure disorder is more common in PD-TBI than in TBI.*
4. *Most persons with PD-TBI improve in presentation, with antipsychotics the most efficacious medications.*
5. *Male gender and family history of schizophrenia are risk factors for developing PD-TBI.*
6. *The most common psychotic symptoms associated with PD-TBI are persecutory delusions and auditory hallucinations. Negative symptoms are less pronounced.*
7. *PD-TBI is associated with cognitive impairments, most commonly in memory and executive functioning.*
8. *PD-TBI is associated with lesions to frontal and temporal areas of the brain as identified by neuroloqical studies.*

Moreover, they analyzed differences between PD-TBI and schizophrenia (Box 8.4).

Box 8.4 PD-TBI Vs Schizophrenia [31]

1. *Less likely to present with negative symptoms*
2. *More likely to demonstrate positive findings on CT/MRI, with the most common findings being focal lesions to temporal and frontal lobes, versus patients with schizophrenia (PWS), who commonly present with enlarged ventricles*
3. *More likely to demonstrated positive findings on EEG, with the most common finding of temporal slowing, versus PWS, who most commonly demonstrate frontal slowing*

However, even if SKZ represents less than one third of psychotic disorders, medical research has mainly focused on it, while less attention has been given to the other 70% of psychotic disorders [36].

Notably, the first classifications of SKZ date back to the nineteenth century: European psychiatrics described cases of male patients with mental deterioration, and different terms were used to identify these situations. Morel used the term "démence précoce," Hecker characterized hebephrenia, and Kahlbaum described catatonic syndrome [37]. Finally, Kraepelin integrated those terms in the unique nosological entity of "dementia praecox" dividing it in nine clinical forms, all of which resulted at the end in cognitive deficit and behavioral decline

[38]. Subsequently, Bleuler replaced the term by introducing "schizophrenia" that "[. . .] *is not a disease in the strict sense, but appears to be a group of diseases* [. . .]. Therefore, we should speak of schizophrenias in the plural." He extended the boundaries of the concept of psychosis to incorporating even less serious cases. Moreover, he introduced the concept of basic and accessory symptoms: the first group included thought and speech derailment, volitional indeterminacy, affective incongruence, and withdrawal from reality, while the second one consisted of delusions and hallucinations [39]. Schneider reformulated this classification, introducing "first rank symptoms," that allowed the diagnosis of schizophrenia. Schneider departed from the classification described by his predecessors and split schizophrenia into "systematic," exogenously determined, and "non-systematic" groups [40]. Recently the concept of psychotic spectrum has been introduced into the literature: psychosis could not be a discrete symptom in the general population, but delusions and hallucinations seem to have a gradual distribution from normal to disease, from schizotypy to schizophrenia [5]. In fact, in the general population, a psychotic experience seems to be frequent, with prevalence about 7%, and most of these individuals meet the criteria for another psychiatric disorder, such as mood and anxiety disorders. Psychosis is also a transdiagnostic link between psychopathological domains (positive, negative, disorganizing, and affective ones) [41]. This variability could be explained by variable interaction between genes and environmental factors, and a graduated clinical phenotype is obtained [40].

When we discuss the concept of schizophrenia, we do not even consider the impact of this diagnosis on the patient and on his social network [42]. The word SKZ was contested by many patients, families, and psychiatrists [36] and seems to have a negative connotation among the general population: it could be associated with "split personalities" or "catastrophic disorganization" [43], and the diagnosis itself seems to have a debilitating impact and stigma, greater than other psychiatric disorders [42, 44]. This may in part be due to the public and professional perception of schizophrenia as irreversible brain disease associated with dangerousness, aggressiveness, incompetence, and dysregulation of emotion and behavior [44]. Howe et al. in 2014 investigated the subjective reactions and stigma related to the diagnostic label of patients when they received the diagnosis of schizophrenia. Many patients sometimes tried to avoid this label, describing symptoms other than those really experienced, thus receiving a wrong diagnosis, which causes delay in setting the correct therapy. Sometimes psychiatrists seem to be reluctant to use the word "schizophrenia," preferring to replace it with synonyms, such as "psychosis," and thus increasing negative perception by patients. Furthermore, patients cannot understand more about their disorder. For these reasons, lots of patients would try to hide the diagnosis, both within the family and with friends [42]. To reduce this stigma, it was proposed to modify the name of the disorder [42, 45], and in 2006 in the United Kingdom, a group of psychiatrists and patients with their families launched *Campaign for the Abolition of Schizophrenia Label* [46]. The first state that modified the label was Japan: in 2002 the term s*eishin bunretsu byo*, which literally means "mind-split-disease," was replaced by *togo-sitcho-syo*, which means "integration disorder" [43, 47]. After this change the stigma associated with the name seems to be reduced, and patients are considered less dangerous, particularly among the young [48], but the evidence is still insufficient [43]. Following Japan's example, a change was achieved in South Korea in 2012, where they introduced the word *johyeonbyung* (attunement disorder). *Johyeon* literally means "to tune a stringed musical instrument," and, referred to schizophrenia, attunement shows a metaphoric meaning of "tuning the strings of the mind" [49].

Lastly, it has been suggested that PD-TBI is a pathophysiological model of primary psychosis [50]. Fuji and Ahmed [18, 19] argued that psychotic symptoms could be similar to neurological symptoms since they represent the result of damage in certain cerebral areas. The disruption of the neurotransmission system in the frontal and temporal lobes implicates an increased activity in the temporal limbic area [18, 19]. All these subjects could therefore be at risk for the development of psychosis, but a genetic predisposition and an exposure to specific environmental risk factors, such as TBI or drug abuse, are required in order to provoke a psychotic disorder [18, 19, 51]. This hypothesis is partially confirmed by the onset of psychotic symptoms in some frontotemporal diseases, such as Alzheimer's disease and temporal lobe epilepsy [16]. However, it seems that TBI could also cause other psychiatric disorders beyond psychosis (0.1–9.8%), such as depression (15.3–33%), mania (9%), obsessive-compulsive disorder (1.6–15%), posttraumatic stress disorder (11.3–24%), alcohol-related disorders (34.9–51%), and personality changes (34.5%) [16].

8.4 Conclusions

The proposed case report remarks the importance of a comprehensive and a complete evaluation of the psychotic patient, without considering only the most frequently observed diagnosis (i.e., schizophrenia). It is important to perform imaging exams to evaluate both the morphological and functional abnormalities of cerebral structures; these two aspects, along with the clinical history, help in drawing a correct diagnosis.

Key Points

- A general transdiagnostic psychosis dimension is complemented by eight specific diagnostic constructs, including hallucinations, delusions, disorganized thought (speech), disorganized or abnormal motor behavior (including catatonia), negative symptoms, cognition impairment, depression, and manic symptoms.
- Different psychotic categories can be viewed as part of the same spectrum syndrome.
- SKZ represents only a minority of psychotic disorders (less than a third).
- Consider the possibility of certain psychiatric disorders, including psychotic disorder, developing after a traumatic brain injury.

Self-Assessment Questionnaire

1. What's the estimated incidence of psychotic disorders due to traumatic brain injury?
   (A) 0.1–10%
   (B) 10–15%
   (C) 15–25%
   (D) 30–40%
2. Which are the main damaged cerebral areas in PD-TBI?
   (A) Bilateral frontal and temporal lobes
   (B) Unilateral occipital and prefrontal lobes
   (C) Bilateral occipital and prefrontal lobes
   (D) Unilateral frontal and temporal lobes
3. How long before half of the patients experienced PD-TBI?
   (A) 2 weeks
   (B) 1 month
   (C) 12 months
   (D) 2 years
4. What is, most likely, the best marker to draw a correct diagnosis of PD-TBI?
   (A) Laboratory tests
   (B) Temporal association
   (C) Cognitive impairments
   (D) Positive psychiatric family history
5. What's the best treatment of PD-TBI?
   (A) Mood stabilizers
   (B) Antidepressants
   (C) Atypical antipsychotics
   (D) Typical antipsychotics

References

1. Kupfer DJ, Regier DA. Neuroscience, clinical evidence, and the future of psychiatric classification in DSM-5. Am J Psychiatr. 2011;168(7):672–4.
2. Heckers S, Barch DM, Bustillo J, Gaebel W, Gur R, Malaspina D, et al. Structure of the psychotic disorders classification in DSM-5. Schizophr Res. 2013;150(1):11–4.
3. Abidi S, et al. CPA guidelines for the pharmacological treatment of schizophrenia Spectrum and other psychotic disorders in children and youth. Can J Psychiatr. 2017;62(9):635–47. PMID: 28764561
4. Barch DM, Bustillo J, Gaebel W, Gur R, Heckers S, Malaspina D, et al. Logic and justification for dimensional assessment of symptoms and related clinical phenomena in psychosis: relevance to DSM-5. Schizophr Res. 2013;150(1):15–20.
5. Allardyce J, Gaebel W, Zielasek J, Van Os J. Deconstructing psychosis conference February 2006: the validity of schizophrenia and alternative approaches to the classification of psychosis. Schizophr Bull. 2007;33(4):863–7.
6. American Psychiatric Association. Diagnostic and statistical manual of mental disorders. 5th ed. Washington: American Psychiatric Association; 2013.

7. Perälä J, Suvisaari J, Saarni SI, Kuoppasalmi K, Isometsä E, Pirkola S, et al. Lifetime prevalence of psychotic and bipolar I disorders in a general population. Arch Gen Psychiatry. 2007;64(1):19–28.
8. Griswold K, Del Regno PA, Berger RC. Recognition and differential diagnosis of psychosis in primary care. Brain. 2015;100(11):18–37.
9. Menon DK, Schwab K, Wright DW, Maas AI. Position statement: definition of traumatic brain injury. Arch Phys Med Rehabil. 2010;91(11):1637–40.
10. Teasdale G, Jennett B. Assessment of coma and impaired consciousness. A practical scale. Lancet. 1974;2:81–4.
11. García-Molina A. Phineas gage and the enigma of the prefrontal cortex. Neurología (English Edition). 2012;27(6):370–5.
12. Harlow JM. Recovery from the passage of an iron bar through the head, vol. 2. Boston: Massachusetts Medical Society; 1868. p. 327–47.
13. Guidotti TL. Phineas gage and his frontal lobe—the "American crowbar case". Arch Environ Occup Health. 2012;67(4):249–50.
14. Thiebaut de Schotten M, Dell'Acqua F, Ratiu P, Leslie A, Howells H, Cabanis E, et al. From Phineas gage and monsieur Leborgne to HM: revisiting disconnection syndromes. Cereb Cortex. 2015;25(12):4812–27.
15. Dukart J, Smieskova R, Harrisberger F, Lenz C, Schmidt A, Walter A, et al. Age-related brain structural alterations as an intermediate phenotype of psychosis. J Psychiatry Neurosci. 2017;42 (4):160179.
16. Schwarzbold M, Diaz A, Martins ET, Rufino A, Amante LN, Thais ME, et al. Psychiatric disorders and traumatic brain injury. Neuropsychiatr Dis Treat. 2008;4(4):797.
17. Kim E, Lauterbach EC, Reeve A, Arciniegas DB, Coburn KL, Mendez MF, et al. Neuropsychiatric complications of traumatic brain injury: a critical review of the literature (a report by the ANPA committee on research). J Neuropsychiatry Clin Neurosci. 2007;19(2):106–27.
18. Fujii D, Ahmed I. Psychotic disorder following traumatic brain injury: a conceptual framework. Cogn Neuropsychiatry. 2002;7(1):41–62.
19. Fujii D, Ahmed I. Characteristics of psychotic disorder due to traumatic brain injury: an analysis of case studies in the literature. J Neuropsychiatry Clin Neurosci. 2002;14(2):130–40.
20. Koponen S, Taiminen T, Portin R, Himanen L, Isoniemi H, Heinonen H, et al. Axis I and II psychiatric disorders after traumatic brain injury: a 30-year follow-up study. Am J Psychiatr. 2002;159(8):1315–21.
21. Nielsen AS, Mortensen PB, O'Callaghan E, Mors O, Ewald H. Is head injury a risk factor for schizophrenia? Schizophr Res. 2002;55(1):93–8.
22. Achté K, Jarho L, Kyykkä T, Vesterinen E. Paranoid disorders following war brain damage. Psychopathology. 1991;24(5):309–15.
23. Fujii D, Fujii DC. Psychotic disorder due to traumatic brain injury: analysis of case studies in the literature. J Neuropsychiatry Clin Neurosci. 2012;24(3):278–89.
24. Sachdev P, Smith JS, Cathcart S. Schizophrenia-like psychosis following traumatic brain injury: a chart-based descriptive and case–control study. Psychol Med. 2001;31(2):231–9.
25. Buckley P, Stack JP, Madigan C, O'Callaghan E, Larkin C, Redmond O, et al. Magnetic resonance imaging of schizophrenia-like psychoses associated with cerebral trauma: clinicopathological correlates. Am J Psychiatry. 1993;150(1):146–8.
26. Maggioni E, Bellani M, Altamura AC, Brambilla P. Neuroanatomical voxel-based profile of schizophrenia and bipolar disorder. Epidemiol Psychiatr Sci. 2016;25(4):312–6.
27. Maggioni E, Altamura AC, Brambilla P. Exploring the neuroanatomical bases of psychotic features in bipolar disorder. Epidemiol Psychiatr Sci. 2017:1–6.
28. Wong A, Lau AY, Yang J, Wang Z, Liu W, Lam BY, et al. Neuropsychiatric symptom clusters in stroke and transient ischemic attack by cognitive status and stroke subtype: frequency and relationships with vascular lesions, brain atrophy and amyloid. PLoS One. 2016;11(9): e0162846.

29. Ismail Z, Nguyen MQ, Fischer CE, Schweizer TA, Mulsant BH. Neuroimaging of delusions in Alzheimer's disease. Psychiatry Res Neuroimaging. 2012;202(2):89–95.
30. Lautenschlager NT, Förstl H. Organic psychosis: insight into the biology of psychosis. Curr Psychiatry Rep. 2001;3(4):319–25.
31. Fujii D, Ahmed I, Hishinuma E. A neuropsychological comparison of psychotic disorder following traumatic brain injury, traumatic brain injury without psychotic disorder, and schizophrenia. J Neuropsychiatry Clin Neurosci. 2004;16(3):306–14.
32. Krivoy A, Fischel T, Weizman A. The cognitive deficit in schizophrenia. Harefuah. 2012;151 (5):277–80.
33. Zhang Q, Sachdev PS. Psychotic disorder and traumatic brain injury. Curr Psychiatry Rep. 2003;5(3):197–201.
34. Saunders K, Brain S, Ebmeier KP. Diagnosing and managing psychosis in primary care. Practitioner. 2011;255(1740):17–21.
35. Levenson JL. Blog archives psychosis in the medically ill. Primary psychiatry. 2005;12 (8):16–8.
36. Van Os J. "Schizophrenia" does not exist. BMJ. 2016;352:i375.
37. Morel BA. *Traité des maladies mentales*. [Treatise on mental diseases]. Paris: Masson; 1860.
38. Kraepelin E. Psychiatrie. *Ein Lehrbuch für Studierende und Aerzte*. 6. Auflage. Leipzig: Barth; 1899. [English translation by Metoui H, Ayed S. Psychiatry, A Textbook for Students and Physicians, 6th edition. Volumes 1 and 2. Canton, MA: Science History Publications; 1990].
39. Bleuler E. *Lehrbuch der Psychiatrie*. Berlin, Germany: Springer; 1920. English translation: *Textbook of Psychiatry*. New York: Arno Press; 1920. p. 1976.
40. Jablensky A. The diagnostic concept of schizophrenia: its history, evolution, and future prospects. Dialogues Clin Neurosci. 2010;12(3):271.
41. Van Os J, Reininghaus U. Psychosis as a transdiagnostic and extended phenotype in the general population. World Psychiatry. 2016;15(2):118–24.
42. Howe L, Tickle A, Brown I. 'Schizophrenia is a dirty word': service users' experiences of receiving a diagnosis of schizophrenia. Psychiatr Bull. 2014;38(4):154–8.
43. Yamaguchi S, Mizuno M, Ojio Y, Sawada U, Matsunaga A, Ando S, Koike S. Associations between renaming schizophrenia and stigma-related outcomes: a systematic review. Psychiatry Clin Neurosci. 2017;71(6):347–62.
44. Brabban A, Morrison T, Read J. Renaming 'schizophrenia': a step too far or not far enough? Psychol Med. 2013;43(7):1558.
45. Van Os J. The transdiagnostic dimension of psychosis: implications for psychiatric nosology and research. Shanghai Arch Psychiatry. 2015;27(2):82.
46. Lasalvia A, Penta E, Sartorius N, Henderson S. Should the label "schizophrenia" be abandoned? Schizophr Res. 2015;162(1):276–84.
47. Sato M. Renaming schizophrenia: a Japanese perspective. World Psychiatry. 2006;5(1):53.
48. Takahashi H, Ideno T, Okubo S, Matsui H, Takemura K, Matsuura M, et al. Impact of changing the Japanese term for "schizophrenia" for reasons of stereotypical beliefs of schizophrenia in Japanese youth. Schizophr Res. 2009;112(1):149–52.
49. Lee YS, Kim JJ, Kwon JS. Renaming schizophrenia in South Korea. Lancet. 2013;382 (9893):683–4.
50. Arciniegas DB, Lauterbach EC, Ginsberg DL, Anderson KE, Chow TW, Flashman LA, et al. The differential diagnosis of pseudobulbar affect (PBA): distinguishing PBA among disorders of mood and affect. CNS Spectr. 2005;10(S5):1–16.
51. Fujii DE, Ahmed I. Risk factors in psychosis secondary to traumatic brain injury. J Neuropsychiatry Clin Neurosci. 2001;13(1):61–9.

Perinatal Depression

9

Marta Serati and Greta Carnevali

Abstract

Perinatal mood disorders are one of the most common complications observed during pregnancy and postpartum, impairing maternal caregiving skills and, in most severe cases, especially in the presence of psychotic symptoms, can lead to suicide and infanticide. In the postpartum period, it is important to detect the possible risk of diagnostic conversion from unipolar to bipolar disorder, with the onset of subtle signs of hypomania or mixed symptoms, till full mania presentation.

In this chapter we present a 42-year-old woman that came to our attention for the onset of a major depressive episode during pregnancy; she previously received a diagnosis of major depressive disorder, when she was 36 years old. Her psychiatric family history was positive for depression and suicide. Four months after delivery, she experienced a recurrence of depressive episode with psychotic symptoms and mixed features. We discuss changes in pharmacological therapy to prevent relapses during pregnancy and breastfeeding. Therapy was modified due to the diagnostic conversion to bipolar II disorder, adding a second generation antipsychotic to SSRI to reach mood stabilization. Furthermore, the patient performed neurocognitive evaluation, brain magnetic resonance imaging, and fluorodeoxyglucose positron emission tomography: results are discussed with respect to the available literature. Finally, we mentioned the main biological markers associated with perinatal depression diagnosis.

In conclusion, it is necessary to screen for perinatal mood disorders during pregnancy and over the first year postpartum, especially to rule out suggestive signs for bipolar disorder onset, with assertive follow-up, to address suicide/infanticide risk.

M. Serati (✉) · G. Carnevali
Department of Neurosciences and Mental Health, Fondazione IRCCS Ca' Granda, Ospedale Maggiore Policlinico, University of Milan, Milan, Italy

A. C. Altamura, P. Brambilla (eds.), *Clinical Cases in Psychiatry: Integrating Translational Neuroscience Approaches*,
https://doi.org/10.1007/978-3-319-91557-9_9

Keywords
Perinatal · Depression · Women mental health · Pregnancy

9.1 Introduction

Perinatal depression is one of the most common complications observed during pregnancy and postpartum, differing from baby blues syndrome, a frequent condition observed in up to 50% of women, with symptoms disappearing spontaneously within few days after delivery [1]. The DSM-5 [2] introduced a specifier for major depressive episode (MDE), "with peripartum onset," considering both the onset during pregnancy and in the 4 weeks following delivery, differently from the previous DSM version. Risk factors predisposing to perinatal depression are a past history of anxiety or depression, psychiatric positive familiar history, single marital status, and inadequate social support [1]. The risk of developing perinatal depression is about 2–5% without a previous history of mood disorder, whereas a history of major depressive disorder in life-span not linked to the peripartum is associated with a 25% risk for a postpartum mood disorder [3, 4]. It is important to underline that most severe perinatal episodes occur within the first month postpartum, with mania or psychosis having an earlier onset than depression; moreover, mood episodes in the postpartum period are significantly more common in bipolar I disorder and recurrent major depression [5]. If depressive symptomatology is untreated, the risk of relapse in a subsequent pregnancy rises to 50%. Moreover, depressed women delay care, and less than 50% attend prenatal clinics; in addition, substance, alcohol, and cigarette use are highly correlated with psychiatric disorders, thus increasing medical risk for both mothers and newborns [4].

From a clinical point of view, it is important to consider the rate of diagnostic conversion (from unipolar to bipolar) in the postpartum period, called "hidden bipolar disorder" by Sharma et al. [6], reporting 11- to 18-fold higher conversion rates, particularly in the first 6 months postpartum [7, 8]. Higher bipolar II disorder diagnosis has been previously reported in patients with depressive symptoms developed in perinatal period by Mandelli et al. [9] and Çelik et al. [10]. Furthermore, we found that a longer duration of untreated illness (DUI) (>1 year) can be associated with a family history of bipolar disorder; hypomanic episodes come to medical attention later with respect to patients with major depressive episodes and suicidal attempts [11].

Beyond genetic loading, perinatal period is at risk for sleep deprivation due to baby care, so women with a risk for mood disorder, in particular BD-I subtype, need to monitor their sleep in order to reduce the possible trigger of mood episodes, as recently reported by Lewis et al. [12] and by Munk-Olsen et al. [13] for patients with previous psychiatric episodes outside postpartum period. In the most severe cases, perinatal depression could lead to suicide (near to 20% of all postpartum deaths) [4]. According to the Australian Department of Health, 73% of suicides by women within 1 year of delivery were conducted by violent means (jumping from a high place, lying in front of moving objects, gunshot, strangulation, and suffocation), compared to suicide in nonpregnant women, and at a higher rate among women with a previous or current mental illness [14]. Another important risk to consider is infanticide, committed to relieve the baby from malignancies or alterations in some body parts, seen by the mother as having psychotic symptoms [4, 15]. We actually know that perinatal depression impairs maternal caregiving skills, with subsequent difficulties in mother-infant relationship, due to lack of maternal emotional and behavioral sensitivity to the infant, leading to reduced breastfeeding, altered neurodevelopment, and future emotional and behavioral problems for children [16, 17]. To evaluate depressive symptomatology in perinatal depression, beyond other diagnostic and evaluation tools, the Edinburgh Postnatal Depression Scale (EPDS) [18] is one of the most well-known and validated instruments across age, languages, and cultural dimensions [19, 20]. It consists of 10 self-report items rated on a 4-point scale; a score of 10 or more (maximum 30) is considered a positive screen. However, some studies have used cutoff scores of 12–13 [18]. Moreover, Clark et al. [21] recently reported that Mood Disorder Questionnaire (MDQ) added to EPDS can improve the distinction of unipolar depression from bipolar depression in a sample of postpartum women. The use of both evaluations is supported also by Merrill et al. [22].

Box 9.1 Differential Diagnosis of Depression in the Peripartum Period

- Baby blues: less than 10 days of duration; onset within 2–3 days postpartum; prevalence up to 50% of women; mild dysfunction; suicidal ideation is usually not present.
- Organic disease: e.g., iron deficiency, thyroid dysfunction, brain tumor, etc.
- Perinatal depression: more than 2 weeks of duration; onset is often within the first month postpartum, but it may occur up to the first year; moderate to severe dysfunction; suicidal ideation may be present.
- Bipolar disorder: subtle signs of emerging hypomania or mixed symptoms, up to full mania presentation.
- Postpartum psychosis: onset is often from the first 48 h to 2 weeks postpartum, but it may occur up to 3 months postpartum; incidence of 1/1000 women; severe dysfunction; high risk for suicide and infanticide ideation.

9.2 Case Presentation

A 42-year-old pregnant woman came to the Department of Psychiatry, University of Milan, in October 2013, sent by the obstetric-gynecological outpatient clinic of our hospital. She was at the 13th week of gestation; the baby's birth was expected at the beginning of April 2014 (last menses July 10, 2013). The patient reported a voluntary interruption of pregnancy when she was 16 years old. With regard to her psychiatric family history, her mother was 61 years old; she suffered from chronic depression (never treated) and hypothyroidism. Her father was 66 years old, described as healthy. Her paternal grandmother died by suicide; her diagnosis is unknown. The patient used to smoke; she reduced her cigarette consumption to one cigarette per day after the beginning of her pregnancy; she didn't consume alcoholic drinks during pregnancy and denied current/previous substance abuse. With regard to medical comorbidity, she suffered from polycystic ovary syndrome, and she had a retroverted uterus.

With regard to patient psychiatric history, psychopathological onset was in April 2007, when she was 36 years old: she experienced difficulties in the workplace (she worked in the State Scientific Police), with quarrels with her colleagues due to persecution ideation. She thought to be unable to cover her working position; she didn't feel to be up to that. At the same time, her mood worsened when she separated from her husband because of his extramarital relationship; she had death thoughts and anti-conservative ideation. She received a diagnosis of MDE, and she was treated by a neurologist with pharmacological treatments (drugs prescribed unknown) and started in association psychotherapeutic sessions, with clinical recovery. The last depressive episode dated back to February 2012, with dissatisfaction in the workplace and persecution ideation, feelings of worthlessness, and difficulties in covering a role of authority in the workplace. Moreover, she described anxiety, psychomotor agitation, feelings of inability, and interpretativity as prodromic symptoms of her psychopathological episodes. At that time she received fluoxetine and olanzapine, taken for 1 year with benefit; she completely reacquired her high work functioning.

At the first psychiatric visit to our clinic, at 13 weeks of gestation, the patient was taking fluoxetine 20 mg for another depressive episode started about 6 months before, with a good psychopathological compensation. There were no ideation alterations, and the mood was stable; she reported mild sensations of anxiety and tension during the interview. Sometimes she experienced guilt feelings and weakness and weariness in starting working activity, with difficulty in paying attention to her tasks. She described a slightly reduced interest in her activities and hobbies. There was no anti-conservative ideation. The sleep-wake rhythm was regular. We reduced fluoxetine till stopping it and introduced sertraline up to 50 mg and started psychotherapeutic supportive sessions.

The patient came to the second appointment at 23 weeks of gestation; she was euthymic, guilt feelings toward herself and her family were reduced, and she experienced a greater willpower in the workplace. She continued psychotherapeutic sessions and maintained the pharmacological treatment; 4 weeks before delivery, the administration of sertraline was gradually reduced to 25 mg. The patient gave birth to her daughter, with a natural delivery with epidural anesthesia, on March 22, 2014. At birth, the baby weighed 2.990 kg and her length was 46 cm. The Apgar score is 10. During hospitalization in the puerperium, the patient decided to breastfeed the baby. Sertraline 50 mg was divided in two separate doses taken just before breastfeeding. Because of the patient's and her family's psychiatric history, to prevent relapses, she was monitored almost once a month during postpartum by a psychiatrist, and twice a month she had psychological supportive sessions with a trained psychologist specialized in the perinatal field [23].

At the first postpartum examination, the patient showed an expressive mimic; sometimes she smiled. There were no ideation alterations with regard to form and content and the mood was good. She was breastfeeding and often went to the counseling center. During the weekdays her mother helped her with the housework and to look after the baby, whereas during the weekend she stayed with her partner. In July 2014 she experienced feelings of worthlessness and incapacity ideas, moreover with guilt feelings, causing difficulties in carrying out daily activities. In addition, she reported increasing anxiety and tension, with thoughts going faster in her head. In the hypothesis of a depressive episode with "mixed features" diagnosis, a dose of olanzapine from 5 mg to 10 mg/day was added to reach mood stabilization. In September 2014 she described a worsening of inadequacy feelings and insecurity; she was critical toward herself, especially with regard to her skills in a mother's role. She experienced these feelings as reflected in the baby's eyes: "my daughter yawns because she gets bored with me, she never yawns with her father." The patient often received disapproving comments from her partner and her mother-in-law, which highlighted her lack of enthusiasm in the couple relationship. We thus augmented sertraline up to 100 mg/day, maintaining olanzapine 10 mg/day. In December 2014 she reported subjective improvement of her symptoms; the therapy was confirmed without changes.

The patient underwent a Structured Clinical Interview for Axis I Disorders (SCID 1) and a Structured Clinical Interview for Axis II Disorders (SCID 2) in postpartum, receiving a diagnosis of severe depressive episode with psychotic features and specifiers: with peripartum onset and with mixed features in patient with bipolar II disorder. No personality disorder was found.

During postpartum, the patient underwent a neurocognitive evaluation, the Brief Assessment of Cognition in Schizophrenia (BACS), showing a deficit in executive functions (frontal efficiency and motor proficiency), as shown in Table 9.1.

Moreover, brain magnetic resonance imaging (MRI) and fluorodeoxyglucose positron emission tomography (FDG-PET) were performed in the postpartum. The MRI resulted in normal ventricular system amplitude and peripheral liquor spaces for the patient's age, except for a minimal extension of the design of the cerebellar

Table 9.1 Neurocognitive evaluation assessed by BACS subtest

Test	Normal score	Score	Comment
Language			
Verbal fluency	v.n. $\geq$ 31.68	48,25	Normal
Memory			
Verbal memory	v.n. $\geq$ 33.01	61	Normal
Motor proficiency			
Token task	v.n. $\geq$ 68.77	57.50	Deficit
Symbol-coding task	v.n. $\geq$ 40.49	48,25	Normal
Frontal efficiency			
Working memory	v.n. $\geq$ 14.93	24,25	Normal
Tower of London	v.n. $\geq$ 12.37	12	Deficit

folia (Fig. 9.1). MRI studies have shown both volume and cortical thickness reduction in gray matter in bipolar disorder, also in the cerebellum [24]. For more information, see paragraph 3.2 (Neuroimaging and Neurocognitive Impairment). The PET did not evidence any significant modification in the normal and symmetric brain metabolism (Fig. 9.2). Moreover psychopathological rating scale have been completed during treatment (Fig. 9.3).

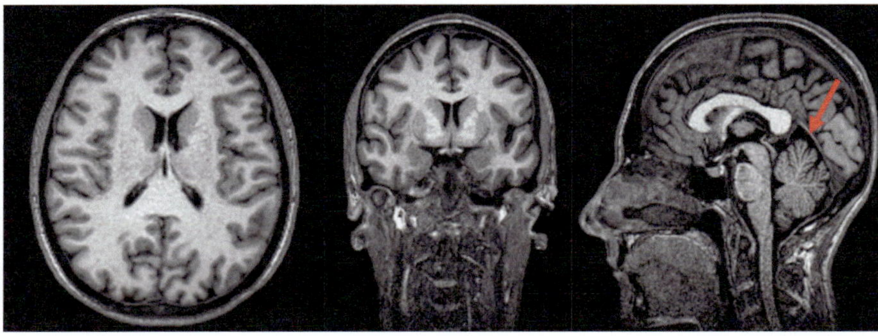

Fig. 9.1 T1-weighted cross, coronal, and sagittal images, showing a normal amplitude of the ventricular system and of the peripheral liquor spaces, except for a minimal extension of the design of the cerebellar folia (red arrow)

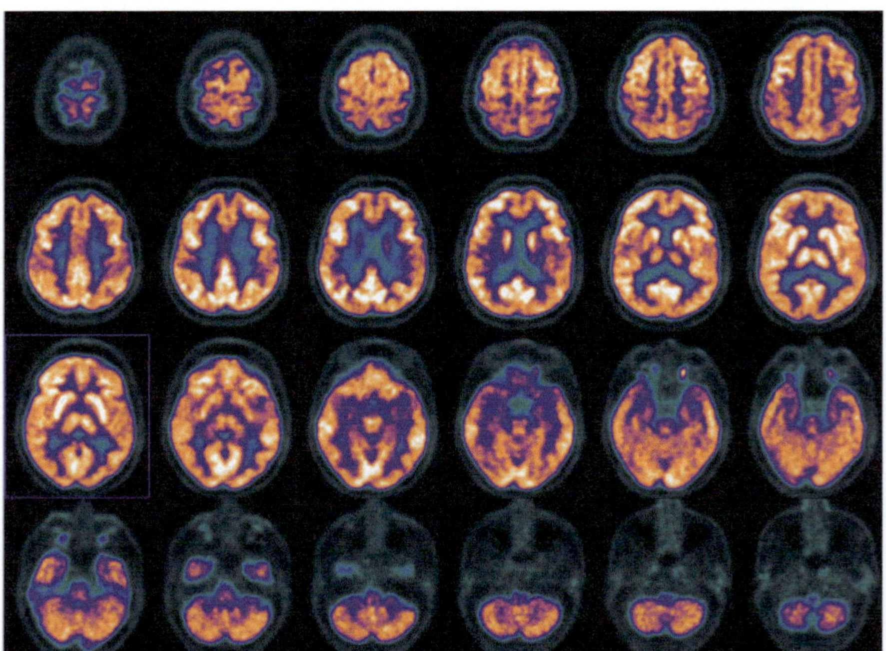

Fig. 9.2 FDG-PET shows no significant alterations of the normal symmetric glucose metabolism of the brain

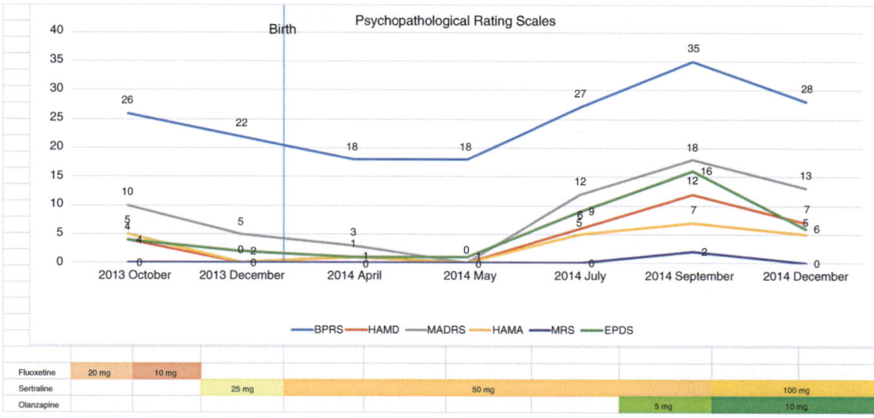

Fig. 9.3 Psychopathological Rating Scale time course. *BPRS* Brief Psychiatric Rating Scale, *HAM-D* Hamilton Depression Rating Scale, *MADRS* Montgomery-Asberg Depression Rating Scale, *HAM-A* Hamilton Anxiety Scale, *MRS* Mania Rating Scale, *EPDS* Edinburgh Postnatal Depression Scale

9.3 Literature Review

9.3.1 Biological Biomarkers

Recently, biomarkers for early perinatal depression detection have been investigated, having for this condition, as for other psychiatric disorders, a genetic loading [25, 26]. Over and above others, BDNF [27, 28], oxytocin system [29, 30], and proinflammatory immune system activation (IL-1, IL-6, and TNF-α) are involved [31, 32]. Such biological alterations probably interfere with nutrient, oxygen, and growth factor transfer to the fetus, thus compromising neurodevelopment [33, 34]. We actually know that prolonged prenatal maternal stress, due to a psychiatric disorder, may alter glucocorticoid feedback, creating a vulnerability to addictive and mood disorders in offspring and contributing to alterations in fetal neurodevelopment [3, 35–37].

9.3.2 Neuroimaging and Neurocognitive Impairment

Neuroimaging studies in bipolar patients have shown an enlargement of the third and lateral ventricles and a reduction in orbital and medial prefrontal cortex gray matter volumes, ventral striatum, and mesotemporal cortex (Severino [38]). Moreover, MRI studies have shown both volume and cortical thickness reduction in gray matter in bipolar disorder, also in the cerebellum [24]. The cerebellum is involved in cognition and affect, beyond motor functioning, with a role in modulating cortical-limbic interconnections: it receives motor and cognitive information through its connections with afferent (cortico-ponto-cerebellar) and efferent (cerebello-thalamo-cortical) pathways and takes part in neural circuitries responsible for superior cognitive functions, usually disrupted in bipolar patients [39]. We actually don't know if cerebellum alterations are due to bipolar disorder itself or to maternal diseases during gestation altering mood regulation pathways and influencing normal brain development, the so-called early neurodevelopmental model [40].

Some data suggest that even normal pregnancy is associated with structural changes in the maternal brain. Using voxel-based morphometry (VBM), Kim et al. [41] reported gray matter volume increases during postpartum period in the superior, middle, and inferior prefrontal cortex, precentral and postcentral gyrus, superior and inferior parietal lobe, insula, and thalamus. These volume changes concern areas involved in maternal behavior and maternal-infant interactions. However, no imaging studies to date have examined structural MRI differences between euthymic peripartum women and those with peripartum mood disorders [42].

Neurodegeneration in mood disorders is the result of several factors, such as inflammation; oxidative stress, increasing with repeated mood episodes; duration of untreated illness; and aging [43–45]. Thus, bipolar patients often present a significant cognitive impairment, during both the acute phase of illness and remission, worsening with cumulative episodes. The main cognitive impairment generally affects executive functions and verbal memory, due to prefrontal and medial temporal cortex impairment, respectively (Severino [38]).

Our patient showed normal ventricular system amplitude and peripheral liquor spaces for age, except for a minimal extension of the design of the cerebellar folia, probably due to atrophy. Her neurocognitive evaluation revealed a deficit in the executive functions (frontal efficiency and motor proficiency; see Table 9.1), without verbal memory impairment.

9.3.3 Treatment Strategies

9.3.3.1 Pregnancy

Ford et al. [46] recently observed that counseling by general practitioners and referrals to specialists were common in the postnatal period, less in pregnancy, with antidepressants as first line of treatment and various SSRIs considered safe and well tolerated [46–49]. No substantial evidence supports the use of one SSRI over another [50]; however, if the patient has a history of response to a particular SSRI, it is reasonable to use that medication initially [51]. Moreover, the use of antidepressants seems to be more favorable compared to exposing the mother and child to untreated depressive illness [52]. An analysis conducted by Langan and colleagues showed that with paroxetine and fluoxetine exposure during pregnancy, there was a small increase in specific birth defects, while the use of sertraline, citalopram, and escitalopram was not associated with severe birth defects [53]. Fluoxetine, with a long half-life and active parent compound and metabolite, is thus associated with neonatal syndrome. Elevated infant fluoxetine levels at birth could result in serotonergic toxicity, so, the high rates of neonatal syndrome (31%) with third trimester exposure, compared to 9% with early gestational exposure, may be explained [54, 55]. Hendrick et al. determined the maternal and umbilical cord blood antidepressant and metabolite concentrations in 38 mother/baby pairs. The lowest ratios of cord to maternal serum concentrations were for the antidepressants sertraline and paroxetine; the highest ratios were for citalopram and fluoxetine. These data suggest that some drugs produce less fetal exposure than others [56]. The recent Council on Patient Safety in Women's Health Care focused on the management of perinatal depression and anxiety [4], with a treatment cascade model, suggesting multiple opportunities to improve perinatal depression management [57].

The administration of drugs during pregnancy calls for special attention due to possible effects on the fetus, since infants exposed to SRI during the third trimester have a threefold increased risk for neonatal behavioral syndrome [58], which consists in respiratory distress, feeding problems, jitteriness, altered muscle tone, agitation, and irritability [54]. Whether the observed adverse fetal effects are related to the mother's medication or her underlying maternal illness remains difficult to determine; weighing the risks of treatment against the risk of untreated depression for both woman and child is warranted. In women in contact with UK psychiatric services, suicides in the perinatal period were more likely to occur in those with a depression diagnosis and no active treatment at the time of death. Assertive follow-up and treatment of perinatal women in contact with psychiatric services are needed to address suicide risk in this group [59]. Moreover, although causality has not been established, SSRI use during pregnancy is associated with increased risk of persistent pulmonary hypertension of the newborn, lower Apgar scores, attention-deficit/hyperactivity disorder, and speech delay [53]. Guidelines recommend that pregnant women exposed to any SSRI in early pregnancy be offered options for prenatal diagnosis through ultrasound examinations and fetal echocardiography to detect the presence of birth defects. Stopping or switching to other therapy in early pregnancy, if appropriate for the individual, may also be considered on a case-by-case basis [60].

With regard to mood stabilizers, women with chronic bipolar illness should continue treatment with lithium during pregnancy, in order to avoid relapses, or starting again at the second or third trimester to prevent postpartum recurrences [61], whereas valproate is banned during pregnancy [62]. Recent studies show promise for second generation antipsychotics in the treatment of bipolar depression with mixed features, due to their mood-stabilizing effects [8, 63]. Antidepressant monotherapy should be avoided, as it may worsen manic symptoms [64]. Electroconvulsive therapy (ECT) is an efficacious alternative choice for treatment-resistant cases or patients requiring emergent treatment who may not tolerate pharmacological therapy.

With regard to antipsychotics, placental transfer varies; typical antipsychotics have been widely used for approximately 50 years, with minimal associated risks during pregnancy. Generally, no increase in the rate of major anomalies or neurodevelopmental problems has been reported with atypical antipsychotics [63].

For a complete and updated drugs used in pregnancy, see http://www.farmaciegravidanza.gov.it.

9.3.3.2 Breastfeeding

If the patient is breastfeeding, the relative transfer of a medication into breast milk should be considered as the main parameter determining the degree of drug penetration into breast milk plasma protein binding: the higher the percentage of protein binding, the less the drug is found in maternal milk. Moreover, drugs with a large volume of distribution are poorly excreted into breast milk, while lipid-soluble drugs readily diffuse across cell membranes by dissolving in the lipid bilayer, compared with water-soluble drugs [61]. In a recent paper, Langan and collaborators observed that fluvoxamine, paroxetine, and sertraline are preferred in breastfeeding women, leading to the lowest serum medication levels in infants [53].

With regard to mood stabilizers, lithium during lactation is generally contraindicated, because of the high variability of the transfer into breast milk; lactation can be permitted through an individualized approach to breastfeeding in women receiving lithium. Valproate is theoretically safe during lactation, due to limited passage into breast milk.

From the available studies, treatment with most antipsychotics does not preclude breastfeeding, as only minimal concentrations of the active components have been found in breast milk. In particular, quetiapine and olanzapine are safe and should be the first-line treatment options when an antipsychotic is indicated, with low infant olanzapine plasma concentrations and low relative infant doses; generally, no adverse events were reported in infants exposed to olanzapine [61].

For updated and complete data on breastfeeding in pregnancy, see https://toxnet.nlm.nih.gov/newtoxnet/lactmed.htm.

9.3.3.3 Other Treatments

Recently, Dama et al. [65] outlined the importance of treating perinatal depression by considering in a new onset episode medical comorbidities above all, performing a full laboratory analysis to exclude, e.g., iron deficiency and thyroid dysfunction. Nutritional medicine could in the future be an important strategy to improve treatment effectiveness, women with perinatal depression having lower levels of total $n-3$ PUFAs and docosahexaenoic acid and significantly increased $n-6/n-3$ ratios [66]. Bright light therapy (BLT) is a well-tested and safe treatment, effective in both depression and circadian/sleep disorders [67]. If the depression is severe, resistant, and with psychotic features and requires rapid treatment, when the risk of complication is lower than the possible benefit, electroconvulsive therapy (ECT) is an excellent option in the peripartum patient [68].

A 10-week Internet-delivered cognitive behavior therapy (ICBT) has been reported to be effective in antenatal depression in a randomized controlled trial [69]. A 9-week CBT in perinatal depression group showed a clinically significant improvement in depressive symptoms [70]. The use of technology is common among perinatal women, and apps aiming to disseminate prevention programs for perinatal depression are under study [71–73].

In conclusion, women with psychiatric disorders, either already affected before pregnancy or with a new psychiatric onset in this period, pose a great challenge for treatment management [14]. Therefore, in several European countries (UK, France,

Holland), in the USA, and in Australia, mother-baby units are available to treat severe psychiatric episodes, to hospitalize both mother and child [74].

Prevention plans from earliest and prodromal depression signs and symptoms (medication prophylaxis, birth plan, and intervention strategies) are fundamental, in order to take care of women and neonatal well-being from pregnancy, including strategies for adequate sleep, maintenance of stable circadian rhythm, stress reduction, and familial support of maternal-newborn bonding [75].

Box 9.2 Treatment

- Second-generation antidepressants are considered to be relatively safe for use during pregnancy. Citalopram, escitalopram, and sertraline do not augment the risk for major malformations, while first trimester exposure to paroxetine, particularly at doses up to 25 mg/day, has been associated with a small increased risk of cardiac defects (see ToxMed for updated info). With regard to breastfeeding, antidepressants are differently excreted in breast milk [76, 77]; routine pediatric care is appropriate monitoring for breastfed infants of women taking SSRI (see LactMed for updated info).
- Lithium prophylaxis immediately postpartum has been shown to decrease relapse in high-risk women with bipolar disorder; however, it is not recommended when breastfeeding, unlike valproic acid which is allowed when breastfeeding and banned in pregnancy.
- No increase in congenital malformations with prenatal antipsychotic exposure has been reported by the National Pregnancy Registry for Antipsychotics [78]. Nevertheless, babies exposed to psychotropic drugs during pregnancy have less optimal neonatal outcome than unexposed and should be considered as a high-risk population [79].

Key Points

- Screen for perinatal depression during pregnancy and over the first year monitoring postpartum patients carefully, to rule out subtle signs of emerging hypomania and mixed symptoms, suggestive of bipolar disorder. Assertive follow-up and treatment of women to address suicide/infanticide risk are needed.
- Women at high risk for perinatal depression: previous history of mood disorders, history of postpartum depression, depressive symptoms in pregnancy, positive psychiatric family history, and poor social/family support.
- SSRI is the gold standard for perinatal depression and anxiety treatment, being considered safe and well tolerated: drug reduction before birth to minimize neonatal behavioral syndrome at birth.
- If the patient is breastfeeding, breast milk transfer should be considered: fluvoxamine, paroxetine, and sertraline lead to the lowest serum medication levels in infants.

- If depression is severe, resistant, and with psychotic features, it is recommended to add antipsychotics/mood stabilizers or consider electroconvulsive therapy (ECT) as an alternative treatment in the peripartum.

Self-Assessment Questionnaire

1. Which is the major assessment tool used to screen perinatal depression?
   (A) HAM-A
   (B) HAM-D
   (C) **EPDS**
   (D) MRS
2. Which biomarker could detect depressive states in pregnancy/postpartum?
   (A) IL-6
   (B) BDNF
   (C) Oxytocin levels
   (D) **All of the above**
3. Which drug has been reported to be unsafe during pregnancy?
   (A) Lamotrigine
   (B) **Valproic acid**
   (C) Quetiapine
   (D) Sertraline
4. Which factors have been reported to augment the risk for a severe perinatal depression?
   (A) Previous depressive episodes
   (B) Inadequate social support
   (C) Family psychiatric history
   (D) **All of the above**
5. Drug penetration into breast milk increases with:
   (A) Increasing plasma protein binding
   (B) **Increasing solubility in lipids**
   (C) Decreasing solubility in lipids
   (D) Increasing volume of distribution

References

1. McCall-Hosenfeld JS, Phiri K, Schaefer E, Zhu J, Kjerulff K. Trajectories of depressive symptoms throughout the peri- and postpartum period: results from the first baby study. J Womens Health. 2016;25(11):1112–21.
2. American Psychiatric Association, editor. Diagnostic and statistical manual of mental disorders. 5th ed. Arlington: American Psychiatric Association; 2013.
3. Apter G, Devouche E, Gratier M. Perinatal mental health. J Nerv Ment Dis. 2011;199:575–7.
4. Kendig S, Keats JP, Hoffman MC, Kay LB, Miller ES, Moore Simas TA, Frieder A, Hackley B, Indman P, Raines C, Semenuk K, Wisner KL, Lemieux LA. Consensus bundle on maternal mental health. The American College of Obstetricians and Gynecologists; 2017.

5. Di Florio A, Forty L, Gordon-Smith K, Heron J, Jones L, Craddock N, Jones I. Perinatal episodes across the mood disorder spectrum. JAMA Psychiat. 2013;70(2):168–75.
6. Sharma V, Xie B, Campbell MK, Penava D, Hampson E, Mazmanian D, Pope CJ. A prospective study of diagnostic conversion of major depressive disorder to bipolar disorder in pregnancy and postpartum. Bipolar Disord. 2014;16(1):16–21.
7. Robakis T, Jernick E, Williams K. Recent advances in understanding maternal perinatal mood disorders. F1000Res. 2017;6.
8. Sharma V, Doobay M, Baczynski C. Bipolar postpartum depression: an update and recommendations. J Affect Disord. 2017;219:105–11.
9. Mandelli L, Souery D, Bartova L, Kasper S, Montgomery S, Zohar J, Mendlewicz J, Serretti A. Bipolar II disorder as a risk factor for postpartum depression. J Affect Disord. 2016;204:54–8.
10. Çelik SB, Bucaktepe GE, Uludağ A, Bulut İU, Erdem Ö, Altınbaş K. Screening mixed depression and bipolarity in the postpartum period at a primary health care center. Compr Psychiatry. 2016;71:57–62.
11. Serati M, Buoli M, Altamura AC. Factors that affect duration of untreated illness in pregnant women with bipolar disorder. Am J Obstet Gynecol. 2015;213(6):876.
12. Lewis KS, Gordon-Smith K, Forty L, Di Florio A, Craddock N, Jones L, Jones I. Sleep loss as a trigger of mood episodes in bipolar disorder: individual differences based on diagnostic subtype and gender. Br J Psychiatry. 2017;211(3):169–74.
13. Munk-Olsen T, Maegbaek ML, Johannsen BM, Liu X, Howard LM, di Florio A, Bergink V, Meltzer-Brody S. Perinatal psychiatric episodes: a population-based study on treatment incidence and prevalence. Transl Psychiatry. 2016;6(10):e919.
14. Orsolini L, Valchera A, Vecchiotti R, Tomasetti C, Iasevoli F, Fornaro M, De Berardis D, Perna G, Pompili M, Bellantuono C. Suicide during perinatal period: epidemiology, risk factors, and clinical correlates. Front Psych. 2016;7:138.
15. Putnam KT, Wilcox M, Robertson-Blackmore E, Sharkey K, Bergink V, Munk-Olsen T, Deligiannidis KM, Payne J, Altemus M, Newport J, Apter G, Devouche E, Viktorin A, Magnusson P, Penninx B, Buist A, Bilszta J, O'Hara M, Stuart S, Brock R, Roza S, Tiemeier H, Guille C, Epperson CN, Kim D, Schmidt P, Martinez P, Di Florio A, Wisner KL, Stowe Z, Jones I, Sullivan PF, Rubinow D, Wildenhaus K, Meltzer-Brody S. Postpartum Depression: Action Towards Causes and Treatment (PACT) Consortium. Clinical phenotypes of perinatal depression and time of symptom onset: analysis of data from an international consortium. Lancet Psychiatry. 2017;4(6):477–85.
16. Barkin JL, Wisner KL, Bromberger JT, Beach SR, Wisniewski SR. Factors associated with postpartum maternal functioning in women with positive screens for depression. J Womens Health (Larchmt). 2016;25(7):707–13.
17. Pawluski JL, Lonstein JS, Fleming AS. The neurobiology of postpartum anxiety and depression. Trends Neurosci. 2017;40(2):106–20.
18. Cox JL, Holden JM, Sagovsky R. Detection of postnatal depression development of the 10-item Edinburgh Postnatal Depression Scale. Br J Psychiatry. 1987;150:782–6.
19. Gibson J, McKenzie-McHarg K, Shakespeare J, Price J, Gray R. A systematic review of studies validating the Edinburgh Postnatal Depression Scale in antepartum and postpartum women. Acta Psychiatr Scand. 2009;119(5):350–64.
20. Zubaran C, Schumacher M, Roxo MR, Foresti K. Screening tools for postpartum depression: validity and cultural dimensions. Afr J Psychiatry (Johannesbg). 2010;13(5):357–65.
21. Clark CT, Sit DK, Driscoll K, Eng HF, Confer AL, Luther JF, Wisniewski SR, Wisner KL. Does screening with the MDQ and EPDS improve identification of bipolar disorder in an obstetrical sample? Depress Anxiety. 2015;32(7):518–26.
22. Merrill L, Mittal L, Nicoloro J, Caiozzo C, Maciejewski PK, Miller LJ. Screening for bipolar disorder during pregnancy. Arch Womens Ment Health. 2015;18(4):579–83.
23. Freeman MP, Davis MF. Supportive psychotherapy for perinatal depression: preliminary data for adherence and response. Depress Anxiety. 2010;27(1):39–45.

24. Maggioni E, Bellani M, Altamura AC, Brambilla P. Neuroanatomical voxel-based profile of schizophrenia and bipolar disorder. Epidemiol Psychiatr Sci. 2016;25:312–6.
25. Miniati M, Callari A, Calugi S, Rucci P, Savino M, Mauri M, Dell'Osso L. Interpersonal psychotherapy for postpartum depression: a systematic review. Arch Womens Ment Health. 2014;17(4):257–68.
26. Serati M, Cattaneo A, Mandolini GM, Orsenigo G, Brambilla P. Il ruolo dei biomarcatori in psichiatria: stato dell'arte. Pneireview n. 2 anno; 2016a.
27. Christian LM, Mitchell AM, Gillespie SL, Palettas M. Serum brain-derived neurotrophic factor (BDNF) across pregnancy and postpartum: associations with race, depressive symptoms, and low birth weight. Psychoneuroendocrinology. 2016;74:69–76.
28. Comasco E, Sylvén SM, Papadopoulos FC, Oreland L, Sundström-Poromaa I, Skalkidou A. Postpartum depressive symptoms and the BDNFVal66Met functional polymorphism: effect of season of delivery. Arch Womens Ment Health. 2011;14(6):453–63.
29. Garfield L, Giurgescu C, Carter CS, Holditch-Davis D, McFarlin BL, Schwertz D, Seng JS, White-Traut R. Depressive symptoms in the second trimester relate to low oxytocin levels in African-American women: a pilot study. Arch Womens Ment Health. 2015;18(1):123–9.
30. Jobst A, Krause D, Maiwald C, Härtl K, Myint AM, Kästner R, Obermeier M, Padberg F, Brücklmeier B, Weidinger E, Kieper S, Schwarz M, Zill P, Müller N. Oxytocin course over pregnancy and postpartum period and the association with postpartum depressive symptoms. Arch Womens Ment Health. 2016;19(4):571–9.
31. Altamura AC, Buoli M, Pozzoli S. Role of immunological factors in the pathophysiology and diagnosis of bipolar disorder: comparison with schizophrenia. Psychiatry Clin Neurosci. 2014;68(1):21–36.
32. Garfield L, Mathews HL, Witek Janusek L. Inflammatory and epigenetic pathways for perinatal depression. Biol Res Nurs. 2016;18(3):331–43.
33. Faa G, Manchia M, Pintus R, Gerosa C, Marcialis MA, Fanos V. Fetal programming of neuropsychiatric disorders. Birth Defects Res C Embryo Today. 2016;108(3):207–23.
34. Lewis AJ, Austin E, Knapp R, Vaiano T, Galbally M. Perinatal maternal mental health, fetal programming and child development. Healthcare (Basel). 2015;3(4):1212–27.
35. Diego MA, Field T, Hernandez-Reif M, Schanberg S, Kuhn C, Gonzalez-Quintero VH. Prenatal depression restricts fetal growth. Early Hum Dev. 2009;85(1):65–70.
36. Gelman PL, Flores-Ramos M, López-Martínez M, Fuentes CC, Grajeda JP. Hypothalamic-pituitary-adrenal axis function during perinatal depression. Neurosci Bull. 2015;31(3):338–50.
37. Gentile S. Untreated depression during pregnancy: short- and long-term effects in offspring. A systematic review. Neuroscience. 2017;342:154–66.
38. Gama CS, Kunz M, Magalhães PVS, Kapczinski F. Staging and neuroprogression in bipolar disorder: a systematic review of the literature. Rev Bras Psiquiatr. 2013;35:070–4.
39. Serati M, Delvecchio G, Orsenigo G, Perlini C, Barillari M, Ruggeri M, Altamura AC, Bellani M, Brambilla P. Potential gender-related aging processes occur earlier and faster in the vermis of patients with bipolar disorder: an MRI study. Neuropsychobiology. 2017;75:32–8.
40. Buoli M, Serati M, Caldiroli A, Cremaschi L, Altamura AC. Neurodevelopmental versus neurodegenerative model of schizophrenia and bipolar disorder: comparison with physiological brain development and aging. Psychiatr Danub. 2017;29(1):24–7.
41. Kim P, Leckman JF, Mayes LC, Feldman R, Wang X, Swain JE. The plasticity of human maternal brain: longitudinal changes in brain anatomy during the early postpartum period. Behav Neurosci. 2010;124(5):695–700.
42. Duan C, Cosgrove J, Deligiannidis KM. Understanding peripartum depression through neuroimaging: a review of structural and functional connectivity and molecular imaging research. Curr Psychiatry Rep. 2017;19:70.
43. Altamura AC, Buoli M, Albano A, Dell'Osso B. Age at onset and latency to treatment (duration of untreated illness) in patients with mood and anxiety disorders: a naturalistic study. Int Clin Psychopharmacol. 2010;25(3):172–9.

44. Buoli M, Caldiroli A, Altamura AC. Psychotic versus non-psychotic major depressive disorder: a comparative naturalistic study. Asian J Psychiatr. 2013;6(4):333–7.
45. Buoli M, Caldiroli A, Caletti E, Zugno E, Altamura AC. The impact of mood episodes and duration of illness on cognition in bipolar disorder. Compr Psychiatry. 2014;55(7):1561–6.
46. Ford E, Shakespeare J, Elias F, Ayers S. Recognition and management of perinatal depression and anxiety by general practitioners: a systematic review. Fam Pract. 2017;34(1):11–9.
47. Latendresse G, Elmore C, Deneris A. Selective serotonin reuptake inhibitors as first-line antidepressant therapy for perinatal depression. J Midwifery Womens Health. 2017;62 (3):317–28.
48. Misri S, Swift E, Abizadeh J, Shankar R. Overcoming functional impairment in postpartum depressed or anxious women: a pilot trial of desvenlafaxine with flexible dosing. Ther Adv Psychopharmacol. 2016;6(4):269–76.
49. Susser LC, Sansone SA, Hermann AD. Selective serotonin reuptake inhibitors for depression in pregnancy. Am J Obstet Gynecol. 2016;215(6):722–30.
50. Molyneaux E, Howard LM, McGeown HR, Karia AM, Trevillion K. Antidepressant treatment for postnatal depression. Cochrane Database Syst Rev. 2014;9.
51. Wisner KL, Parry BL, Piontek CM. Clinical practice. Postpartum depression. N Engl J Med. 2002;347(3):194–9.
52. Muzik M, Hamilton SE. Use of antidepressants during pregnancy? What to consider when weighing treatment with antidepressants against untreated depression. Matern Child Health J. 2016;20(11):2268–79.
53. Langan RC, Goodbred AJ. Identification and management of peripartum depression. Am Fam Physician. 2016;93(10):852–8.
54. Sit D, Perel JM, Wisniewski SR, Helsel JC, Luther JF, Wisner KL. Mother-infant antidepressant levels, maternal depression and perinatal events. J Clin Psychiatry. 2011;72(7):994–1001.
55. Yang A, Ciolino JD, Pinheiro E, Rasmussen-Torvik LJ, Sit DKY, Wisner KL. Neonatal discontinuation syndrome in serotonergic antidepressant-exposed neonates. J Clin Psychiatry. 2017;78(5):605–11.
56. Hendrick V, Stowe Z, Altshuler L, Hwang S, Lee E, Haynes D. Placental passage of antidepressant medications. Am J Psychiatr. 2003;160:993–6.
57. Cox EQ, Sowa NA, Meltzer-Brody SE, Gaynes BN. The perinatal depression treatment cascade: baby steps toward improving outcomes. J Clin Psychiatry. 2016;77(9):1189–200.
58. Moses-Kolko E, Bogen D, Perel J, Bregar A, Uhl K, Levin B, et al. Neonatal signs after late in utero exposure to serotonin reuptake inhibitors: literature review and implications for clinical applications. JAMA. 2005;293(19):2372–83.
59. Khalifeh H, Hunt IM, Appleby L, Howard LM. Suicide in perinatal and non-perinatal women in contact with psychiatric services: 15 year findings from a UK national inquiry. Lancet Psychiatry. 2016;3(3):233–42.
60. Alwan S, Friedman JM, Chambers C. Safety of selective serotonin reuptake inhibitors in pregnancy: a review of current evidence. CNS Drugs. 2016;30(6):499–515.
61. Pacchiarotti I, León-Caballero J, Murru A, Verdolini N, Furio MA, Pancheri C, Valentí M, Samalin L, Roigé ES, González-Pinto A, Montes JM, Benabarre A, Crespo JM, Perrino C d D, Goikolea JM, Gutiérrez-Rojas L, André FC, Vieta E. Mood stabilizers and antipsychotics during breastfeeding: Focus on bipolar disorder. Eur Neuropsychopharmacol. 2016;26:1562–78.
62. Anniverno R, Bramante A, Petrilli G, Mencacci C. Prevenzione, diagnosi e trattamento della psicopatologia perinatale. Indicazione di buona pratica clinica per la gestione della depressione perinatale. ONDa; 2010.
63. Ornoy A, Weinstein-Fudim L, Ergaz Z. Antidepressants, antipsychotics, and mood stabilizers in pregnancy: what do we know and how should we treat pregnant women with depression. Birth Defects Res. 2017;109:933–56.
64. Rosenblat JD, McIntyre RS. Treatment recommendations for DSM-5-defined mixed features. CNS Spectr. 2017;22:147–54.

65. Dama M, Van Lieshout RJ, Steiner M. A potential gap in the perinatal depression treatment cascade. J Clin Psychiatry. 2017;78(5):612.
66. Lin PY, Chang CH, Chong MF, Chen H, Su KP. Polyunsaturated fatty acids in perinatal depression: a systematic review and meta-analysis. Biol Psychiatry. 2017;82(8):560–9.
67. Baiardi S, Cirignotta F, Cicolin A, Garbazza C, D'Agostino A, Gambini O, Giordano A, Canevini M, Zambrelli E, Marconi AM, Mondini S, Borgwardt S, Cajochen C, Rizzo N, Manconi M. Chronobiology, sleep-related risk factors and light therapy in perinatal depression: the "Life-ON" project. BMC Psychiatry. 2016;16(1):374.
68. Robakis TK, Williams KE. Biologically based treatment approaches to the patient with resistant perinatal depression. Arch Womens Ment Health. 2013;16:343–35.
69. Forsell E, Bendix M, Holländare F, Szymanska von Schultz B, Nasiell J, Blomdahl-Wetterholm M, Eriksson C, Kvarned S, Lindau van der Linden J, Söderberg E, Jokinen J, Wide K, Kaldo V. Internet delivered cognitive behavior therapy for antenatal depression: a randomized controlled trial. J Affect Disord. 2017;221:56–64.
70. Van Lieshout RJ, Yang L, Haber E, Ferro MA. Evaluating the effectiveness of a brief group cognitive behavioral therapy intervention for perinatal depression. Arch Womens Ment Health. 2017;20(1):225–8.
71. Moore D, Ayers S, Drey N. A thematic analysis of stigma and disclosure for perinatal depression on an online forum. JMIR Ment Health. 2016;3(2):e18.
72. Osma J, Barrera AZ, Ramphos E. Are pregnant and postpartum women interested in health-related apps? Implications for the prevention of perinatal depression. Cyberpsychol Behav Soc Netw. 2016;19(6):412–5.
73. Ride J, Lancsar E. Women's preferences for treatment of perinatal depression and anxiety: a discrete choice experiment. PLoS One. 2016;11(6):e0156629.
74. Kimmel MC, Lara-Cinisomo S, Melvin K, Di Florio A, Brandon A, Meltzer-Brody S. Treatment of severe perinatal mood disorders on a specialized perinatal psychiatry inpatient unit. Arch Womens Ment Health. 2016;19(4):645–53.
75. Wesseloo R, Kamperman AM, Munk-Olsen T, Pop VJ, Kushner SA, Bergink V. Risk of postpartum relapse in bipolar disorder and postpartum psychosis: a systematic review and meta-analysis. Am J Psychiatry. 2015;173:117–27.
76. Lanza di Scalea T, Wisner KL. Antidepressant medication use during breastfeeding. Clin Obstet Gynecol. 2009;52(3):483–97.
77. Pinheiro E, Bogen DL, Hoxha D, Ciolino JD, Wisner KL. Sertraline and breastfeeding: review and meta-analysis. Arch Womens Ment Health. 2015;18(2):139–46.
78. Cohen LS, Viguera AC, McInerney KA, Freeman MP, Sosinsky AZ, Moustafa D, Marfurt SP, Kwiatkowski MA, Murphy SK, Farrell AM, Chitayat D, Hernández-Díaz S. Reproductive safety of second-generation antipsychotics: current data from the Massachusetts General Hospital National Pregnancy Registry for atypical antipsychotics. Am J Psychiatry. 2016;173 (3):263–70.
79. Sutter-Dallay AL, Bales M, Pambrun E, Glangeaud-Freudenthal NM, Wisner KL, Verdoux H. Impact of prenatal exposure to psychotropic drugs on neonatal outcome in infants of mothers with serious psychiatric illnesses. J Clin Psychiatry. 2015;76(7):967–73.

Depression, Dementia, and Pseudodementia

10

Sara Pozzoli, Vera De Carlo, and Domenico Madonna

Abstract

Depression and dementia represent frequent clinical presentations in the elderly population. Both diseases are interlinked. Indeed, depression in the elderly may reflect an increased risk for the later development of dementia. Worldwide, about 47 million people are affected by dementia, which represents about 10% of the general population over 65 years of age. On the other hand, among elderly people, the prevalence of depressive symptoms and major depressive disorder (MDD) is 15% and 1–3%, respectively. Female gender, alcohol and substance or drug abuse, family history, and medical conditions are factors associated with depression in the elderly.

In the present chapter we specifically focus on discrimination between elderly MDD and major neurocognitive disorder. Concerning this point, two steps have to be performed: (1) observation of clinical response to antidepressants' treatment, taking into consideration a possible longer latency of antidepressants' efficacy in elderly/lower response to antidepressants in older age, and (2) cross-clinical assessments with neuropsychological evaluation, imaging, and laboratory tests.

Considering the background of the present literature, we pointed out the case of a woman, 72 years old, with a long history of MDD and a recent diagnosis of mild cognitive impairment (MCI). She was referred to our outpatient clinic for a relapsing depressive episode. In the present case presentation, we investigated the differential diagnostic process, observing clinical course, using neuropsychological evaluation, blood and urine laboratory exams, computed tomography, magnetic resonance imaging, positron emission tomography, cerebrospinal fluid analysis, and electroencephalogram.

S. Pozzoli (✉) · V. De Carlo · D. Madonna
Department of Neurosciences and Mental Health, Fondazione IRCCS Ca' Granda Ospedale Maggiore Policlinico, University of Milan, Milan, Italy

A. C. Altamura, P. Brambilla (eds.), *Clinical Cases in Psychiatry: Integrating Translational Neuroscience Approaches*,
https://doi.org/10.1007/978-3-319-91557-9_10

Keywords
Depression · Dementia · Pseudodementia · Elderly

10.1 Introduction

Depression and dementia are frequent in the elderly [1] and seem not to be completely independent: many studies have shown how history of elderly depression may increase risk for later development of dementia [2, 3]. Therefore assessing late-life depression could be relevant in relation to possible onset of dementia [4].

Some epidemiological studies have shown that in about 15 years the elderly population is expected to reach 20% of the general population [5], with an increase of specific health problems, including depression and progressive cognitive deficits.

10.1.1 Epidemiology

Worldwide about 47 million people are affected by dementia, with nearly 9.9 million new cases per year [6], which represent about 10% of the general population over 65 years of age [7]. Both incidence and prevalence increase with older age [8]. The most frequent kind of dementia is Alzheimer's disease (42%), followed by vascular dementia (26%), mixed dementia (caused by both Alzheimer's and vascular disease—12%), frontotemporal dementia (FTD—9%), Lewy body dementia (8%), and Creutzfeldt-Jakob's disease (3%) [9].

75% of people suffering from Alzheimer are affected by a sporadic form of Alzheimer's disease with a late onset, the remaining 25% by a familial one, which is due to genetic factors and is characterized by earlier onset (around 55 years). Vascular dementia characterizes elderly population with cardiovascular risk factors such as hypertension, heart disease, hypercholesterolemia, and thrombosis [9].

Frontotemporal dementia has the earliest onset: it usually peaks at 58 years of age, although large variability is possible, with cases where it arises in the second or in the ninth decade of life [10, 11].

On the other hand, mood disorders are widespread: it has been estimated that 1/7 of the general population is affected over the whole lifetime [12]. Among elderly people, the prevalence of depressive symptoms and major depressive disorder (MDD) is 15% and 1–3%, respectively [13]. Female gender, alcohol and substance or drug abuse, family history, and medical conditions are factors associated with depression in the elderly. British and American studies reported a prevalence of a depressive clinical presentation in 14.7% to 20% of elderly people, whereas in the Canadian community, depressive symptoms are prevalent in 10% to 15% of the elderly population [14]; finally, European epidemiological data shows that, in subjects aged 65 years, prevalence of depressive diagnoses ranges from 9% to 24%, with a mean value of 12% [15].

10.1.2 Diagnosis of Dementia

Global clinical examination is essential to recognize whether patients' symptoms and signs could be ascribed to a degenerative cognitive impairment and, more precisely, to dementia. A wide range of diagnostic tools, i.e., psychiatric, neurologic, and medical assessments, as well as blood and cerebrospinal fluid (CSF) and imaging studies, should be performed [16, 17]. Detailed anamnestic interview should be performed as a first approach to investigate psychiatric and/or neurological history, previous/ongoing medical conditions, and pharmacological history. The help of the patient's relatives or caregivers could be useful for clinicians [18]. Anamnestic results should be compatible with the diagnostic criteria of the *Diagnostic and Statistical Manual of Mental Disorders*, fifth version (DSM-5) [19], where the term "dementia," previously reported in the DSM-IV [20], has been redefined as "major neurocognitive disorder." To make a diagnosis, a significant cognitive decline from a previous level of performance should be investigated. Moreover, assessment of one or more cognitive domains (memory, language, executive functions, complex attention, perceptual-motor abilities, and social cognition) should be performed. Furthermore, cognitive deficits should interfere with one's independence in daily activities, and assistance should be required with complex instrumental daily living activities, such as paying bills or managing medications. Cognitive deficits should not occur exclusively in the context of a delirium and cannot be better explained by another mental disorder (e.g., MDD, schizophrenia).

After an accurate clinical assessment, impairment in cognitive performance has to be documented by standardized global functioning rating scale and neuropsychological testing, for example, IADL (instrumental activities of daily living) [21, 22], which investigates eight daily life activities. Other tests, such as Mini-Mental State Examination (MMSE) [23, 24] or the Mental Deterioration Battery (MDB) [25], can measure global cognitive performance and can be used as screening tests. Furthermore, blood and urine exams should be performed as soon as possible, both in hospitalized and in outpatients, to determine whether hydro-electrolyte alterations, thyroid dysfunction, or any infections have occurred. These medical conditions can be responsible for a delirium, which is characterized by confusion, loss of orientation, and memory impairment. In addition, plasma levels of daily medications should be monitored, because many psychiatric and nonpsychiatric drugs may cause hydro-electrolyte alterations or imitate symptoms or signs of the dementia spectrum when plasma levels are over the standard range [20].

Neuroimaging exams should be also taken into consideration. In particular, as a first step, a computed tomography (CT) is useful to exclude acute neurological conditions, such as stroke and intracranial hemorrhage. After that, magnetic resonance imaging (MRI) could show possible vascular neurodegeneration. Also positron emission tomography (PET), in addition to the previously listed imaging exams, may help clinicians in the diagnostic processes of dementia [26]. The cerebrospinal fluid (CSF) examination can confirm the clinical and/or imaging suspect of dementia, and it could point to different forms of dementia [27, 28]. Lastly, the electroencephalogram (EEG) is considered a diagnostic completion exam which is able to exclude epileptic syndromes in elderly patients and can help to discriminate between healthy elderly subjects, Alzheimer's disease, and vascular dementia [29].

Box 10.1 Diagnosis of Dementia

- The presence of a significant cognitive decline from a previous level of performance.
- The domains concerned are complex attention, executive function, learning and memory, language, and social cognition.
- All this must be documented by standardized neuropsychological tests (IADL, MMSE, MDB).
- Cognitive deficits interfere with autonomy in daily activities.
- Patients must not be tested in the context of a delirium or other mental disturbance, in order to be able to provide a primary diagnosis.
- To differentiate the etiologic subtypes, additional diagnostic markers, in particular neuroimaging studies such as magnetic resonance imaging or positron emission tomography, may come into play.

10.1.3 Differential Diagnosis of Depression and Dementia

Geriatric pathologies are often characterized by an acute change in mental status, a clinical condition called delirium, which could be due to thyroid dysfunction, B12 vitamin deficiency, hydro-electrolyte alterations, and epileptic or vascular syndrome. This represents a possible confusing condition; nevertheless, differently from dementia, whose course is more progressive, delirium is characterized by an acute and reversible change in mental status, and this relevant feature allows the differential diagnosis.

However, discriminating between elderly major depression and major neurocognitive disorder is not easy. Concerning this point, it might be useful to perform the following steps [19]:

- Observation of clinical response to antidepressant treatment, taking into consideration a possible longer latency of antidepressant efficacy in elderly/lower response to antidepressants in older age [30]
- Cross-clinical assessments with imaging and laboratory tests

Box 10.2 Differential Diagnosis Between Depression and Dementia

Depression	*Dementia*
• Onset can be accurately datable • Symptoms, often, are acute and progress quickly • Frequently there is a positive psychiatric history • Patients express strong discomfort and complain of the loss of cognitive efficiency, highlighting mistakes and engaging in little simple tasks • Affectivity is altered; there is loss of social-relational abilities • Cognitive disorders do not worsen overnight • Attention and concentration remain preserved • Subjects often answer "I don't know" to questions asked • Memory loss for both recent and past events • Sleep disorders • Frequent vegetative symptoms • High risk of suicide	• Onset cannot be precisely dated • Onset is sneaky with subtle symptoms • Symptom progression is very slow • Typically negative psychiatric history • Patients complain little about cognitive disorders but tend to hide their shortfalls • Affectivity is evanescent with emotional disinhibition • Social skills are conserved normally • Frequently, symptoms worsen late in the afternoon or early evening • Compromised memory and attention • The answers are often imprecise • In orientation tests, patients confuse familiar and unknown things • Severe memory loss for recent events (anterograde memory) • Fatigue without drowsiness but with restlessness or stiffness or cramps • Absence of vegetative symptoms • Lower risk of suicide

10.2 Case Presentation

10.2.1 Clinical History and Presentation

Ms. S.C., 72 years old, was referred to our outpatient clinic for a relapsing depressive episode. She had a long history of major depressive disorder (MDD) and a family history of depression (mother). She was treated for hypertension, but did not have other diseases.

The patient has been known to the psychiatric services since 1973, when, aged 28, she was admitted to the psychiatric ward for the first time. On that occasion she presented symptoms of a major depressive episode (MDE) and was treated with non-referred antidepressants. The treatment resulted in full recovery and she did not receive neither maintenance treatment nor psychotherapy. After this episode, she described a few years of well-being. She relapsed three times from 1999 to 2007, each one following a stressful life event (e.g., divorce, economic collapse, second marriage). Every time, she fully recovered and interrupted the pharmacological treatment. The episodes were always treated in outpatient services and never became so severe as to require admission to the psychiatric ward. From 2000 to 2007 the patient had a weekly session of cognitive psychotherapy.

In 2007, aged 62 years, her second husband died. After the mourning, depressive symptoms progressively worsened, and the patient contacted our psychiatric day hospital. During this episode, she referred slight memory loss. After a year and a half, neuropsychological tests resulted in range. Her mood worsened and she underwent two further hospitalizations in 2009 and 2010.

From 2010 to the present, the patient has been treated with several medicaments (antidepressants in association with atypical antipsychotics or mood stabilizers) and with psychological therapy. Mood deflection, hypersomnia, apathy, ablution, lack of initiative, and memory deficits remained rather constant.

Timeline of depression is reported in Fig. 10.1.

In January 2017, at the time of our first evaluation, the patient presented mood deflection, emotional liability, apathy, anhedonia, memory deficits, and reduction of daily functioning.

The patient had already been at our unit a few times in the previous 2 years. However, the day she came to our building, she got lost and went to the dentist's next door, believing that she was sent there to treat some teeth problems. She had a paper with the address, but was unable to reach the right place. She appeared confused and not oriented, so the dentist called our service to understand what was going on. She was then led to our service by a nurse and started to cry. She reported to have lost her keys several times in the last few months and to forget very simple daily tasks. She could not exactly refer her daily pharmacotherapy. She took 75 mg of clomipramine every day, in addition to random antidepressants. She reported that she helped herself to remember her daily medications by some notes she kept in the kitchen. Hamilton Depression (HAM-D) Rating Scale showed a mild-moderate depression (score = 18), with prevalent deficits of mood deflection and anhedonic and anxious sphere.

In day hospital she was treated with 50 mg of intravenous trazodone daily and 10 mg of vortioxetine daily. Her depressive symptoms partially decreased within 8 weeks and her HAM-D became 10. However, the patient remained sad, persistent in her speech, amnesic, and unable to cope with her daily activities. She seemed to have a global cognitive deficit.

10.2.2 Neuropsychological History and Previous Neuroimaging Results

The onset of the first symptoms of memory loss is 2007. The result of the neuropsychological evaluation was a "minimum cognitive deficit in amnesic-attentive capacities that appears to be consequent to an affective disorder rather than a primary sign of neuro-progressive disease." A second evaluation performed in 2008 did not show cognitive deficits.

From 2009 to 2011, annual neuropsychological assessments revealed slight fluctuating deficits, defined as "minimal cognitive deficits consequent to anxiety and depression."

In 2012, the patient was diagnosed with *minimal cognitive impairment* (MCI), and the diagnosis was confirmed in the following 2 years.

The last neuropsychological evaluation was performed in 2014 and showed mild diffused cognitive deficits, with a worsening in speech tasks. Results have been ascribed to the patient's depressive disease.

Timeline of neuropsychological evaluation is shown in Fig. 10.2.

Neuropsychological tests have shown over time a decrease in tests that relate to selective attention, visual-spatial search capability, and memory. In this chart, we have reported the most important deficitary tests

In association with neuropsychological evaluations, the patient underwent several neuroimaging scans from 2007 to 2017.

In 2007 the brain RMN showed an expansion of the convexity liquor spaces, as well as micro-involutive lesions of the periventricular white matter (WM).

Another brain RMN carried out in 2014 confirmed the same result. In 2009 the brain SPECT resulted overall normal. In 2011 the brain RMN showed subcortical and periventricular WM point intensities, especially in the bilateral frontal lobes.

In February 2017 the patient underwent neuropsychological tests again, which showed "a multifocal cognitive decay." This condition was again considered secondary to a depressive disorder. However, the neuropsychological performance seemed to be more severe and spread to several other cognitive domains, so we

Fig. 10.1 Depression timeline

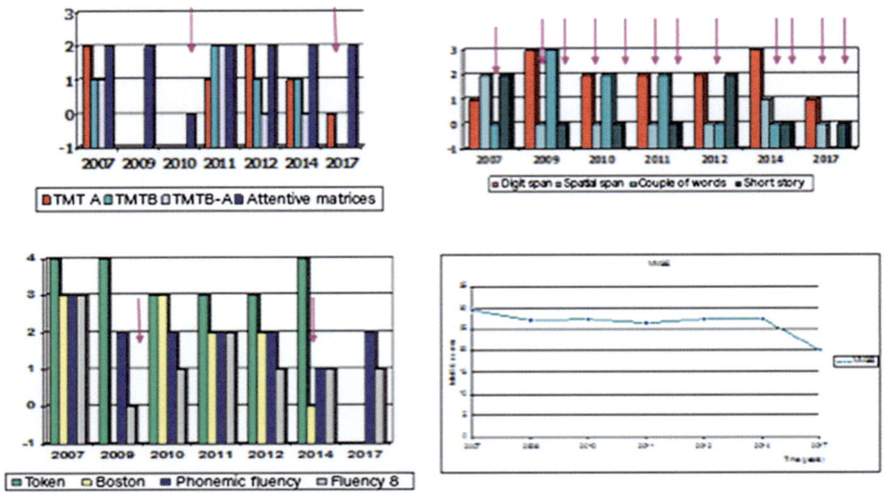

Fig. 10.2 Neuropsychological test results

decided to perform a magnetic resonance imaging (MRI), a positron emission tomography (PET), and a liquor examination. Results and pictures are reported in images (Panel a and Panel b, Fig. 10.3).

10.2.3 Conclusions

In conclusion, the patient was diagnosed with "a moderate cognitive deficit of probable primary degenerative nature, such as Alzheimer's disease." Moreover, the neurologist suggested to introduce 5 mg of donepezil daily and to repeat neuropsychological tests after 6 months.

This case describes the situation of a well-known patient with MDD who presents, at some point of her life, different clinical features in her depression. Until 2007 depressive episodes were characterized by mood deflection, apathy, and anhedonia, but she responded to the antidepressant treatment, reaching full recovery. The patient also experienced several years of well-being, without any pharmacological treatment. In 2007, however, she experienced the onset of her memory deficits, which, at the beginning, seemed to be ascribed to memory domain and connected to mood deflection (depressive "pseudodementia"). These deficits, however, did never remit and worsened over time. At this point, the neuropsychologist supposed a cognitive deficit independent from depression (MCI) (Fig. 10.4).

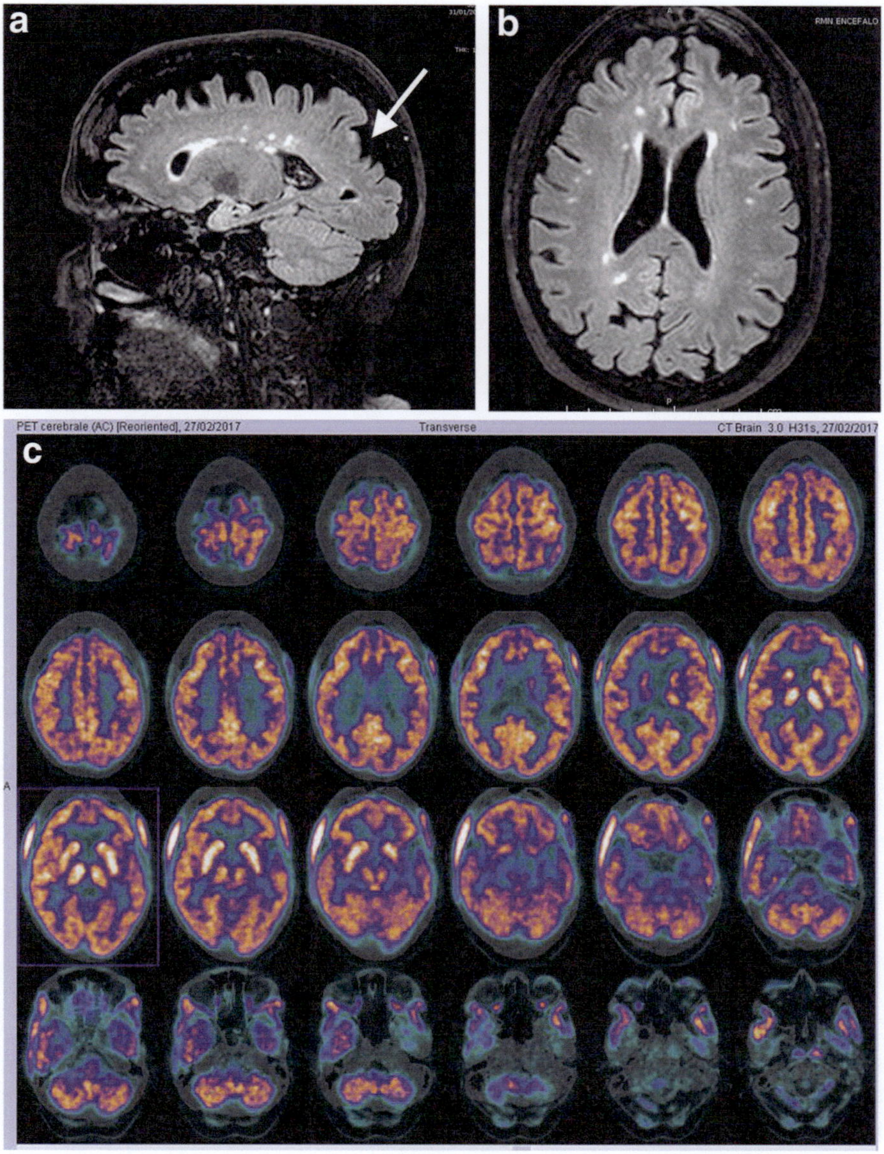

Fig. 10.3 (Panel **a**) T1-weighted sagittal image shows a moderate enlargement of the sovratentorial cortical liquoral spaces. (Panel **b**) T1-weighted sagittal image shows white matter hyperintensities, probably on vascular bases most expressed in the periventricular white matter. (Panel **c**) FDG-PET shows areas of modest reduction of the fixation of the analogue of radiolabeled glucose at the mesial regions of the temporal lobes of both sides, predominantly in the left one

MOLECULAR GENETICS

MUTATION C9ORF72	**Absent**

NEUERODEGENERATIVE DISEASES

AMYLOIDE, TAU and PHOSPHO-TAU PROTEINS LIQUOR					ELISA
AMYLOIDE PROTEIN	**525**	*	pg/ml	>600	
TAU PROTEIN	**678**	*	pg/ml	<500	
PHOSPHO-TAU PROTEIN	**64**	*	pg/ml	<61	
PROGRANULIN PROTEIN PLASMA	**525**	*	ng/mL	>61.5	ELISA

Fig. 10.4 Liquor examination shows an increase in the abovementioned proteins

In the following years, the patient became amnesic and also unable to cope with her daily activities. She experienced partial disorientation, and she easily lost objects such as keys or forgot to take her medications correctly.

The doubt that the patient showed features of the onset of a degenerative disorder (e.g., Alzheimer's disease) was investigated, and the diagnosis of Alzheimer's dementia was made.

10.3 Literature Review

The described case focuses on the overlap between depression and major neurocognitive disorder in the elderly population. Recent studies tried to determine the relationship between these two diagnoses and comorbidity [31]. Evidence generated different hypotheses. For example, Butters et al. found that depression severity could impact on the threshold of the dementia onset [32], whereas other authors considered dementia/cognitive impairment as features of depression [33] or elderly depression as a prodromal phase of dementia [34]. On the other hand, depression could be a reactive mood disorder due to cognitive decay [35], or the two disorders could share the same risk factors and consequently present common aspects [36–38]. This could justify the increased prevalence of both disorders in the elderly population and their frequent comorbidity. Other potential links between late-onset depression and dementia may be common vascular factors [39–41], because depression has been associated with vascular risk factors and with cerebrovascular lesions on neuroimaging [1, 42]. Nevertheless, the same vascular lesions are observed in MRI/TC images of patients affected by mild cognitive impairment.

The term "pseudodementia" was coined for the first time by Kiloh in 1961 [43] to describe cases in which depressed patients closely mimed aspects typical of the dementia spectrum. It underlines the overlap between elderly depression and dementia. Dementia and "pseudodementia" share cognitive deficits (e.g., memory, language, speech, executive functions), particularly during the acute phase or after depression, when patients respond to treatment but not fully remit.

Even though performances in neuropsychological tests of depressed patients show some specific characteristics (e.g., long reaction time, deficits spread in different domains with variable severity, time fluctuation of performance), they tend to underestimate their cognitive functioning; it is due to a lack of self-esteem. On the other hand, patients with dementia have no insight of bad performances; they tend to worsen in the same domains, to answer fast and spontaneously; and they overestimate their memory and cognitive functioning. Awareness of memory deficits also seems to be useful to distinguish between depression and mild cognitive impairment (MCI) [44].

However, the main difference is the reversible nature of depressive pseudodementia (the so-called "pseudo" component), whereas the cognitive pattern in dementia tends to become increasingly worse during the following months and years.

In any case, to date, pseudodementia has no diagnostic value and remains a descriptive denomination to describe cognitive deficits observed in psychiatric disorders, in particular in depressive disorders [45].

MCI is characterized by the presence of cognitive concerns, objective evidence of impairment in one or more cognitive domains, preservation of independence in functional abilities, and no dementia [46]. Substantially, the construct of MCI has been considered as a transitional step between normal aging and dementia, for which early preventive interventions may be possible [47]. Over time, this concept has been widely revised; in fact MCI was introduced for the first time in 1988 by Reisberg and colleagues, and it corresponded to stage 3 of the Global Deterioration Scale (GDS), an evaluation scale which rates the dementia state [48]. Then, also the Clinical Dementia Rating (CDR) scale became an instrument to quantify and to describe both mild impairment and very early dementia. However, it has been realized that none of the two rating scales alone could adequately characterize the subtle differences between MCI and the early stage of dementia; indeed, in a study conducted by Brodaty et al. subjects affected by MCI (as currently defined) could be classified as GDS stages 2–3 and as showing a CDR of 0–0.5 [49].

Regarding the diagnostic criteria of MCI, they have been revised in a conference of international experts on MCI [50, 51]. In fact patients with MCI have been observed to be more memory impaired than the cognitively normal subjects, but to present a lower memory impairment than subjects with dementia. However, other (non-memory) cognitive performances only showed minimal alterations in MCI spectrum [52].

In any case, keeping in mind that MCI and dementia are different diagnostic entities, it has been observed that only 16.5% of patients affected by MCI develop dementia in about 12 months [52]. The remaining 83.5% do not progress to a dementia status; so we should consider MCI and dementia as two different diagnoses, in a possible context of continuum, where MCI may imply a further development to dementia or it might imply stability or, unfortunately less commonly, improvement in their clinical symptoms. Consequent relevant clinical implication consists of the possibility of intervention only in the MCI phase in those patients, who are destined to a progressive cognitive decline, with the medical aim of decelerating or arresting further development.

Furthermore, many studies investigated this topic and showed that depression significantly increases the risk for dementia and for prodromal phase of MCI [46, 48, 53]. Consequently, given this correlation, it may be important to understand whether antidepressant treatment alone or combined with other regimens could positively impact on cognition [54]. Finally, a close monitoring of elderly depression may be a possible therapeutic tool for preventing or decelerating the development of dementia.

Regarding the clinical-diagnostic gray area of pseudodementia, many years ago Kaszniak [55] provided an important summary, which clarifies the difficult differentiation between reversible cognitive deficits due to depression and dementia:

- Early signs of a pervasive developmental disorder can easily be confused with cognitive alterations due to old age.
- Cognitive impairment represents a frequent sign of depressive clinical spectrum. It can be severe enough to play a confusing role in the differential diagnosis between dementia and depression.
- Signs of neurologic diseases, associated with progressive decline (i.e., Alzheimer's or Parkinson's disease), show many clinical aspects, which overlap with depression.
- Dementia and depression can be in comorbidity.

Thus, we are not able to exactly address the mechanisms linking depression and dementia, but all previous evidence supports the hypothesis that elderly depression often accompanies cognitive impairment. An interesting study showed how late-life depression quite often co-occurs with cognitive decline, but does not precede it [56].

Regarding the diagnostic pathway, according to recent literature, we underlined the importance of instrumental diagnostic tests, such as CSF laboratory analysis [27], imaging [26], or EEG [29]. In fact, since many clinical aspects could overlap, these second-/third-line procedures could provide objective data, which could help the diagnostic process. Thus imaging, CFS analysis and EEG are useful not only to exclude other pathological presentations but also to recognize the diagnosis of dementia, depression, or both.

Opinions about antidepressant pharmacotherapy in elderly people are not univocal. Antidepressants (ADs) have been found to be protective against dementia [57]. Moreover, AD therapy decreases the risk of developing depressive episodes associated with cognitive impairment. This pattern was found for both older and last generations' ADs [57]. This concept has been further supported in a study with paroxetine, in which interaction between paroxetine and mitochondrial proteins could play a neuroprotective role [58]. Also fluoxetine has shown to promote both the proliferation and neuronal differentiation of neural stem cells and to have protective effects against various neurodegenerative diseases [59]. Conversely, some authors showed how antidepressant drugs were associated with an increased risk of developing cognitive impairment [60]. Elsewhere, no significant differences between AD treatment and posttrial cognitive functioning suggested cognitive impairment resistance to ADs [61].

New therapeutic strategies for dementia have been widely investigated: they are based on several blood peripheral markers in patients who have cognitive deficits, which could play a causal role. For example, circadian rhythm alterations, immune dysfunction, and/or oxidative stress, which are key pathologic processes involved in depressive cognitive dysfunction and neurodegeneration, could be potential targets for novel treatments [12]. Moreover, interventions on the monoaminergic system (protagonist of the oxidative stress mechanism) could prevent neurodegeneration. Obviously, these are only initial steps for a new treatment, but further research is required to better understand how the cited pathways could impact on the physiopathology of a depressive cognitive impairment/neurodegenerative cognitive decline.

Key Points

- Numerous epidemiological studies demonstrate how, in the next few years, ageing population will reach a higher percentage (approximately 20%). This will result in an increase of incidence and prevalence of progressive cognitive deficits and geriatric depression.
- With the new DSM-5, the term "dementia" was replaced by the concept of "major neurocognitive disorder." The term dementia is still used, especially to classify the various subtypes (i.e., dementia with Lewy bodies).
- Differential diagnosis is crucial with other disorders, which impair mental status (i.e., delirium). Delirium symptoms could mirror those of dementia—often causing misdiagnosis, such as in the case of confusion, loss of orientation, and memory impairment. However, many organic conditions typical of the elderly (hydro-electrolyte alterations, thyroid dysfunction, or any infections) can simulate cognitive degeneration.
- In the differential diagnosis of the geriatric population, it is essential to monitor blood tests and assumption of psychiatric and nonpsychiatric drugs. It is also important to perform a computed tomography or a magnetic resonance imaging, an electroencephalogram, and other second-level tests such as positron emission tomography and cerebrospinal fluid analysis.
- Neurocognitive tests and history of symptoms progression can be useful to make a differential diagnosis between dementia and cognitive deficits secondary to depression.
- Dementia and depression can easily overlap and appear in comorbidity.

- An element that differentiates depression from dementia is the rapid onset of cognitive declines, associated with depressed mood. The depressed subject appears aware of his cognitive symptoms and their severity. On the other side, Alzheimer's dementia progresses slowly, and the patient is unaware of the severity of his deficits.

Self-Assessment Questionnaire

1. What is the most common type of dementia that shows an early onset, compared to the others?
   (A) Alzheimer's disease
   (B) Dementia with Lewy body
   (C) Frontotemporal dementia
   (D) Parkinson's disease
2. Who has been coined for the first ever term "pseudodementia"?
   (A) Kraepelin, 1880
   (B) Kohl, 1961
   (C) Babinski, 1900
   (D) Perusini, 1930
3. What is the main difference between dementia and pseudodementia?
   (A) The reversibility of pseudodementia after treatment.
   (B) The reversibility of memory deficits in dementia patients.
   (C) Mood disorders are more often associated with dementia than pseudodementia.
   (D) Antidepressants don't have any effect in pseudodementia patients, compared to dementia.
4. What, among these, is the cause that most frequently can simulate a delirium in the geriatric population?
   (A) Celiac disease
   (B) Thyroid dysfunction
   (C) Iron deficiency
   (D) Cardiac alterations

References

1. Steffens DC, Otey E, Alexopoulos GS, Butters MA, Cuthbert B, Ganguli M, et al. Perspectives on depression, mild cognitive impairment, and cognitive decline. Arch Gen Psychiatry. 2006;63(2):130. https://doi.org/10.1001/archpsyc.63.2.130.
2. Heser K, Tebarth F, Wiese B, Eisele M, Bickel H, Köhler M, et al. Age of major depression onset, depressive symptoms, and risk for subsequent dementia: results of the German study on Ageing, Cognition, and Dementia in Primary Care Patients (AgeCoDe). Psychol Med. 2013;43(8):1597–610. https://doi.org/10.1017/S0033291712002449.

3. Ownby RL, Crocco E, Acevedo A, John V, Loewenstein D. Depression and risk for Alzheimer disease: systematic review, meta-analysis, and meta-regression analysis. Arch Gen Psychiatry. 2006;63(5):530–8. https://doi.org/10.1001/archpsyc.63.5.530.
4. Diniz BS, Butters MA, Albert SM, Dew MA, Reynolds CF. Late-life depression and risk of vascular dementia and Alzheimer's disease: systematic review and meta-analysis of community-based cohort studies. Br J Psychiatry. 2013;202(5):329–35. https://doi.org/10.1192/bjp.bp.112.118307.
5. Mintzer J, Targum SD. Psychosis in elderly patients: classification and pharmacotherapy. J Geriatr Psychiatry Neurol. 2003;16(4):199–206. https://doi.org/10.1177/0891988703258658.
6. WHO-Dementia. World Health Organization endorses global action plan on rising incidence of dementia. Nurs Older People. 2017;29(6):7–7. https://doi.org/10.7748/nop.29.6.7.s6.
7. Espinoza RT. Improving the recognition and management of dementia in long-term care: obstacles and opportunities. J Am Med Dir Assoc. 2006;7(2):128–30. https://doi.org/10.1016/j.jamda.2005.10.002.
8. Fillit HM, Doody RS, Binaso K, Crooks GM, Ferris SH, Farlow MR, et al. Recommendations for best practices in the treatment of Alzheimer's disease in managed care. Am J Geriatr Pharmacother. 2006;4(Suppl A):S9–S28. https://doi.org/10.1016/j.amjopharm.2006.10.001.
9. Pazzaglia P. Clinica Neurologica. 7 edizione. Capitolo "Demenze"; 2012.
10. Roberson ED, Hesse JH, Rose KD, Slama H, Johnson JK, Yaffe K, et al. Frontotemporal dementia progresses to death faster than Alzheimer disease. Neurology. 2005;65(5):719–25. https://doi.org/10.1212/01.wnl.0000173837.82820.9f.
11. Rosso SM, Donker Kaat L, Baks T, Joosse M, de Koning I, Pijnenburg Y, et al. Frontotemporal dementia in The Netherlands: patient characteristics and prevalence estimates from a population-based study. Brain J Neurol. 2003;126(Pt 9):2016–22. https://doi.org/10.1093/brain/awg204.
12. Brown GM, McIntyre RS, Rosenblat J, Hardeland R. Depressive disorders: processes leading to neurogeneration and potential novel treatments. Prog Neuro-Psychopharmacol Biol Psychiatry. 2017;80(Pt C):189–204. https://doi.org/10.1016/j.pnpbp.2017.04.023.
13. Mulsant BH, Ganguli M. Epidemiology and diagnosis of depression in late life. J Clin Psychiatry. 1999;60(Suppl 20):9–15. Retrieved from http://www.ncbi.nlm.nih.gov/pubmed/10513852
14. CADTH-Depression. Diagnosing, screening, and monitoring depression in the elderly: a review of guidelines – PubMed – NCBI. Canadian Agency for Drugs and Technologies in Health; 2015. Retrieved from https://www.ncbi.nlm.nih.gov/pubmed/26468558.
15. Copeland JRM, Beekman ATF, Braam AW, Dewey ME, Delespaul P, Fuhrer R, ... Wilson KCM. Depression among older people in Europe: the EURODEP studies. World Psychiatry. 2004;3(1):45–9. Retrieved from http://www.ncbi.nlm.nih.gov/pubmed/16633454.
16. Fereshtehnejad S-M, Johannsen P, Waldemar G, Eriksdotter M. Dementia diagnosis, treatment, and care in specialist clinics in two Scandinavian countries: a data comparison between the Swedish Dementia Registry (SveDem) and the Danish Dementia Registry. J Alzheimers Dis. 2015;48(1):229–39. https://doi.org/10.3233/JAD-150144.
17. Miller AM, Begley E, Coen R, Doyle M, Dunne J, Hutchinson S, ... Lawlor BA. Clinical consensus guidelines on the application of cerebrospinal fluid biomarkers for Alzheimer's disease diagnosis: recommendations of the Irish network for biomarkers in neurodegeneration. Ir Med J. 2016;109(10):483. Retrieved from http://www.ncbi.nlm.nih.gov/pubmed/28644588.
18. Hofmann W. Guideline-based diagnosis of dementia etiology. Zeitschrift Fur Gerontologie Und Geriatrie. 2012;45(8):761–71–3. https://doi.org/10.1007/s00391-012-0399-y
19. APA American Psychiatric Association. Diagnostic and statistical manual of mental disorders: DSM-V; 2013.
20. APA American Psychiatric Association. Diagnostic and Statistical Manual of Mental Disorders: DSM-IV; 1994.

21. Gold DA. An examination of instrumental activities of daily living assessment in older adults and mild cognitive impairment. J Clin Exp Neuropsychol. 2012;34(1):11–34. https://doi.org/10.1080/13803395.2011.614598.
22. Lawton MP, Brody EM. Assessment of older people: self-maintaining and instrumental activities of daily living. The Gerontologist. 1969;9(3):179–86. Retrieved from http://www.ncbi.nlm.nih.gov/pubmed/5349366.
23. Ozer S, Young J, Champ C, Burke M. A systematic review of the diagnostic test accuracy of brief cognitive tests to detect amnestic mild cognitive impairment. Int J Geriatr Psychiatry. 2016;31(11):1139–50. https://doi.org/10.1002/gps.4444.
24. Tsoi KKF, Chan JYC, Hirai HW, Wong SYS, Kwok TCY. Cognitive tests to detect dementia: a systematic review and meta-analysis. JAMA Intern Med. 2015;175(9):1450–8. https://doi.org/10.1001/jamainternmed.2015.2152.
25. Carlesimo GA, Caltagirone C, Gainotti G. The mental deterioration battery: normative data, diagnostic reliability and qualitative analyses of cognitive impairment. The group for the standardization of the mental deterioration battery. Eur Neurol. 1996;36(6):378–84. Retrieved from http://www.ncbi.nlm.nih.gov/pubmed/8954307.
26. Barthel H, Schroeter ML, Hoffmann K-T, Sabri O. PET/MR in dementia and other neurodegenerative diseases. Semin Nucl Med. 2015;45(3):224–33. https://doi.org/10.1053/j.semnuclmed.2014.12.003.
27. Lista S, Toschi N, Baldacci F, Zetterberg H, Blennow K, Kilimann I, et al. Cerebrospinal fluid neurogranin as a biomarker of neurodegenerative diseases: a cross-sectional study. J Alzheimers Dis. 2017;59(4):1327–34. https://doi.org/10.3233/JAD-170368.
28. Società Italiana di Neuroscienze. Expert Panel Alzheimer. Malattia di Alzheimer. Documento di consenso, DSM-V, MMSE, RMN, Liquor, es. ematici per escludere infezioni disdratazione, squilibri elettrolitici, disfunzioni tiroidee; 1999.
29. Neto E, Biessmann F, Aurlien H, Nordby H, Eichele T. Regularized linear discriminant analysis of EEG features in dementia patients. Front Aging Neurosci. 2016;8:273. https://doi.org/10.3389/fnagi.2016.00273.
30. De Carlo V, Calati R, Serretti A (2016). Socio-demographic and clinical predictors of non-response/non-remission in treatment resistant depressed patients: a systematic review. Psychiatry Research. Elsevier Ireland Ltd.
31. Bennett S, Thomas AJ. Depression and dementia: cause, consequence or coincidence? Maturitas. 2014;79(2):184–90. https://doi.org/10.1016/j.maturitas.2014.05.009.
32. Butters MA, Young JB, Lopez O, Aizenstein HJ, Mulsant BH, Reynolds CF, ... Becker JT. Pathways linking late-life depression to persistent cognitive impairment and dementia. Dialogues Clin Neurosci. 2008;10(3):345–57. Retrieved from http://www.ncbi.nlm.nih.gov/pubmed/18979948.
33. Sheline YI, Barch DM, Garcia K, Gersing K, Pieper C, Welsh-Bohmer K, et al. Cognitive function in late life depression: relationships to depression severity, cerebrovascular risk factors and processing speed. Biol Psychiatry. 2006;60(1):58–65. https://doi.org/10.1016/j.biopsych.2005.09.019.
34. Panza F, Frisardi V, Capurso C, D'Introno A, Colacicco AM, Imbimbo BP, et al. Late-life depression, mild cognitive impairment, and dementia: possible continuum? Am J Geriatr Psychiatry. 2010;18(2):98–116. https://doi.org/10.1097/JGP.0b013e3181b0fa13.
35. Huang C-Q, Wang Z-R, Li Y-H, Xie Y-Z, Liu Q-X. Cognitive function and risk for depression in old age: a meta-analysis of published literature. Int Psychogeriatr. 2011;23(4):516–25. https://doi.org/10.1017/S1041610210000049.
36. Djernes JK. Prevalence and predictors of depression in populations of elderly: a review. Acta Psychiatr Scand. 2006;113(5):372–87. https://doi.org/10.1111/j.1600-0447.2006.00770.x.
37. Enache D, Winblad B, Aarsland D. Depression in dementia: epidemiology, mechanisms, and treatment. Curr Opin Psychiatry. 2011;24(6):461–72. https://doi.org/10.1097/YCO.0b013e32834bb9d4.

38. Ferri CP, Prince M, Brayne C, Brodaty H, Fratiglioni L, Ganguli M, et al. Global prevalence of dementia: a Delphi consensus study. Lancet. 2005;366(9503):2112–7. https://doi.org/10.1016/S0140-6736(05)67889-0.
39. Alexopoulos GS, Meyers BS, Young RC, Campbell S, Silbersweig D, Charlson M. "Vascular depression" hypothesis. Arch Gen Psychiatry. 1997;54(10):915–22. Retrieved from http://www.ncbi.nlm.nih.gov/pubmed/9337771
40. Flicker L. Vascular factors in geriatric psychiatry: time to take a serious look. Curr Opin Psychiatry. 2008;21(6):551–4. https://doi.org/10.1097/YCO.0b013e32830ef736.
41. Teper E, O'Brien JT. Vascular factors and depression. Int J Geriatr Psychiatry. 2008;23 (10):993–1000. https://doi.org/10.1002/gps.2020.
42. de Groot JC, de Leeuw FE, Oudkerk M, Hofman A, Jolles J, Breteler MM. Cerebral white matter lesions and depressive symptoms in elderly adults. Arch Gen Psychiatry. 2000;57 (11):1071–6. Retrieved from http://www.ncbi.nlm.nih.gov/pubmed/11074873.
43. Kiloh LG. Pseudo-dementia. Acta Psychiatr Scand. 1961;37:336–51.
44. Coutinho G, Drummond C, Teldeschi A, Mattos P. Awareness of memory deficits is useful to distinguish between depression and mild cognitive impairment in the elderly. Rev Bras Psiquiatr. 2016;38(3):231-234. doi: https://doi.org/10.1590/1516-4446-2015-1772. Epub 2016 May 17
45. Kang H, Zhao F, You L, Giorgetta C, Venkatesh D, Sarkhel S, Prakash R. Pseudo-dementia: A neuropsychological review. Ann Indian Acad Neurol. 2014 Apr-Jun; 17(2):147–54. PMID: 25024563.
46. Lara E, Koyanagi A, Domènech-Abella J, Miret M, Ayuso-Mateos JL, Haro JM. The impact of depression on the development of mild cognitive impairment over 3 years of follow-up: a population-based study. Dement Geriatr Cogn Disord. 2017;43(3–4):155–69. https://doi.org/10.1159/000455227.
47. Petersen RC, Roberts RO, Knopman DS, Boeve BF, Geda YE, Ivnik RJ, et al. Mild Cognitive Impairment. Arch Neurol. 2009;66(12):1447–55. https://doi.org/10.1001/archneurol.2009.266.
48. Apostolova LG, Cummings JL. Neuropsychiatric manifestations in mild cognitive impairment: a systematic review of the literature. Dement Geriatr Cogn Disord. 2007;25(2):115–26. https://doi.org/10.1159/000112509.
49. Brodaty H, Heffernan M, Kochan NA, Draper B, Trollor JN, Reppermund S, et al. Mild cognitive impairment in a community sample: The Sydney Memory and Ageing Study. Alzheimers Dement. 2013;9(3):310–317.e1. https://doi.org/10.1016/j.jalz.2011.11.010.
50. Petersen RC. Mild cognitive impairment as a diagnostic entity. J Intern Med. 2004;256 (3):183–94. https://doi.org/10.1111/j.1365-2796.2004.01388.x.
51. Winblad B, Palmer K, Kivipelto M, Jelic V, Fratiglioni L, Wahlund L-O, et al. Mild cognitive impairment–beyond controversies, towards a consensus: report of the International Working Group on Mild Cognitive Impairment. J Intern Med. 2004;256(3):240–6. https://doi.org/10.1111/j.1365-2796.2004.01380.x.
52. Petersen RC, Aisen PS, Beckett LA, Donohue MC, Gamst AC, Harvey DJ, et al. Alzheimer's Disease Neuroimaging Initiative (ADNI): clinical characterization. Neurology. 2010;74 (3):201–9. https://doi.org/10.1212/WNL.0b013e3181cb3e25.
53. Xu W, Tan L, Wang H-F, Jiang T, Tan M-S, Tan L, et al. Meta-analysis of modifiable risk factors for Alzheimer's disease. J Neurol Neurosurg Psychiatry. 2015;86(12):1299–306. https://doi.org/10.1136/jnnp-2015-310548.
54. Byers AL, Yaffe K. Depression and risk of developing dementia. Nat Rev Neurol. 2011;7 (6):323–31. https://doi.org/10.1038/nrneurol.2011.60.
55. Kaszniak AW, Fox J, Gandell DL, Garron DC, Huckman MS, Ramsey RG. Predictors of mortality in presenile and senile dementia. Ann Neurol. 1978;3(3):246–52.
56. Richard E, Reitz C, Honig LH, Schupf N, Tang MX, Manly JJ, et al. Late-Life Depression, Mild Cognitive Impairment, and Dementia. JAMA Neurol. 2013;70(3):383. https://doi.org/10.1001/jamaneurol.2013.603.
57. Kessing LV, Søndergård L, Forman JL, Andersen PK. Antidepressants and dementia. J Affect Disord. 2009;117(1–2):24–9. https://doi.org/10.1016/j.jad.2008.11.020.

58. Steiner JP, Bachani M, Wolfson-Stofko B, Lee M-H, Wang T, Wang T, et al. Interaction of paroxetine with mitochondrial proteins mediates neuroprotection. Neurotherapeutics. 2015;12 (1):200–16. https://doi.org/10.1007/s13311-014-0315-9.
59. Chang K-A, Kim JA, Kim S, Joo Y, Shin KY, Kim S, et al. Therapeutic potentials of neural stem cells treated with fluoxetine in Alzheimer's disease. Neurochem Int. 2012;61(6):885–91. https://doi.org/10.1016/j.neuint.2012.03.017.
60. Tifratene K, Robert P, Metelkina A, Pradier C, Dartigues JF. Progression of mild cognitive impairment to dementia due to AD in clinical settings. Neurology. 2015;85(4):331–8. https://doi.org/10.1212/WNL.0000000000001788.
61. Halahakoon DC, Roiser JP. Cognitive impairment in depression and its (non-)response to antidepressant treatment. Evid Based Ment Health. 2016;19(4):e23–e23. https://doi.org/10.1136/eb-2016-102438.

Catatonia and Cotard's Syndrome

11

Andrea Botturi, Lucio Oldani, and Francesca Bottinelli

Abstract

Catatonia is a complex neuropsychiatric syndrome characterized by motor signs, disturbances of volition, inability to suppress complex motor activities, stereotypies, and autonomic instability. It has been reported to occur in more than 10% among hospitalized acute psychiatric patients. In DSM-5, catatonia syndrome is defined as a specifier to major mood, psychotic disorders, and general medical conditions.

Cotard's syndrome is a particular cluster of psychotic symptoms characterized by nihilistic delusions. It has been described among depressive psychotic disorders and in bipolar depression. Even though the concurrent presence of these two psychopathological conditions is rare, they can be associated with medical internal complications as dehydration, pneumonia, and thromboembolic disease, due to the lack of movements (immobilization). These syndromes are complex and difficult to treat, needing multiple psychopharmacological choices (antidepressants, mood stabilizers, antipsychotics).

Here we describe the case of a young woman with bipolar disorder, who presented catatonia and Cotard's syndrome during a depressive psychotic episode; she required hospitalization because of the severity of symptoms. The recovery from the acute phase was obtained only after several months of treatment and rehabilitation.

A. Botturi (✉)
Department of Neurosciences and Mental Health, Fondazione IRCCS Ca' Granda Ospedale Maggiore Policlinico, University of Milan, Milan, Italy

Fondazione IRCCS Istituto Neurologico Carlo Besta, Milan, Italy
e-mail: andrea.botturi@istituto-besta.it

L. Oldani · F. Bottinelli
Department of Neurosciences and Mental Health, Fondazione IRCCS Ca' Granda Ospedale Maggiore Policlinico, University of Milan, Milan, Italy

A. C. Altamura, P. Brambilla (eds.), *Clinical Cases in Psychiatry: Integrating Translational Neuroscience Approaches*,
https://doi.org/10.1007/978-3-319-91557-9_11

Keywords
Catatonia · Cotard's syndrome · Major depressive disorder · Bipolar disorder · Psychosis · Pharmacological treatment

11.1 Introduction

Psychosis, generally defined as the occurrence of hallucinations or delusions, is a common feature across numerous psychiatric disorders [1].

Major depressive episodes with psychotic features are more common in bipolar disorder (BD) than in major depressive disorder (MDD), and a recent study has reported that depressed bipolar patients with psychotic features have worse illness outcomes, lower rates of recovery, and shorter time to first recurrence, including first-episode patients [2, 3].

Most studies of psychotic symptoms in BD focus on psychotic mania, because psychotic symptoms are more common in manic episodes [4, 5]. However, psychosis is also frequent in bipolar depressive episodes [6].

The prevalence of psychosis in bipolar depression could be as high as 66% in historical descriptions [7, 8]. More recent studies report lower but still significant prevalence: 10.4% for a current episode [9] and 10–28% for lifetime bipolar depressive episodes.

Moreover, major depressive episodes with psychosis are more frequent and more recurrent in BD than in unipolar depression [10, 11].

About 75% of patients with an acute manic episode may present psychotic symptoms [12].

DSM-5 includes specifiers to describe the nature of a person's mood disorder [13]. Two syndromes can be distinguished among depressive psychotic disorders but can also characterize bipolar depression: catatonia with neuropsychiatric facets and Cotard's (a particular cluster of psychotic symptoms characterized by nihilistic delusions).

These are complex disorders and may be complex to treat, also due to their need of psychopharmacological polytherapy. In catatonia the rate of response to benzodiazepine is 70–90%, but, when they are ineffective, second-generation antipsychotics as well as GABAergic drugs can be employed [14].

Psychotic symptoms in Cotard's syndrome (CS) require mood-stabilizing treatments and (one or more) antipsychotic drugs. Both syndromes can also have medical internal complications (thrombosis, pneumonia) due to bedridden status as a result of psychotic delusional depression.

The contemporary presence of catatonia and CS is rare [15].

11.2 Case Report

Y.X. is a 26-year-old girl, who moved to Italy in 2013. She was born in China, where her parents are currently living. In a short amount of time after she moved to Italy, she found a job in a fashion company, work that she kept up until hospitalization. Even though she has been living in Italy for 3 years, she cannot speak Italian fluently, so it was hard to collect anamnestic information; all the details were gathered with the help of her parents, her boyfriend, and some friends of hers.

Psychopathological onset dates back to 2009, with a manic episode which lasted about 1 month and subsequently shifted to a depressive picture. Because of this symptomatology, she was treated in China with quetiapine and lamotrigine (dosages were not provided), in addition to six electroconvulsive therapy (ECT) sessions implemented during a 4-month hospitalization period.

Without medical advice, she stopped her psychopharmacological treatment in 2016, which facilitated the appearance of a new psychotic episode. This last manic period did not undergo any medical assessment or therapy because of its rapidity (it lasted only a few days).

In April 2017 she was taken to the emergency room (ER) of our hospital by ambulance, called by two friends of hers. They were worried about her health status, because in the last 3 days, she neither had gone to work nor answered the phone. When visited at home by such friends, she was found motionless on her sofa, probably after some hours of sleep deprivation (as subsequently referred by her friends and colleagues).

The psychiatric examination in ER described Y.X. as generally uncooperative, negative, confused, and not oriented in time and space. She showed poor eye contact, with blank and blunted facial expression. She did not answer to any question and repeated, with flat voice, lacking intonations, or modulation: "I don't remember—I don't know." Her mood seemed unstable, alternating tears and laughter with no apparent reason.

General physical examination resulted within the range of normality:

- Blood pressure was 115/70, regular pulse at 110 beats per minute.
- Cardiovascular and respiratory examinations were normal.
- No abdominal guarding or rigidity.
- Head TC scan: unremarkable.
- Central nervous system examination: no focal signs, except for motor slowing retardation and immobility.
- EEG: no seizures or cerebral dysfunction.
- No signs of drugs when her urine was screened for substance of abuse (testing alcohol, opioids, and cannabinoids).

The patient was therefore admitted to the psychiatric ward with a provisional diagnosis of psychotic picture.

During hospitalization a clinical screening was assessed, based on the administration of the Structured Clinical Interview for DSM-IV (SCID-I and SCID-II) [16]

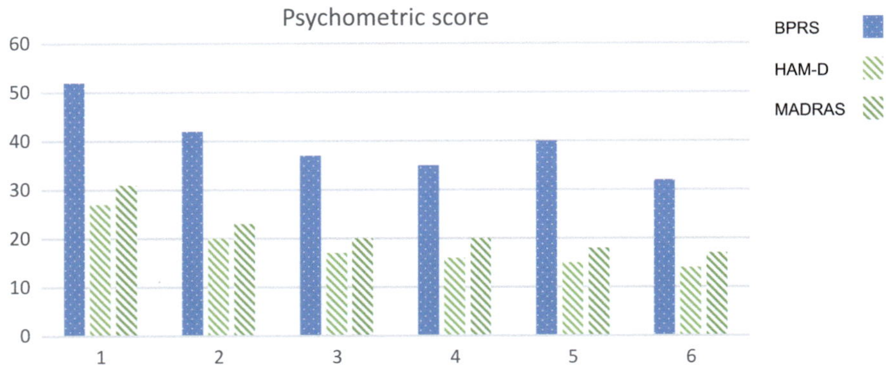

Fig. 11.1 Psychometric evaluation grouped by treatment weeks

and psychometric scales, such as Brief Psychiatric Rating Scale (BPRS) [17], Hamilton Depression Rating Scale (HAM-D) [18], Montgomery-Asberg Depression Rating Scale (MADRS) [19], and Young Mania Rating Scale (YMRS) [20] (Fig. 11.1).

The SCID-I returned a diagnosis of bipolar disorder type I, current depressive episode with catatonic features.

During the following days, the psychotic symptoms emerged clearly, especially as ideas of guilt and ruin and delusions with persecutory features. In addition, she referred auditory hallucinations concerning negative voices and noises which the patient could not specifically describe, expressing a severe degree of anguish. She showed delusional somatic ideas that appeared as inability to walk or to eat. She looked perplexed and afraid of her body movements; in fact she could not move her limbs, and she spent a lot of time looking at her legs.

Her mood was mostly depressed with anhedonia. In the inpatient unit of our hospital, she used to spend most of the time on her own, in her room, laying on her bed, but was sporadically seen walking around for her physiological needs (eating, using the toilet).

Y. X. was treated with haloperidol 5 mg and delorazepam 4 mg per day IM for 1 week, with slight improvement of the catatonic state; indeed she started answering questions with mild and little head movements; nonetheless she still spent most of her time in bed, presented poor physical activity and gesture, and showed avolition and insomnia (Fig. 11.2). For these reasons, the pharmacological treatment was switched to aripiprazole 5 mg per day IM, quetiapine 200 mg RP, and lamotrigine 25 mg (Fig. 11.3).

During hospitalization, a magnetic resonance imaging (MRI) was performed; it did not show any pathological findings. An 18F-fluorodeoxyglucose positron emission tomography (PET) showed a bilateral relative increasing in the fixation of 18F-fluorodeoxyglucose both on the frontal cortex and cingulate gyrus, as well as hypofixation on the mesial portion of the cortex of both temporal lobes (Fig. 11.4).

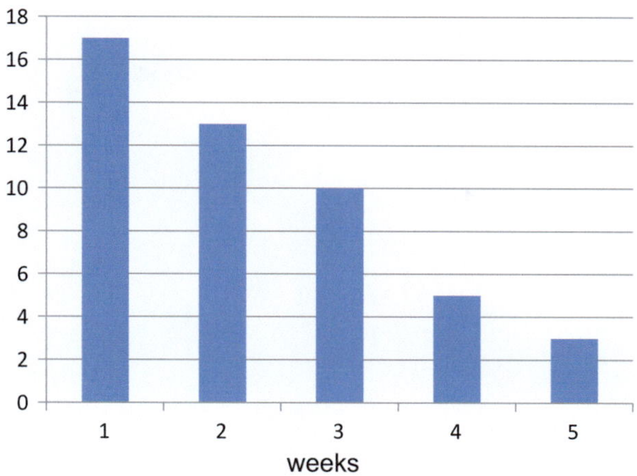

Fig. 11.2 Psychometric score Bush Catatonia Rating Scale

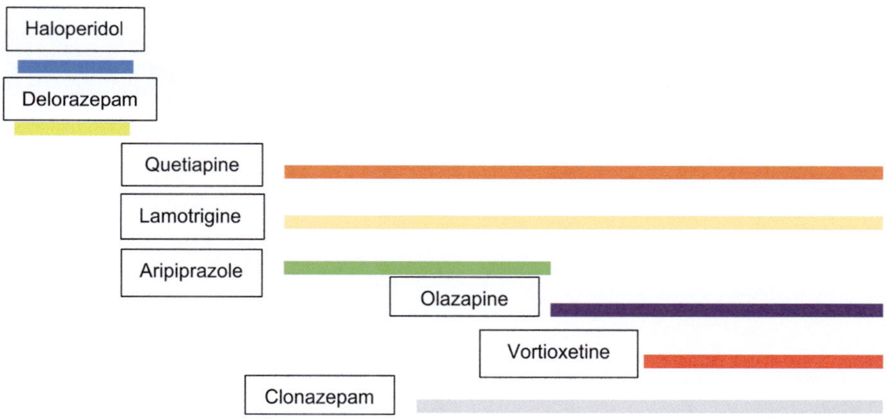

Fig. 11.3 Time course visual representation of pharmacological treatments

After 42 days of hospitalization, catatonic features regressed. In fact she was able to talk, to walk through the corridor, and to take a shower by her own. Y. X. was still experiencing a depressed mood, showing static facial gestures, a slow way of speaking and anguish. However, formal thought disorder, characterized by auditory hallucinations, was not present. She was finally transferred to a rehabilitation center, with the aim of stabilizing the psychopharmacological treatment in a safe environment.

Before discharge, some clinical scales were assessed one more time.

The progress of results through hospitalization are shown in Table 11.1.

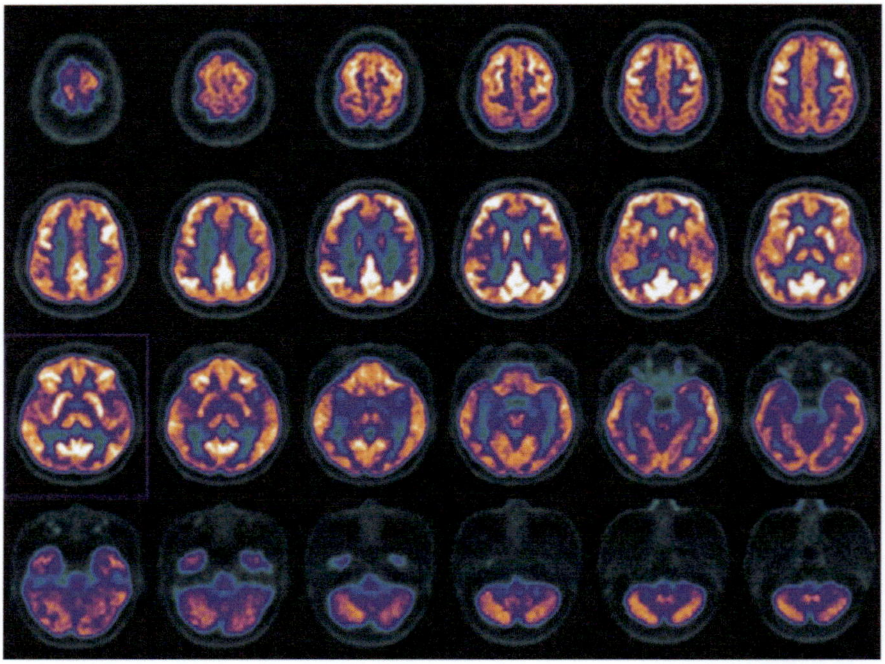

Fig. 11.4 Metabolic study of the patient's brain

Table 11.1 Psychometric tests scores at baseline (T0) and ever 7 days; from hospitalization to discharge

	(T0)	(T1)	(T2)	(T3)	(T4)	(T5)
BPRS	52	42	37	35	40	32
HAM-D	27	20	17	15	13	12
MADRS	31	24	20	20	18	17
YMRS	8	/	/	/	/	/

11.3 Catatonia

11.3.1 Epidemiology and Clinical Presentation

Catatonia is a complex neuropsychiatric syndrome that has been reported to occur in more than 10% of patients with acute psychiatric illnesses (hospital admitted) [21, 22].

In the main epidemiological studies, the prevalence of catatonia in psychiatric patients varies from 7% to 31%. It appears to be more frequent in hospitalized patients, and it can be present both in adults and adolescents, as well as infants: in the USA, each year, 90,000 individuals are hospitalized for catatonia [23].

It was named as "catatonia" for the first time by Karl Kahlbaum in the 1870s, and he described it as "tension and insanity"; (*katatonia*) [24]. Other authors described it with a large number of different motor signs and even vegetative instability; in particular it has been associated with up to 40 different signs and symptoms.

These signs may be summarized in four groups: *pure motor signs* (e.g., posturing, rigor, immobility), *disturbances of volition* (e.g., ambitendency, negativism, automatic obedience), *inability to suppress complex motor activities* (e.g., stereotypies, rituals, echophenomena), and *autonomic instability* (e.g., tachycardia, hyperthermia).

Symptoms usually wax and wane within 1 h.

Some catatonia patients experience a complete remission within 24 h. But there could be acute and chronic forms of catatonia that share the same symptoms.

Finally, catatonic syndrome may become malignant with increased mortality, particularly when autonomic instability is included.

Two subtypes have been described. Catatonia of the *retarded type* is associated with signs reflecting a paucity of movement, including immobility, staring, mutism, rigidity, withdrawal, and refusal to eat, along with more bizarre features such as posturing, grimacing, negativism, waxy flexibility, echolalia or echopraxia, stereotypy, verbigeration, and automatic obedience. *Excited catatonia*, on the other hand, is characterized by severe psychomotor agitation, potentially leading to life-threatening complications such as hyperthermia, altered consciousness, and autonomic dysfunction [22, 25].

In the 1970s, when multiple reports indicated that catatonia is more closely associated with affective disorders than schizophrenia, it was finally separated (Box 11.1).

In DSM-5, catatonia syndrome is defined on the basis of 3 or more than 12 (Box 11.1) symptoms and may be diagnosed as a specifier to major mood disorders, psychotic disorders, general medical conditions, and catatonia not otherwise specified. This allows diagnosing the syndrome in a large variety of psychiatric disorders [13] (Box 11.2).

In two surveys including several patients with psychiatric catatonia syndrome, 30% had diagnosis of mood disorder (17–47% in mania and 0–20% with a depressive episode) and 43% of schizophrenia [26, 27].

Between patients with a unipolar disorder, 20% of patients meet the criteria for catatonia [28].

Box 11.1 Catatonia Criteria in DSM-5 [13]

Three or more of the following:

1. Catalepsy (i.e., passive induction of a posture held against gravity)
2. Waxy flexibility (i.e., slight and even resistance to positioning by examiner)
3. Stupor (no psychomotor activity; not actively relating to environment)
4. Agitation, not induced by external stimuli
5. Mutism (i.e., no or very little verbal response)
6. Negativism (i.e., opposing or not responding to instructions or external stimuli)
7. Posturing (i.e., spontaneous and active maintenance of a posture against gravity)
8. Mannerisms (i.e., odd caricature of normal actions)
9. Stereotypies (i.e., repetitive, abnormally frequent, non-goal-directed movements)
10. Grimacing
11. Echolalia (i.e., mimicking another's speech)
12. Echopraxia (i.e., mimicking another's movements)

Patients with psychiatric catatonia often present with an association of catatonic symptoms and signs, usually more than five; the most frequent symptoms are mutacism (68% of cases) and negativism or psychomotor arrest (62% of cases) [29].

Box 11.2 Catatonia Diagnosis in DSM-5 [13]

1. Catatonic disorder due to a general medical condition
2. Catatonia specifier for
   (a) Schizophrenia
   (b) Schizoaffective disorder
   (c) Schizophreniform disorder
   (d) Brief psychotic disorder
   (e) Substance-induced psychotic disorder
3. Catatonia specifier for affective disorders
   (a) Major depressive disorder
   (b) Bipolar I disorder
   (c) Bipolar II disorder
4. Catatonic disorder NOS

11.3.2 Clinical Assessment: Differential Diagnosis

There are no available neuroimaging exams or laboratory tests helping physician to make diagnosis.

Once catatonia syndrome is recognized on clinical signs and symptoms, the clinician should consider psychiatric form as a matter of exclusion.

A number of neurological conditions may appear similar to catatonia and may even have substantial overlap with respect to pathophysiological mechanisms (Box 11.3).

Box 11.3 Differential Diagnosis

Extrapyramidal side effects
Neuroleptic malignant syndrome
Nonconvulsive status epilepticus
Abulia or akinetic mutism
Stiff person syndrome

11.3.3 Measuring Catatonia

The first rating scale designed for a systematic and standardized evaluation of catatonia is the Bush-Francis Catatonia Rating Scale (BFCRS). It was created in 1996 by Bush and collaborators [30].

Since its publication, it has become widely used and has been translated into several languages.

The examination consists of 23 items. Seventeen items are scored on a 0–3 scale, while the remaining six are rated as either absent ("0") or present ("3"). Selection of the items was based on the classical description by Kahlbaum, Kraepelin, Taylor, and contemporary literature [31] (Box 11.4).

Box 11.4 Bush-Francis Catatonia Rating Scale Items

Screening items	Full-scale items
Excitement	Impulsivity
Immobility/stupor	Automatic obedience
Mutism	Mitgehen
Staring	Gegenhalten
Posturing/catalepsy	Ambitendency
Grimacing	Grasp reflex
Echopraxia/echolalia	Perseveration
Stereotypy	Combativeness
Mannerisms	Autonomic abnormality
Verbigeration	
Rigidity	
Negativism	
Waxy flexibility	
Withdrawal	

11.3.4 Complications

If catatonia remains unrecognized and untreated, it may become chronic and associated with several complications (malnutrition, immobility) that eventually will lead the patient to die [32].

Box 11.5 Complications Associated with Patient's Clinical Features

Clinical features	Potential complications
Autonomic instability	Hypotension, hypertension, arrhythmias, myocardial ischemia, rhabdomyolysis, hemodynamic instability
Decreased movement	Thromboembolic disease, pressure ulcers, rhabdomyolysis
Decreased oral intake	Dehydration, hypovolemia, electrolyte derangements, inability to take oral medications
Catalepsy	Decreased airway clearance, aspiration pneumonia, hypoxemia, pneumonia already

11.3.5 Neurobiology

Several neurotransmitter disturbances have been implicated as putative causes for catatonia, including dopamine, gamma-aminobutyric acid (GABA), and glutamate.

The functional status of dopaminergic system in catatonia remains unclear. Indeed neuroleptics with high anti-D2 potency can induce catatonia-like status or worsen preexisting form, but cannot explain all the signs or symptoms of catatonia. Dopamine agonists, however, failed to alleviate catatonia [33].

The GABAergic system seems to play a central role, with a decreased GABAA receptor density in sensorimotor cortex catatonia patients, correlating to severity of disease. Furthermore GABAergic drugs are the most effective treatment.

Glutamatergic system may contribute to the pathogenesis of catatonia due to hyperactivity of the NMDA receptor on the GABAergic system. The NMDA receptor antagonist (e.g., ketamine) can cause catatonia-like symptoms in healthy subject. On the contrary, glutamatergic antagonists (e.g., amantadine) are useful in treating catatonia [34].

Alterations of brain function or structure due to catatonia are found within the cerebral motor circuit. The majority of studies reported hypoactivity in cortical motor areas of the frontal and parietal cortex. Early work on regional cerebral blood flow (rCBF) indicated frontal and parietal hypoperfusion in mixed groups of predominantly akinetic catatonia. Furthermore, some reports noted an increase of frontal motor and parietal rCBF during the improvement of catatonia by ECT. Moreover, akinetic catatonia patients had delayed onset of movement-related potentials and readiness potentials in motor areas, which correlated with catatonia severity. Likewise, a few studies consistently found reduced neural activation in cortical motor and premotor areas, as well as in the parietal cortex in catatonia during finger-tapping or finger-opposition tasks [35, 36].

11.3.6 Treatment

Correct management of catatonia requires, first of all, identification and treatment of any underlying medical conditions (internal, neurologic, toxicological) that are responsible for clinical symptoms [37].

A characteristic feature of catatonia consists of its striking responsiveness to benzodiazepine treatment. The most commonly used treatment is intravenous lorazepam, with a reported remission rate of catatonic manifestations of about 70%.

Elective treatment consists of intravenous benzodiazepine. Lorazepam (1 mg) is administered; if there are no changes in symptoms after 5 min, another intravenous dose of 1 mg is given [24]. A remission rate of about 70% is reported.

A negative result, even if it does not exclude a future response to lorazepam (at doses higher than those normally used), suggests that ECT should be preferred (with 85% remission rate) [38–40].

11.4 Cotard's Syndrome

11.4.1 Epidemiology and Clinical Presentation

In 1880, Jules Cotard (1840–1889) described a psychopathological condition characterized by a constellation of "false nihilistic beliefs" concerning own body, the non-existence of the self, and severe depressive symptoms.

It affects middle-aged or older people, and the typical picture also includes depressive mood, which is present in approximately 90% of cases described in the literature [41, 42]. Although there is historical controversy over the precise clinical picture of CS, it is most often an eponym for *délire des négations*, translated to English as "nihilistic delusion." There are several proposed mechanisms for CS. It may begin with anomalous perceptual experiences or their disconnection from emotional or limbic processes, and there may be a failure in belief evaluation or a tendency to negative self-attribution.

It is possible to divide the syndrome according to prevailing psychopathological aspects.

In Cotard type I, patients suffer hypochondriacal delusions, characterized by denial of self-related aspects (body or personal space); altered self-movement/capacity to walk or eat; uncertain existence or function of the heart, head/brain, and liver/stomach/intestine/bowels; doubts regarding aspects of the external world, existence of the world, doctors, patient's father, mother, or children; and delusional thoughts of being pregnant.

In Cotard type II, anxiety, delusions of immortality, auditory hallucinations, nihilistic delusions concerning existence, and suicidal behavior prevail.

Many patients with CS have psychiatric disease with psychotic features.

G. E. Berrios et al. [43] reported depression in 89%, anxiety in 65%, and guilt in 63%, delusions concerning the body in 86%, and existence in 69% in schizophrenic patients.

11.4.2 Psychiatric Disorder Related to Cotard's Syndrome

CS may be seen in both neurological and psychiatric illnesses. It is not listed as a specific disorder in DSM 5 but as a symptom of an underlying disorder.

The most frequently associated diseases are indicated in the table (Box 11.6), and a specific diagnostic work-up must be done before starting pharmacological treatments.

Box 11.6
Psychiatric disease

Depressive disorder
Schizophrenic spectrum
Bipolar disorder
Catatonia
Capgras syndrome
Lycanthropy

Neurological disease

Cerebrovascular disease
Subdural hemorrhage
Parkinson's disease
Traumatic brain injury
Multiple sclerosis
Arteriovenous malformation
Epilepsy
Semantic dementia
Atrophy of the insular cortex

11.4.3 Treatment Strategies

It still remains to be elucidated, however, if symptom resolution is possible after treatment of their underlying neurological and psychiatric disturbances.

Treatment for CS focuses on its clinical origin.

Although a definite agreement on the best treatment protocol for catatonia does not exist, antidepressants seem to be the most effective compounds, including TCAs and SSRIs.

Thus, antidepressants can be useful in patients with affective disorders, and there are studies which show that (ECT) was successful, leading to proposing it as the treatment of choice. Antipsychotic drug treatment or combination strategies (antipsychotics plus antidepressants) are also used.

Formal internationally recognized guidelines to treat Cotard's syndrome still lack; in fact, such clinical condition is not listed as a univocal psychiatric disorder in DSM 5, but it is described as a number of symptoms deriving from an underlying disorder.

In a case report, Chou et al. [44] observed a dramatic improvement of symptoms in a patient with CS after 2 months of treatment with fluoxetine 40 mg/day and risperidone 6 mg/day. Another reported combination was venlafaxine 225 mg/day and quetiapine 600 mg/day, which produced relief of depressive symptoms and delusions of negation in a 68-year-old patient after 2 weeks of treatment [45]. In yet another case report, clozapine 50 mg/day, fluvoxamine 200 mg/day, and imipramine 50 mg/day led to a complete remission of delusions in 5 days.

Monotherapy treatment with sulpiride 300 mg/day was reported as successful in a 33-year-old patient with schizophrenia who developed CS [46], and Shiraishi et al. concluded that treatment with sulpiride is the first line for CS in schizophrenia.

11.4.4 Neuroimaging

The application of neuroimaging techniques, such as PET, to the study of depressive disorders can contribute to the development of pathophysiologic models and to the understanding of the neurobiology of the disease [47].

PET uses radioactive tracers to measure changes in the composition of brain metabolism and to study functional imaging of the brain.

Some PET studies showed that depressed subjects have reduced lateral prefrontal metabolism and increased medial prefrontal and subgenual cingulate metabolism [48].

Meta-analysis results demonstrated that MDD patients exhibit decreased brain activation on bilateral insula, particularly in the right insula.

The results also verify that the limbic system including the cingulate gyrus and the basal ganglia including caudate head, lentiform nucleus putamen, and had abnormal brain metabolism in MDD patients [49, 50].

Results from a previous study indicated that in the context of untreated symptoms of depression, these regions showed decreased activation and that limbic right posterior cingulate and right basal ganglia activation were seen after antidepressant treatment.

Few studies concerning CS are available in literature and are discordant. Some showed hypometabolism in the frontal and parietal cortex [51, 52], whereas Lee et al. showed hypermetabolism in the same region [53] in catatonic patients.

11.5 Bipolar and Psychotic Depression

The World Health Organization has ranked BP among the leading causes of disability globally, affecting 4% of the population.

The morbidity associated with BP is increasingly recognized as not the result of mania but depression [54].

Depressive phase compared to manic is more pervasive, with functional impairment, working disability, and a wide range of medical comorbid disorders. Often patients attempt suicide [55].

Depressive symptoms dominate the long-term course of the illness.

Psychotic features are highly prevalent in the course of BP. Several studies have shown that patients with psychotic symptoms during episodes of BP represent a form more severe of illness, associated with lower rates of recovery [56].

The prevalence of psychosis in bipolar depression has significant prevalence (10.4%).

Psychotic depression is associated with a high suicidality compared to psychotic manic or mixed states. The heavy course of the disease is quite similar in patients with unipolar depression with psychotic features [57].

11.6 Discussion

11.6.1 Evidence-Based Therapy

In order to offer clinicians the best guidance regarding optimal pharmacological treatment of BP, several expert guidelines have been developed. However, despite the significant clinical differences between psychotic and nonpsychotic episodes of BP, guidelines do not give specific advice for the treatment of the psychotic episodes, reflecting the presence of few studies of pharmacotherapy for bipolar depression with psychotic features [58, 59].

A recent study, [56] which investigated the differences in psychopharmacological treatment between psychotic and nonpsychotic episodes in BD, reported that psychotic bipolar depression was associated with treatment with antipsychotics and the combination of an antipsychotic and an antidepressant.

However, the efficacy of antidepressants for the treatment of bipolar depression has not been well established. Paroxetine is less effectiveness to improving depression than in unipolar disorder [60].

Over the past decade, however, there have been only the following treatments approved by the FDA for bipolar depression: olanzapine, fluoxetine (2003), quetiapine monotherapy (2006 and 2008), and lurasidone monotherapy and adjunctive therapy (2013) [61–63].

Lamotrigine, a drug approved by the FDA for the maintenance phase (i.e., delaying manic and depressive recurrence) of bipolar I disorder, is not FDA-approved for acute bipolar depression, but a meta-analysis (five studies, one positive, four negative) and one controlled LamLit study suggest possible efficacy in acute bipolar depression [64].

In a case report by Kajiya 2017, lamotrigine was given in add-on to a bipolar-depressed patient [65].

Aripiprazole is an atypical antipsychotic with a novel mechanism of action.

Aripiprazole is approved by the FDA for the treatment of acute bipolar mania and maintenance treatment of bipolar I disorder. Further, aripiprazole adjunctive therapy has shown some benefits in the treatment of bipolar depression and in MDD [66–71].

However, aripiprazole seems to have less effect by week 8.

The evidence available thus far does not support the efficacy of aripiprazole for the treatment of acute bipolar depression and prevention of depressive relapse.

11.7 Conclusion

While the remarkable number of drugs is available, effective therapeutic options for bipolar (psychotic) depression are seldom inconclusive.

Although the literature has not well established the exact criteria for diagnosing CS (including patient with affective and nonaffective psychosis), treatment strategies remain to be elucidated, especially in those patients who have a psychotic feature.

Catatonia is a neuropsychiatric syndrome with movement disorder. It should be considered when a patient shows a heavy psychomotor retardation, in order to prevent fatal complication.

The association between catatonia and CS is very rare and potentially harmful.

Personalized medicine approach may establish efficacy and effectiveness of the pharmacological treatment, together with the ability to recognize different psychopathological pictures.

Key Points

- Cotard is a syndrome typically described as a constellation of "false nihilistic beliefs," often in the form of self-negation. It affects middle-aged or older people, and the peculiar manifestation includes depressive mood, which is present in approximately 90% of cases described in the literature.
- Catatonia is a complex neuropsychiatric syndrome that has been reported to occur in more than 10% of patients with acute psychiatric illnesses (hospital admitted).

- In the main epidemiological studies, prevalence of catatonia in psychiatric patients varies from 7% to 31% and can be present both in adults and adolescents, as well as in infants.
- In DSM-5 catatonia syndrome is defined on the basis of the presence of three or more symptoms; this condition can be a specifier to major mood disorders, psychotic disorders, general medical conditions, or as catatonia not otherwise specified.
- Management of catatonia in psychiatric conditions consists of intravenous lorazepam, with a reported remission of catatonic manifestations of about 70%. Anyway pharmacological treatment is complex, and it needs multiple psychotropic drugs.
- Almost half of patients with bipolar disorder show psychotic symptoms during illness' phases, and their presence is associated with worse prognosis.

  Psychosis is more frequent in manic episode than in bipolar depressive episodes, in which the lifetime prevalence reported is 10–28%. Indeed, during psychotic depressive phase, catatonia and nihilistic delusion are rarely present.

Self-Assessment Questionnaire

1. Which of the following statements about catatonia treatment is correct?
   (A) **It is responsive to benzodiazepine treatment (lorazepam 1–2 mg) IM or EV.**
   (B) High doses of dopaminergic agents are useful.
   (C) No therapy is currently effective.
   (D) Electroconvulsive therapy is the only treatment approved.
2. Which of the following statements about catatonia treatment is incorrect?
   (A) Catatonia was described by Karl Kahlbaum in the 1870s for the first time.
   (B) Catatonia is a transnosographic syndrome, so it may be diagnosed as a specifier to major mood disorders, psychotic disorders, and general medical conditions.
   (C) Two subtypes have been described: retarded and excited.
   (D) **It is only present in schizophrenic disorder.**
3. Which of the following statements about Cotard's syndrome is correct?
   (A) **It is a nihilistic delusion that can occur in psychiatric and neurological conditions.**
   (B) It occurs only in elderly patients.
   (C) The only therapy to which the patients respond is ECT.
   (D) In schizophrenic patients, it is two times more common.
4. Psychotic bipolar depression can be psychopharmacologically treated with the combination of an antipsychotic and an antidepressant:
   (A) **Yes**
   (B) No
   (C) Never
   (D) Only in young patients

5. Major depressive episodes with psychotic features are more common in BD than MDD:
   (A) **Yes.**
   (B) No.
   (C) It depends on psychometric scale.
   (D) Only in women.

References

1. Arciniegas DB. Psychosis. Continuum (Minneap Minn). 2015;21(3 Behavioral Neurology and Neuropsychiatry):715–36.
2. Goghari VM, Harrow M. Twenty year multi-follow-up of different types of hallucinations in schizophrenia, schizoaffective disorder, bipolar disorder, and depression. Schizophr Res. 2016;176:371–7.
3. Pallaskorpi S, Suominen K, Ketokivi M, et al. Five-year outcome of bipolar I and II disorders: findings of the Jorvi Bipolar Study. Bipolar Disord. 2015;17:363–74.
4. Altamura AC, Buoli M, Caldiroli A, et al. Misdiagnosis, duration of untreated illness (DUI) and outcome in bipolar patients with psychotic symptoms: a naturalistic study. J Affect Disord. 2015;182:70–5.
5. Mantere O, Suominen K, Leppämäki S, et al. The clinical characteristics of DSM-IV bipolar I and II disorders: baseline findings from the Jorvi Bipolar Study (JoBS). Bipolar Disord. 2004;6:395–405.
6. Caldieraro MA, Sylvia LG, Dufour S, et al. Clinical correlates of acute bipolar depressive episode with psychosis. J Affect Disord. 2017;217:29–33.
7. Fountoulakis KN, Vieta E, Young A, et al. The International College of Neuropsychopharmacology (CINP) treatment guidelines for bipolar disorder in adults (CINP-BD-2017), Part 4: Unmet needs in the treatment of bipolar disorder and recommendations for future research. Int J Neuropsychopharmacol. 2017;20(2):196–205.
8. Frankland A, Cerrillo E, Hadzi-Pavlovic D, et al. Comparing the phenomenology of depressive episodes in bipolar I and II disorder and major depressive disorder within bipolar disorder pedigrees. J Clin Psychiatry. 2015;76:32–8.
9. Goes FS, Sadler B, Toolan J, et al. Psychotic features in bipolar and unipolar depression. Bipolar Disord. 2007;9:901–6.
10. Zaninotto L, Souery D, Calati R, et al. Dimensions of delusions in major depression: socio-demographic and clinical correlates in an unipolar-bipolar sample. Clin Psychopharmacol Neurosci. 2015;13:48–52.
11. Mitchell PB, Wilhelm K, Parker G, et al. The clinical features of bipolar depression: a comparison with matched major depressive disorder patients. J Clin Psychiatry. 2001;62:212–6.
12. Grande I, Berk M, Birmaher B, et al. Bipolar disorder. Lancet. 2016;387:1561–72.
13. DSM-5. Diagnostic and statistical manual of mental disorders. 5th ed. American Psychiatric Publishing; 2013.
14. Sienaert P, Dhossche DM, Vancampfort D, De Hert M, Gazdag G. A clinical review of the treatment of catatonia. Front Psychiatry. 2014;5:181.
15. Simpson P, Kaul E, Quinn D. Cotard's syndrome with catatonia: a case presentation and discussion. Psychosomatics. 2013;54(2):196–9.
16. First MB, Williams JB, Karg RS, Spitzer R. SCID-5-CV: structured clinical interview for DSM-5 disorders, clinician version. Washington, DC: American Psychiatric Association; 2016.
17. Overall J, Gorham D. The brief psychiatric rating scale. Psychol Rep. 1962;10:799–812.
18. Hamilton M. A rating scale for depression. J Neurol Neurosurg Psychiatry. 1960;23:56–62.
19. Montgomery SA, Asberg M. A new depression scale designed to be sensitive to change. Br J Psychiatry. 1979;134:382–9.

20. Young RC, Biggs JT, Ziegler VE, Meyer DA. A rating scale for mania: reliability, validity and sensitivity. Br J Psychiatry. 1978;133:429–35.
21. Rosebush PI, Hildebrand AM, Furlong BG, Mazurek MF. Catatonic syndrome in a general psychiatric inpatient population: frequency, clinical presentation, and response to lorazepam. J Clin Psychiatry. 1990;51:357–62.
22. Peralta V, Campos MS, de Jalon EG, Cuesta MJ. DSM-IV catatonia signs and criteria in first-episode, drug-naive, psychotic patients: psychometric validity and response to antipsychotic medication. Schizophr Res. 2010;118:168–75.
23. Taylor MA, Fink M. Catatonia in psychiatric classification: a home of its own. Am J Psychiatry. 2003;160(7):1233–41.
24. Kahlbaum KL. Catatonia (originally published in 1874). Baltimore: John Hopkins University Press; 1973.
25. Mann SC, Caroff SN, Bleier HR, Welz WK, Kling MA, Hayashida M. Lethal catatonia. Am J Psychiatry. 1986;143:1374–81.
26. Pommepuy N, Januel D. Catatonia: resurgence of a concept. A review of the international literature. Encéphale. 2002;28(6 Pt 1):481–92.
27. Krüger S, Bagby RM, Höffler J, Bräunig P. Factor analysis of the catatonia rating scale and catatonic symptom distribution across four diagnostic groups. Compr Psychiatry. 2003;44 (6):472–82.
28. Starkstein SE, Petracca G, Teson A. Catatonia in depression: prevalence, clinical correlates, and validation of a scale. J Neurol Neurosurg Psychiatry. 1996;60:326–32.
29. Francis A, Fink M, Appiani F, et al. Catatonia in diagnostic and statistical manual of mental disorders, fifth edition. J ECT. 2011;26:246–7.
30. Bush G, Fink M, Petrides G, Dowling F, Francis A. Catatonia rating scale and standardized examination. Acta Psychiatr Scand. 1996;93:129–36.
31. Sienaert P, Rooseleer J, De Fruyt J. Measuring catatonia: a systematic review of rating scales. J Affect Disord. 2011;135(1–3):1–9.
32. Carrol BT. Complications of catatonia. J Clin Psychiatry. 1996;57(2):95.
33. Daniels J. Catatonia: clinical aspects and neurobiological correlates. J Neuropsychiatry Clin Neurosci. 2009;21:371–80.
34. Carroll BT, Goforth HW, Thomas C, Ahuja N, McDaniel WW, Kraus MF. Review of adjunctive glutamate antagonist therapy in the treatment of catatonic syndromes. J Neuropsychiatry Clin Neurosci. 2007;19(4):406–12.
35. Dhossche DM, Stoppelbein L, Rout UK. Etiopathogenesis of catatonia: generalizations and working hypotheses. J ECT. 2010;26:253–8.
36. Nisijima K, Ishiguro T. Cerebrospinal fluid levels of monoamine metabolites and gamma-aminobutyric acid in NMS. J Psychiatr Res. 1995;29:233–44.
37. Ungvari GS, Kau LS, Wai-Kwong T, Shing NF. The pharmacological treatment of catatonia: an overview. Eur Arch Psychiatry Clin Neurosci. 2001;251(Suppl 1):I31–4.
38. Fink M, Taylor MA. Catatonia: a clinician's guide to diagnosis and treatment. Cambridge: Cambridge University Press; 2003.
39. Goodchild CS. GABA receptors and benzodiazepines. Br J Anaesth. 1993;71(1):127–33.
40. Smith M, Lebegue B. Lorazepam in the treatment of catatonia. Am J Psychiatry. 1991;148(9): 1265.
41. Cotard J. Du délire hypocondriaque dans une forme grave de la mélancolie anxieuse. Ann Med Psychol. 1880;38:168–14.
42. Cotard J. Du délire des negations. Arch Neurol. 1882;4:152–70.
43. Berrios GE, Luque R. Cotard's syndrome: analysis of 100 cases. Acta Psychiatr Scand. 1995; 91(3):185–8.
44. Chou PH, Lin BT, Lan TH, Chan CH. Chronic Cotard's syndrome: recovery from 2 years' bed-ridden status. Psychiatry Clin Neurosci. 2011;65(3):301.
45. Chan JH, Chen CH, Robson D, Tan HK. Case report: effective treatment of Cotard's syndrome: quetiapine in combination with venlafaxine. Psychiatry Clin Neurosci. 2009;63(1):125–6.

46. Shiraishi H, Ito M, Hayashi H, Otani K. Sulpiride treatment of Cotard's syndrome in schizophrenia. Prog Neuro-Psychopharmacol Biol Psychiatry. 2004;28(3):607–9.
47. McGuire PK. Functional neuroimaging in mental disorders. World Psychiatry. 2004;3(1):6–11.
48. Fusar-Poli P, Allen P, McGuire P. Neuroimaging studies of the early stages of psychosis: a critical review. Eur Psychiatry. 2008;23:237–44.
49. Kimbrell TA, Ketter TA, George MS. Regional cerebral glucose utilization in patients with a range of severities of unipolar depression. Biol Psychiatry. 2002;51:237–52.
50. Su L, Cai Y, Xu Y, Dutt A, Shi S, Bramon E. Cerebral metabolism in major depressive disorder: a voxel-based meta-analysis of positron emission tomography studies. BMC Psychiatry. 2014; 14:321.
51. Charland-Verville V, Bruno MA, Bahri MA, Demertzi A, Desseilles M. Brain dead yet mind alive: a positron emission tomography case study of brain metabolism in Cotard's syndrome. Cortex. 2013;49(7):1997–9.
52. Ozkan N, Caliyurt O. Brain metabolism changes with 18F-fluoro-deoxy-glucosepositron emission tomography in a patient with Cotard's syndrome. Aust N Z J Psychiatry. 2016;50(6): 600–1.
53. Lee EM, Kang JK, Oh JS, et al. 18F-fluorodeoxyglucose positron-emission tomography findings with anti-N-methyl-D-aspartate receptor encephalitis that showed variable degrees of Catatonia: three cases report. J Epilepsy Res. 2014;4(2):69–73.
54. World Health Organization. The global burden of disease: 2004 update. Geneva, Switzerland, World Health Organization, http://www.who.int/healthinfo/global_burden_disease/GBD report 2004update; 2008 Accessed October 17, 2014.
55. Wijkstra J, Lijmer J, Burger H, Geddes J, Nolen WA. Pharmacological treatment for psychotic depression. Cochrane Database Syst Rev. 2013;(11).
56. Bjørklund LB, Horsdal HT, Mors O, Gasse C, Østergaard SD. Psychopharmacological treatment of psychotic mania and psychotic bipolar depression compared to non-psychotic mania and non-psychotic bipolar depression. Bipolar Disord. 2017;19(6):505–12.
57. Fryea MA, Prieto ML, Bobo WV, et al. Current landscape, unmet needs, and future directions for treatment of bipolar depression. J Affect Disord. 2014;169(S1):17–23.
58. Grunze H, Vieta E, Goodwin GM, et al. The World Federation of Societies of Biological Psychiatry (WFSBP) guidelines for the biological treatment of bipolar disorders: update 2009 on the treatment of acute mania. World J Biol Psychiatry. 2009;10:85–116.
59. Grunze H, Vieta E, Goodwin GM, et al. The World Federation of Societies of Biological Psychiatry (WFSBP) guidelines for the biological treatment of bipolar disorders: update 2010 on the treatment of acute bipolar depression. World J Biol Psychiatry. 2010;11:81–109.
60. Pacchiarotti I, Bond DJ, Baldessarini RJ, et al. The International Society for Bipolar Disorders (ISBD) task force report on antidepressant use in bipolar disorders. Am J Psychiatry. 2013; 170(11):1249–62.
61. Tohen M, Vieta E, Calabrese J, et al. Efficacy of olanzapine and olanzapine-fluoxetine combination in the treatment of bipolar I depression. Arch Gen Psychiatry. 2003;60(11):1079–88.
62. Calabrese JR, Keck PE Jr, Macfadden W, et al. A randomized, double-blind, placebo-controlled trial of quetiapine in the treatment of bipolar I or II depression. Am J Psychiatry. 2005;162(7): 1351–60.
63. Loebel A, Cucchiaro J, Silva R, et al. Lurasidone monotherapy in the treatment of bipolar I depression: a randomized, double-blind, placebo-controlled study. Am J Psychiatry. 2014; 171(2):160–8.
64. Geddes JR, Calabrese JR, Goodwin GM. Lamotrigine for treatment of bipolar depression: independent meta-analysis and meta-regression of individual patient data from five randomized trials. Br J Psychiatry. 2009;194(1):4–9.
65. Kajiya T, Sugawara H, Kajio Y, Morieda S. Effect of lamotrigine in the treatment of bipolar depression with psychotic features: a case report. Ann General Psychiatry. 2017;16:31.

66. Dunn RT, Stan VA, Chriki LS, Filkowski MM, Ghaemi SN. A prospective, open-label study of aripiprazole mono- and adjunctive treatment in acute bipolar depression. J Affect Disord. 2008; 110:70–4.
67. Kemp DE, Gilmer WS, Fleck J, Straus JL, Dago PL, Karaffa M. Aripiprazole augmentation in treatment-resistant bipolar depression: early response and development of akathisia. Prog Neuro-Psychopharmacol Biol Psychiatry. 2007;31:574–7.
68. Ketter TA, Wang PW, Chandler RA, Culver JL, Alarcon AM. Adjunctive aripiprazole in treatment-resistant bipolar depression. Ann Clin Psychiatry. 2006;18:169–72.
69. McElroy SL, Suppes T, Frye MA, et al. Open-label aripiprazole in the treatment of acute bipolar depression: a prospective pilot trial. J Affect Disord. 2007;101:275–81.
70. Quante A, Zeugmann S, Luborzewski A, et al. Aripiprazole as adjunct to a mood stabilizer and citalopram in bipolar depression: a randomized placebo controlled pilot study. Hum Psychopharmacol. 2010;25:126–32.
71. Sokolski KN. Adjunctive aripiprazole in bipolar I depression. Ann Pharmacother. 2007;41: 35–40.

Generalized Anxiety Disorder, Somatization, and Emotional Dysregulation: A Possible Link

12

Gianluigi Tacchini and Matteo Vismara

Abstract

In clinical practice, it is quite common to deal with patients who primarily express somatic symptoms. They can be distinctive features of depressive disorders, of anxiety disorders, in particular general anxiety disorder, and, less commonly, of somatic symptom disorders. Nonetheless, in several patients, these three conditions could coexist and delineate a clinical picture driven by emotional dysregulation (ED). ED is an emotional response to external stimuli that is poorly modulated and does not fall within the conventionally accepted range of emotive response, which can be characterized by marked and rapid fluctuation of mood, mood lability, weeping crisis, eating problems, and up to behavior outbursts. In our clinical case, a female patient came to our attention reporting headache, gastrointestinal disturbance, hyporexia, and leg restlessness. The diagnostic approach is pictured, and a correct pharmacological treatment is shown in this chapter.

Keywords

Generalized anxiety disorder · Somatization · Emotional dysregulation · Magnetic resonance

G. Tacchini (✉) · M. Vismara
Department of Neurosciences and Mental Health, Fondazione IRCCS Ca' Granda Ospedale Maggiore Policlinico, University of Milan, Milan, Italy
e-mail: gianluigi.tacchini@policlinico.mi.it

A. C. Altamura, P. Brambilla (eds.), *Clinical Cases in Psychiatry: Integrating Translational Neuroscience Approaches*,
https://doi.org/10.1007/978-3-319-91557-9_12

12.1 Introduction

12.1.1 Generalized Anxiety Disorder

Generalized anxiety disorder (GAD) is a common mental health condition characterized by an excessive and persistent worry about everyday life with no obvious reasons and not restricted to particular circumstances. Physical anxiety symptoms (such as tachycardia, tremors, headaches, sweating, nausea) and psychological symptoms (including restlessness, fatigue, difficulty in concentrating, irritability, and altered sleep) represent characteristic key points [1]. A recent cross-sectional study conducted around the globe reported a 3.7% lifetime prevalence, being higher in high-income countries [2]. These data make GAD a common public health concern, particularly for females and older subjects that show an even higher prevalence. Moreover, associated functional impairment is reported to be similar to patients suffering from major depressive disorder. The course of the disorder is chronic, with poor family relationships, high rates of comorbid axis I disorders, and cluster C personality disorders that can worsen its evolution [3].

Box 12.1 Diagnostic and Statistical Manual of Mental Disorders 5 (DSM 5) Criteria for GAD

(A) Excessive anxiety and worry (apprehensive expectation), occurring more days than not for at least 6 months, about a number of events or activities (such as work or school performance).

(B) The individual finds it difficult to control the worry.

(C) The anxiety and worry are associated with three (or more) of the following six symptoms (with at least some symptoms having been present for more days than not for the past 6 months) (Note: Only one item is required in children):

1. Restlessness, feeling keyed up, or on edge
2. Being easily fatigued
3. Difficulty concentrating or mind going blank
4. Irritability
5. Muscle tension
6. Sleep disturbance (difficulty falling or staying asleep, or restless, unsatisfying sleep)

(D) The anxiety, worry, or physical symptoms cause clinically significant distress or impairment in social, occupational, or other important areas of functioning.

(E) The disturbance is not attributable to the physiological effects of a substance (e.g., a drug of abuse, a medication) or another medical condition (e.g., hyperthyroidism).

(continued)

Box 12.1 (continued)

(F) The disturbance is not better explained by another medical disorder (e.g., anxiety or worry about having panic attacks in panic disorder, negative evaluation in social anxiety disorder [social phobia], contamination or other obsessions in obsessive-compulsive disorder, separation from attachment figures in separation anxiety disorder, reminders of traumatic events in posttraumatic stress disorder, gaining weight in anorexia nervosa, physical complaints in somatic symptom disorder, perceived appearance flaws in body dysmorphic disorder, having a serious illness in illness anxiety disorder, or the content of delusional beliefs in schizophrenia or delusional disorder).

As for other psychiatric disorders, GAD etiology is multifactorial with genetic and environmental factors playing an important role in its development. Among genetic factors, literature studies have reported that GAD tends to cluster in families, where a positive family history plays a role in increasing the likelihood to develop GAD. Twin studies have shown a moderate hereditary influence, which is, however, less intense than in other anxiety disorders, e.g., panic disorder [4]. Some genes associated with GAD are key genes implicated in the adrenergic nervous system, GABA, and serotonin pathways [5]. These genes might contribute to abnormal neuronal pathways, involved in cognitive functions like thinking and emotion. Functional MRI (fMRI) studies in GAD patients have identified abnormal amygdala and prefrontal cortex activation and decreased functional connectivity between these areas. Furthermore, additional studies showed increased gray matter volume and decreased structural connectivity between these brain structures [6]. Additionally, a recent review analyzed the role of proton magnetic resonance spectroscopy (^{1}H MRS) investigations in patients with GAD. This new technique has the aim of identifying differences in metabolite levels between conditions in key brain areas. As results, the majority of studies reviewed showed altered metabolite levels in the dorsolateral prefrontal cortex and hippocampus, suggesting regional specificity [7].

Among environmental factors, trauma and stressful life events, such as abuse, loved one's death, divorce, and changing jobs or schools, may contribute to the onset. Addictive substance use/withdrawal, including alcohol, can worsen anxiety symptoms and facilitate GAD development. Additionally, an association between smoking and anxiety symptoms has been reported, with a risk of GAD 5–6 times higher among adolescents who smoke heavily compared to nonsmokers [8].

12.1.2 Somatization, Somatic Symptom, and Related Disorders

Somatization is defined as a patient's concern about physical symptoms, occurring when someone tends to communicate psychological distress in the form of a somatic symptom. This tendency is common in the general population and can be considered a worldwide phenomenon, more common in women and older ages [9]. When a significant functional impairment is associated, specific psychiatric disorders can be delineated.

DSM 5 collects these disorders under the diagnostic group of "somatic symptom and related disorders" that constitutes a new diagnostic category compared to DSM IV-TR. This chapter includes:

- Somatic symptom disorder (SSD), when patient symptom concern worries him/her constantly and/or drives him/her to see doctors frequently
- Illness anxiety disorder, when patients are excessively preoccupied and worried about the possibility of having or getting a serious illness
- Functional neurological symptom disorder (formerly known as conversion disorder), when physical symptoms resemble a nervous system disorder
- Psychological factors affecting other medical conditions
- Factitious disorder, when people pretend to have symptoms for no apparent external reason (such as to get time off from work)
- Other specified somatic symptom and related disorders
- Unspecified somatic symptom and related disorder

Patients suffering from somatic symptom disorder (SSD) report a significant concern about physical symptoms, with abnormal thoughts, feelings, and behaviors. Although medical conditions are excluded, excessive worry leads to frequent doctor visits, experiencing major emotional distress and difficulties in overall daily functioning. The SSD key point is that the patient feels and behaves in response to physical sensations and not only the symptom itself [1]. Studies report widely variable lifetime prevalence of SSD, varying from 0.2% to 2% for females and less than 0.2% for males [10]. However, SSD is particularly more common in primary care and other medical settings, with an estimated rate reported of up to 15.1% [11].

In relation to functional neurological symptom disorder, patients report different neurological symptoms (i.e. hemiparesis, paraparesis, monoparesis, alteration of consciousness, visual loss, seizure-like activity, pseudocoma, abnormal gait disturbance, aphonia or dysphonia, lack of coordination, or a bizarre movement disorder) with no medical explanation. Conversion symptoms are seen also in other clinical conditions like affective disorders, antisocial personality disorder, alcohol or drug abuse, or organic, neurologic, or medical illnesses. Other functional conditions such as irritable bowel syndrome, fibromyalgia, chronic pelvic pain, and multiple chemical sensitivity syndrome also have strong associations with conversion disorders.

This clinical picture was classically called "hysteria" derived from Hippocrates' "wandering" uterus. Freud developed this term, reconsidering the psychoanalytic concept of repression through "conversion" of psychological distress into a physical symptom. Recently, DSM 5 focused on brain function abnormalities and reconsidered the diagnostic category of conversion disorder into functional neurological symptom disorders. In this respect, recent studies have shown a possible neurobiological basis, with significantly smaller mean left and right amygdala volumes, without any differences in total gray/white matter or hippocampus volumes in patients with conversion disorders [12]. Additionally, stronger connectivity values in the insula, inferior frontal gyrus, and parietal cortex and precentral sulcus were shown in patients with psychogenic non-epileptic seizures, a quite common presentation of this disorder [13]. Environmental factors also play a role in its pathogenesis; for instance, an immediate precipitating source of stress, such as divorce or loss of employment, may be associated with the development of the disorder. Moreover, a history of sexual or physical abuse is common among these patients and is reported in up to one half of patients suffering functional neurological symptom disorders [14].

Box 12.2 DSM 5 Criteria for Functional Neurological Symptom Disorder

(A) One or more symptoms of altered voluntary motor or sensory function.
(B) Physical findings provide evidence of incompatibility between the symptom and recognized neurological or medical conditions.
(C) The symptom or deficit is not better explained by another medical or mental disorder.
(D) The symptom or deficit causes clinically significant distress or impairment in social, occupational, or other important areas of functioning or warrants medical evaluation.

To correctly diagnose functional neurological symptom disorders, teamwork between a neurologist and a psychiatrist is mandatory. The psychiatric assessment can differentiate functional neurological symptom disorder from other somatoform disorders and factitious disorders and can elucidate the psychodynamic basis. On the other hand, the neurologist must recognize the nonorganic process and avoid potentially dangerous diagnostic or therapeutic interventions.

12.1.3 Emotional Dysregulation

Emotional dysregulation (ED) refers to an emotional response to external stimuli that is poorly modulated and does not fall within the conventionally accepted range of emotive response. ED might express itself with marked and sudden fluctuation of mood, mood lability, weeping crisis, eating problems, right down to angry outbursts or behavior outbursts such as destroying or throwing objects, and aggression toward self or others [15]. Differently from mood changes described in depressive or manic/hypomanic episodes, these variations usually occur rapidly, in seconds to minutes or hours. ED must be considered a psychiatric concern when its extent interferes with personal social interactions and relationships and causes a marked impairment in daily life [16].

Over 85% of diagnoses in the DSM involve excesses or deficits of emotions or a lack of coherence among emotional features [17]. In particular, ED is a core psychopathological factor in psychiatric disorders such as bipolar disorder, major depressive disorder, borderline personality disorder, attention deficit hyperactivity disorder, posttraumatic stress disorder, and autism spectrum disorders. ED is common also in patients with somatoform disorder, alcohol and substance abuse disorder, and anorexia and bulimia nervosa [18].

Emotion regulation involves a set of processes and systems (e.g., attentional, cognitive, behavioral, social, and biological) that act to modulate, manage, and organize emotions in order to help individuals to meet environment demands and achieve their goals [19]. Different factors play a role in the genesis of emotions and the limbic system is one of the most important. Emotional experience involves the integration of visceral signals from the limbic cortices through the cognitive appraisal of these signals in the prefrontal cortex (PFC) [20]. In healthy individuals, cognitive reappraisal of emotion can occur through the top-down prefrontal regulation of limbic activity, using elaboration to modulate initially negative appraisals as being less negative. In mood disorders, however, this cognitive control seems to be impaired, as evidenced by altered connectivity between the PFC and limbic regions, in particular the amygdala. An important clinical implication of these findings is that in patients with a mood disorder, the process of regulating negative emotions may be maladaptive, as emotional information from the limbic system is not regulated by the prefrontal cortices. In confirmation of these hypotheses, neuroimaging studies reported atypically diffuse prefrontal recruitment in depressed patients both during cognitive task performance and during task-free protocols [21].

12.2 Clinical Case

12.2.1 Case Report

A.P. is a middle-aged woman, born in Central Italy, who came for the first time to our department in June 2015. She was brought to the ER of our hospital by her mother. At the psychiatric evaluation, she appeared cooperative toward the examiner. She had a normal state of consciousness, oriented in time, space, and person. Attention was maintained. She looked very agitated and tense and was crying and showing involuntary movements of legs and arms. She said that she had lately been excessively anxious and worried during most of the day at work and in her family life. Speech rate was increased and elevated in tone. Her mood was depressed and her affect was labile. She described the inability to express her feelings completely and clearly, with words such as “an interior chaos” dominating her emotions. Formally the thought was correct, with content focused on worries related to her depressed mood state and obsessions related to her body. She reported altered sleep, with approximately 4 hours per night in the last week, and her food intake was poor with nearly 5 kg weight loss in the last months. No alteration in perceptions was reported. She had no insight into her clinical condition. Due to the clinical picture, she was admitted to the psychiatric ward of our department.

Collecting patient's history, she reported a positive family history for psychiatric illness: a cousin was diagnosed with a major depressive disorder and committed suicide at the age of 70. Her mother, at that time aged 70, was described as a hyper-protective and anxious person, never treated by a psychiatrist. Her father, who left the family when the patient was 6 years old, was described as a violent person who committed crimes. In relation to patient medical history, no complications were described during pregnancy; however forceps was applied due to a delay in delivery procedure. She received regular breastfeeding, and no problems in somatopsychic development are reported. Menarche occurred at the age of 13 and menstruation was regular. She had neither abortions nor pregnancies. She never smoked and never took any recreational drugs/alcohol in her life. With regard to medical diseases, she suffered hypothyroidism, treated with levothyroxine. No allergies were reported. A low food intake was described, with several occasions of voluntary weight loss, without meeting the diagnostic criteria for anorexia nervosa. Personality was characterized by shyness, with few friends since she was prone to solitary activities. The mother has been a determining figure in the patient's life: described as hyper-protective, used to controlling every aspect of patient's life, from education to the relationship with her husband. During childhood the patient was frequently sick with episodes of fever and vomiting, and her mother was worried about her and therefore tended to limit her autonomy. As to education, she got a high school diploma with good results. She wanted to go to university, but her mother dissuaded her because she considered her daughter "too weak." So she began working in a call center which is the only employment she has had in her life, remaining till the age of 46. She got married at the age of 28, but the relationship was described as unstable with some violent episodes, and they separated when the patient was 46; they did not have any children.

From the age of 33, the patient started to report different concerns about physical symptoms: occasionally, she suffered from headache and nausea, with an associated decrease in food intake, and during sleep, she had the feeling of being unable to keep her legs still. For these concerns, she went to her general practitioner who prescribed symptomatic drugs. At that time, she could hold down her job and maintain an overall good functioning. A worsening of the clinical picture occurred when the patient was 36, with episodes characterized by extreme anxiety, anhedonia with clinophilia, altered sleep patterns with periods of hypersomnia and terminal insomnia, and akathisia. At that time, she had her first contact with a psychiatrist and started the first proper oral pharmacological treatment with amisulpride 100 mg per day, valproic acid 750 mg per day, and quetiapine 50 mg per day. Pramipexole was specifically prescribed for "restless leg syndrome."

Box 12.3 Restless Leg Syndrome

Restless leg syndrome (RLS) is a neurologic movement disorder of the lower limbs associated with sleep distress. Patients with RLS report an irresistible urge to move the legs that are not painful but are distinctly annoying. This happens typically during the night and may be associated with daytime tiredness and physical and emotional disability.

(continued)

Box 12.3 (continued)

The diagnostic criteria from the International RLS Study Group are:

- An urge to move the legs usually but not always accompanied by or felt to be caused by uncomfortable and unpleasant sensations in the legs.
- The urge to move the legs and any accompanying unpleasant sensations that begin or worsen during periods of rest or inactivity such as lying down or sitting.
- The urge to move the legs and any accompanying unpleasant sensations that are partially or totally relieved by movement, such as walking or stretching, at least as long as the activity continues.
- The urge to move the legs and any accompanying unpleasant sensations during rest or inactivity that only occur or are worse in the evening or at night, rather than during the day.
- The occurrence of the preceding features is not solely accounted for as symptoms primarily due to another medical or behavioral condition such as myalgia, venous stasis, leg edema, arthritis, leg cramps, positional discomfort, or habitual foot tapping.

All patients with symptoms of RLS should be tested for iron deficiency. If a secondary cause of RLS is suspected on the basis of history, abnormal findings on neurologic examination, or poor response to treatment, other laboratory tests should be performed. In-depth analysis includes needle electromyography/nerve conduction studies to exclude a polyneuropathy or radiculopathy and a polysomnography.

Drug therapy for primary RLS is mainly symptomatic, since cure is possible only in secondary disease. Medications used in the treatment of RLS include dopaminergic agents, benzodiazepines, opioids, anticonvulsants, presynaptic alpha 2-adrenergic agonists, and iron salt.

Nonpharmacological treatment, like sleep hygiene measures, avoidance of stimulants (caffeine, alcohol, and nicotine), and discontinuation of medications that cause or exacerbate RLS (such as selective serotonin reuptake inhibitors, serotonin-norepinephrine reuptake inhibitors, diphenhydramine, and dopamine antagonists), proved to have some benefits in RLS treatment.

With a proper pharmacological treatment, she reported a partial symptoms improvement; however she started to be focused on several side effects that she attributed to the medications, and for this reason, she voluntarily interrupted all drugs. A worsening in the anxious symptoms led to a first hospitalization in a private clinic at the age of 38. A treatment based on valproic acid 750 mg per day, paroxetine 10 mg per day, and clonazepam 2 mg per day was set. After discharge, a long period of wellness with only a few mild relapses followed till she was 43 years old. At that time, a second hospitalization occurred due to increased psychomotor agitation and restless leg syndrome that the patient could not control, despite the correct consumption of pharmacological treatment. During hospitalization, she was treated orally with valproic acid 550 mg per day, trazodone 100 mg per day in combination with paroxetine 10 mg per day, and flurazepam 30 mg per day. She received a diagnosis of major depressive disorder comorbid with histrionic personality disorder. After discharge, the patient did not show stability in her clinical picture, a decrease in overall functioning occurred, with loss of employment and gradual social isolation. She used to stay at home and she began to be hyporexic, with lack of appetite. Moreover, she started to have bowel problems; her diet was based only on squashed fruit. A weight loss of 5 kg in the past months was noted. An increase in anxiety symptoms with distress, restlessness, and depressed mood without suicidal ideas occurred. The symptoms worsened till the patient came to the attention of our clinic in July 2015 and, due to her overall state, was admitted to our psychiatric unit, at the age of 46.

During the first hospitalization in our department, gabapentin 900 mg per day was added as second mood stabilizer, in addition to valproic acid, which was increased to 900 mg per day. Zuclopenthixol 20 mg per day was initially started, later substituted with aripiprazole 10 mg per day and also interrupted later on due to an increase of agitation and restlessness. Intravenous valproic acid 400 mg/day was then introduced. A second antipsychotic, chlorpromazine 75 mg per day, was added to control akathisia and the lower limb movements. Dothiepin 75 mg per day, a tricyclic antidepressant, was started to control the depressed mood. Levothyroxine was continued because of patient hypothyroidism.

To evaluate global neurological functioning, a brain magnetic resonance imaging (MRI) was performed (Fig. 12.1a) which reported: "Sulci and ventricles within the limits. No acute intraparenchymal alterations are observed. Non-specific minor signal alterations in the subcortical frontal region. Minimal cortical atrophy in frontal and parietal lobe. Hypoplastic trajectory of the left vertebral artery. Mucous cysts in the left maxillary sinus." A positron emission tomography (PET) (Fig. 12.2) was also performed: "Mild globally diffused reduction of the concentration of the tracer throughout the cortical hemispheres, most likely on a functional basis, but there

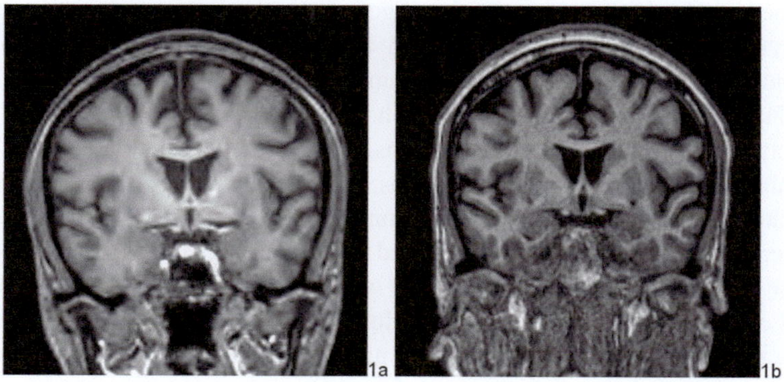

Fig. 12.1 (**a** and **b**) MRI performed in 2015 (**a**) and in 2016 (**b**)

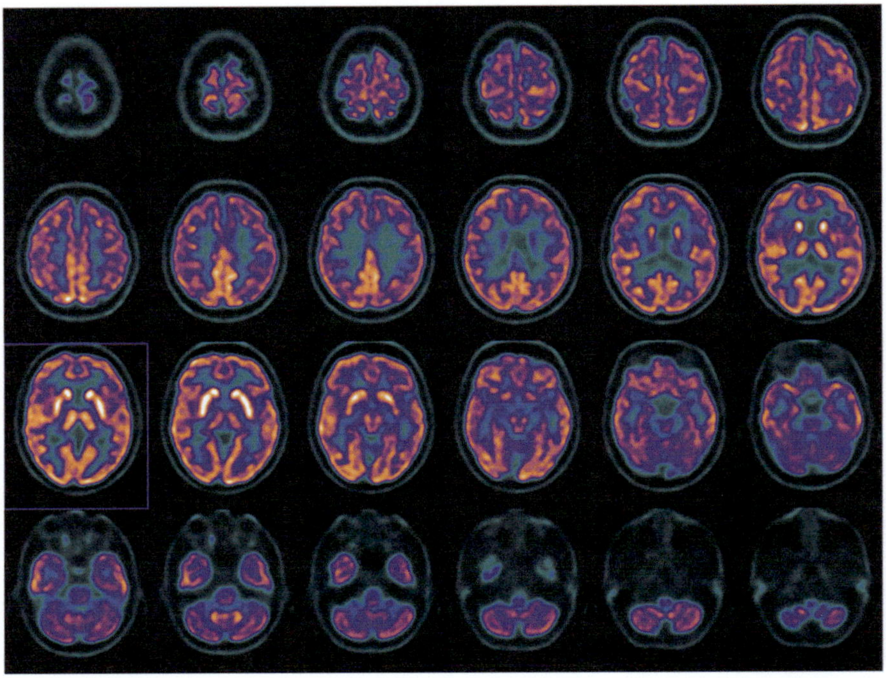

Fig. 12.2 PET performed in 2015

are no clear hypometabolic areas in examined cortical structures. The fixation of the tracer at the level of the examined subcortical structures (thalamus and basal nuclei bilaterally) is normal. In conclusion, PET study reported a glucidic metabolism globally maintained and therefore not suggestive of neurodegenerative aspects." Furthermore, an electroencephalogram showed "9–10 Hz background rhythm, good consistency, symmetrical, reagent, HP and SLI inactivated. In conclusion, EEG shows no significant specific abnormalities."

To evaluate global cognitive functioning, neuropsychological tests were performed and showed an 88 Intelligence Quotient. Speech and memory performances were within the limit, with a motor proficiency value at the lower level and a frontal proficiency evaluation under the lower limit. Lastly, a neurological evaluation was performed; the patient showed hypomimia and bradykinesia, with involuntary movement of the lower limbs. During the evaluation, an episode of involuntary movements of the lower limbs, with choreoathetosis characteristics, of the duration of a few seconds, without disturbance of consciousness, was described. No other neurological signs were depicted. The neurologist confirmed the overall decrease in the bilateral frontal functions with minimal atrophy in the frontal and parietal cortical regions seen in the MRI. In relation to the involuntary movements, the specialist concluded for a picture of ambiguous interpretation that requires a follow-up evaluation due to the effects of ongoing therapy.

To evaluate weight loss and bowel distress, a complete abdomen echography was performed and any organic diseases were excluded, as confirmed also by the internal medicine consultation. A cardiologist examination was requested due to hypotension arising during hospitalization: non-critical illness was documented, and a supportive therapy with midodrine was indicated.

With a proper pharmacological treatment, the symptoms decreased. The psychometric scale of evaluation showed a marked decrease in the overall evaluation (BPRS decrease 48%) with an improvement in the mood state (HAM-D decrease 74%) and reduction of anxiety (HAM-A decrease 78%). After 30 days of hospitalization, the patient was discharged from the psychiatric ward with the indication to continue the treatment in the day hospital (DH) of our department. At discharge the therapy comprised valproic acid 400 mg intravenous in addition to 900 mg per day orally, gabapentin 900 mg per day, chlorpromazine 75 mg per day, dothiepin 75 mg per day, and ketazolam 60 mg per day, with specific therapy for hypotension (dihydergot 10 gtt per day), hypothyroidism, and gastroprotection. According to SCID-I for DSM-IV-TR, the diagnosis at discharge was mood disorder due to a medical condition and organic affective disorder according to ICD-9.

For approximately 10 days the patient was seen daily in DH, with intravenous therapy with valproic acid 400 mg. Gabapentin was slightly increased to 1200 mg/day, and oral valproic acid decreased to 300 mg/day and chlorpromazine to 50 mg/day. The patient showed a fair improvement in her global functioning that permitted her to go back to her hometown, improve social relations, and go back to work; she maintained it till February 2016. She started a family psychoeducation aimed at resolving her conflict with her mother and husband, reporting partial benefits. In February 2016, she reported a worsening of her psychopathological state characterized by depressed mood, clinophilia, social isolation, anxious symptoms, and somatizations (muscle tension, tremors, difficulty in swallowing). This relapse led her to quit her job and to several arguments in her family, increasing her distress. She came back to DH, and an oral therapy based on levomepromazine 50 mg/day and chlordesmethyldiazepam 2 mg intravenous was started, in addition to the oral therapy that she reported to have taken regularly in the last months. Only a partial symptomatology stabilization occurred; in May 2016 a further worsening of her symptoms brought her again to hospitalization. She was treated with gabapentin 900 mg per day and intravenous valproic acid 200 mg per day, as a second mood stabilizer. Dothiepin 75 mg/day was continued as antidepressant medication. Clonazepam up to 6 mg/day, in association with delorazepam 4 mg/day and ketazolam 60 mg per day, was introduced to control anxiety symptoms. During hospitalization, blood pressure reported low values (mean value around 90/60 mmHg), and dihydroergotamine mesylate 40 gtt per day in addition to hydrocortisone intravenous 100 mg per day permitted a good control of blood pressure. The patient showed a decrease in anxiety with a reduction of lower limb distress and somatization in general. The patient was discharged after 27 days, and a mood disorder due to a medical condition diagnosis (DSM IV-TR)/organic affective disorder (ICD-9) was confirmed. Patient's mood and anxiety levels reached a normalization that permitted her to go back to her hometown where she was later followed at the local psychiatric services. A wellness period occurred till November 2016 when she again experienced an increase of the anxiety symptoms with depressed mood and increased concerns on somatic symptoms. In the same period, she divorced her husband, perceived as a huge stressor. During hospitalization fluvoxamine 150 mg per day and mirtazapine 30 mg per day were added to her usual therapy. She underwent additional imaging study with a MRI (Fig. 12.1b) that reported a slight increase of the sulci and gyrus, confirming a minimal diffuse cortical atrophy and minor signal alterations in the subcortical frontal lobe. Neuropsychological tests were carried out a second time and reported a worsening in overall functions with speech, memory, motor proficiency, and frontal proficiency all above the limit. Receiving benefit from the hospitalization, the patient reported a partial remission of her symptoms that permitted, after 14 days, discharge from the psychiatric ward. She was followed daily at the DH, but, after few days of follow-up, the patient reported a further worsening with restlessness, anxiety, somatizations (she described the feeling of "snakes" in her lower limbs and tremors also in the upper limbs), and side effects attributed to the medicine. Due to the impossibility of treating her in a DH setting, she was admitted for the 5th time to the psychiatric ward of our clinic. She was treated with fluvoxamine 150 mg per day, mirtazapine 30 mg/day, and dothiepin 75 mg/day, as she was before admission. Gabapentin 900 mg per day and valproic acid 300 mg per day orally in addition to intravenous valproate 400 mg were continued. Selegiline 5 mg per day was introduced, and chlorpromazine 150 mg/day was started to control involuntary movements. During hospitalization 24 hours of blood pressure monitoring and cardiologic evaluation excluded any pathological reasons responsible for the hypotension. After 30 days of hospitalization, her mood state reached a stable level, with a reduction of the anxiety symptoms, but without reaching a complete normalization. The patient still reported some somatic symptoms (especially a sensation of restlessness in the lower limbs) that she could better control with a reduction of the stressful situations in her daily life. She was followed regularly for 20 days in the DH, and later she went back to her hometown to live with her mother, followed at the local psychiatric services, reporting a temporary stabilization of the symptoms.

Figure 12.3 shows the timeline of patient's history.

12.3 Discussion

The symptoms described in this clinical case report cannot be framed into a unique diagnostic category, due to different psychopathological aspects that need to be underlined. Anxiety symptoms, somatization, and emotion dysregulation are central features in patient's psychopathology, and a link between these can be traced.

As mentioned in the introduction, GAD is a quite common disorder, and referring to Box 12.1, our patient reported feelings of restlessness, muscle tension, sleep disturbance, difficulty in concentrating, and irritability. These physical and psychological symptoms, associated with general worries and concerns perceived most of the day, were present for a long period, without a proper pharmacological treatment

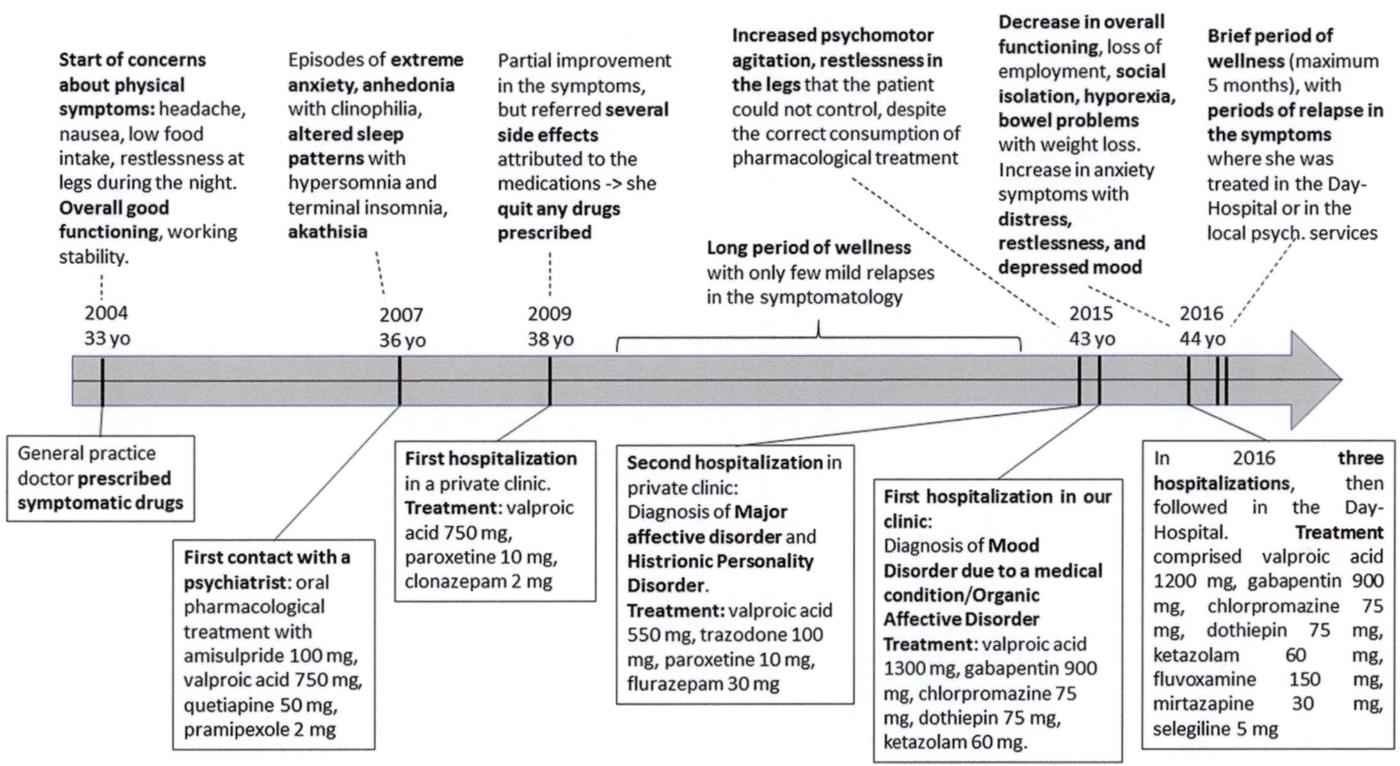

Fig. 12.3 Patient's history timeline

set. Moreover, the symptomatology caused a marked functional deterioration with inability to work and problems with her husband.

If we consider the functional neurological symptom disorder criteria listed in Box 12.2, our patient reported different neurological symptoms, in particular physical complaints in the lower limbs. This concern started with restlessness during the night and turned into involuntary movement in the legs with tremors and shakes also during the day, also depicted during the first neurologist examination she received. During the last hospitalization, the patient also referred sensory alterations, with feelings of "snakes moving" in the lower limbs and tremors also in the arms. No neurological/medical disease could explain the symptoms reported, as indicated by the neurologist's response. Overall, an overlap between a GAD and functional neurological symptom disorder may be underlined. In this respect, tremors and shaking of the limbs in particular can be considered as either linked with anxious burden or secondary to functional aspects.

If we consider the third central feature of patient's psychopathology, emotion dysregulation, a keystone may be found. As noted above, ED is a psychopathological trait associated with different psychiatric disorders and, in particular, with a variety of anxiety disorders [22]. Deficit in emotion regulation may be due to ineffective coping with conditioned fear responses, leading to fear reactions that seem aversive and uncontrollable, leading to a reconditioning in the reactions and to avoidance behaviors that could be chronic. In this respect, when compared with non-anxious controls, individuals with GAD reported poorer understanding of emotions, greater negative reactivity to emotions, and less ability to self-soothe after experiencing negative emotions [23, 24]. Additionally, emotion regulation has long been thought to play a central role in the development of somatoform symptoms. The concept of alexithymia, referring to the difficulty in identifying and describing emotions, can also be introduced in our clinical presentation. Individuals unable to detect, name, and express emotions are more prone to have difficulties using cognitive resources to regulate emotions and, thus, have an increased likelihood of misrepresenting bodily sensations accompanying emotions [25]. In recent years, different studies have shown substantial evidence that somatoform disorders are associated with deficits in the abilities to consciously experience and tolerate emotions, correctly identify emotions, and accurately link emotions to body sensations [18]. In our clinical case, ED could be responsible for the huge distress perceived by the patient, which is then expressed with an increased level of anxiety and somatization, being these the only ways she knows to express emotions, due to her inner inability to control and modulate them.

As mentioned in the introduction, GAD, functional neurological symptom disorder, and ED share alterations in neurological patterns, especially those associated with the limbic system. Imaging studies performed during hospitalization might reveal an organic cause of the symptoms reported. In particular, MRI studies reported a cortical atrophy in the parietal and frontal lobe, while PET studies depicted a globally diffused reduction of the concentration of the tracer. As reported in literature, the frontal lobe and its connectivity with the amygdala are important factors implicated in emotions, social life expression, and behavior regulation. PET has been widely used in the study of psychiatric disorders, and several alterations have been reported in literature. Our patient showed several depressive features like depressed mood, social isolation, and suicidal ideation. A global dysfunction demonstrated by lower cerebral blood flow and decreased cerebral metabolism are the most common findings in depressed patients [26], and this aspect was found also in ours. Instead, in relation to GAD, lower tracer concentration in the basal ganglia and in the white matter and relatively increased metabolism in the left inferior occipital lobe, right posterior temporal lobe, and right precentral frontal gyrus were reported in literature [27], but not confirmed in our case. Overall, the results of imaging study reported in our clinical case do not address a causality issue between symptoms expressed and the alterations found, but could help to correctly characterize the disorder and make a proper differential diagnosis. Furthermore, in clinical practice, an additional evaluation with functional MRI should be encouraged, due to its importance in depicting altered pattern in brain circuits and its correlation with brain disorders, and may yield therapeutic benefits.

The treatment approach used in this case, yet another time, was not based on a unique diagnosis. Antidepressants have become first-line pharmacological treatments in patients with GAD, based on efficacy and tolerability in different randomized controlled trials. As GAD tends to manifest with a chronic course, long-term treatment is usually required. Relapse prevention studies support the long-term efficacy of a range of pharmacological treatments, including some SSRIs (escitalopram, paroxetine) and SNRIs (duloxetine and venlafaxine) [28]. However, the SSRIs and SNRIs have some efficacy limitations, such as lack of response in many cases, a 2- to 4-week delay before the onset of symptom relief, lack of full remission, and risk of relapse. Evidence from early clinical studies of atypical antipsychotics indicates that they may have a potential role in the treatment of anxiety and GAD. In this respect, low-dose augmentative quetiapine may be a useful treatment option for patients with GAD and partial/no response to SSRIs, as reported in a recent randomized clinical study [29]. Lastly, the antianxiety drug pregabalin provides some benefits in GAD treatment [30].

In relation to functional neurological symptom disorder, to date there are no official guidelines for the treatment. Different therapeutic approaches, including pharmacotherapy (mainly antidepressants), psychological therapies (both cognitive-behavioral and psychodynamic), hypnotherapy, and physical rehabilitation, have been considered helpful in a variable proportion of patients [31]. Transcranial magnetic stimulation has been a subject of recent interest as a potential treatment for functional movement disorders [32].

If we agree with the concept of ED, we must evaluate whether ED is the main psychopathological domain or a common trait in different comorbid psychiatric disorders affecting our patient. The therapeutic target of an altered emotion regulation is still controversial. Different behavior therapies have been proposed to treat ED in different disorders, but no pharmacological treatments have been specifically targeted to control ED itself [18]. In the end, a treatment should be direct to control symptoms, like mood fluctuation, anxiety, and somatization depicted in our clinical case.

According to patient pharmacology history, she was initially treated with a mood stabilizer (valproic acid) and atypical antipsychotics (quetiapine and amisulpride). During the first hospitalization in our department, a second mood stabilizer (gabapentin) was added. She was also treated with antidepressant medication. In particular, she received SSRIs (paroxetine, fluvoxamine), SARI (trazodone), tricyclic antidepressant (dothiepin, amitriptyline), and NaSSA (mirtazapine). All antidepressant compounds were used in association with a mood stabilizer, in order to ameliorate her depressed mood without increasing anxiety symptoms.

Additionally, it is worth noticing the long period of illness after the onset of the symptomatology without a proper pharmacological intervention set, clinically defined as duration of untreated illness (DUI). In this respect, our patient stayed at least 3 years without receiving the appropriate medications. A long DUI seems to be the rule rather than the exception in anxiety disorders, as shown in a clinical study where DUI was approximately 6 years in patients suffering GAD. Moreover, different clinical variables were reported to play a role in determining DUI, which, in turn, has therapeutic and prognostic outcomes [33].

Looking back over the patient's timeline, a progressive worsening of symptomatology and clinical response is clear: at the beginning, she completely recovered after the first hospitalization, and she could go back to her job and reconstitute her social life. After the last committal, however, a subthreshold symptomatology was observed, and two hospitalizations occurred one close to the other. This clinical presentation reflects the course of an organic affective disorder, in which a pharmacological treatment fights against a global worsening in the neuronal pathways. In this respect, as mentioned above, our patient showed, 1 year after the first hospitalization, a marked deterioration in cognitive functions, especially in the frontal ones, as outlined also in MRI and PET studies; the specific cause of this progressive worsening is not known and cannot be determined easily. One possible associated factor might be birth complications, as described, due to the use of the forceps. Literature data are concordant on the association between some perinatal complications and mental illness onset [34]. Our patient suffered from hypothyroidism, which could be associated with altered mood, psychosis, and cognitive deficits [35], but a good control with levothyroxine was always guaranteed and not related to changes in symptomatology. Treatment partial efficacy in controlling symptoms could be another proof that an organic cause could have played an important role in the patient's illness.

According to our clinical interpretation, the patient does not show the ability to adaptively cope with the external world, and any external event is seen as impossible to confront, causing her to express her discomfort with extreme anxiety and with physical symptoms: both can recall GAD and functional neurological symptom disorder. Looking back over the patient's history, she also experienced the inability to deal with emotional distress: her difficulties with her mother and her husband could be seen as secondary to her psychopathology and not vice versa as a possible cause of the condition.

Key Points

- General anxiety disorder (GAD) is a common mental health condition characterized by an excessive and persistent worry about everyday life with no obvious reasons and not restricted to circumstances.
- Patient's concern about physical symptoms is known as somatization that, when associated to a worsening in the global functioning, DSM-5 describes as specific diseases in the diagnostic group of "somatic symptom and related disorders."
- Emotional dysregulation is an emotional response to external stimuli that is poorly modulated that can be characterized by marked and rapid fluctuation of mood and behaviors.
- If a patient comes to your attention with a symptomatology overlapping between anxiety disorders, somatic disorders, and mood disorders, a possible link between these conditions can be pointed out.

Self-Assessment Questionnaire

1. Which of the following sentences about the epidemiology of GAD is true?
   (A) **It is more common in female and in the elderly.**
   (B) It is less common in high-income countries.
   (C) It is associated with smoking.
   (D) It is a rare disorder, with a prevalence $< 1\%$ in the general population.
2. Which of the following diseases is not included in the DSM 5 diagnostic group of "somatic symptom and related disorders"?
   (A) Somatic symptoms disorder
   (B) Illness anxiety disorder
   (C) **Body dysmorphic disorder**
   (D) Functional neurological symptom disorder
3. Which is one of the main brain structures implicated in the regulation of emotional response?
   (A) **The prefrontal cortex**
   (B) The basal nuclei
   (C) The corpus callosum
   (D) The postcentral gyrus
4. Which is the most common alteration in positron emission tomography study associated with patients affected by depression?
   (A) Decreased diffuse captation of the tracer
   (B) Augmented captation of the tracer in the frontal lobe
   (C) **Decreased captation of the tracer in the frontal lobe**
   (D) Decreased captation of the tracer in the limbic system
5. Which could be the best therapeutic approach for this clinical case?
   (A) Antidepressants
   (B) **Antidepressants + mood stabilizers**
   (C) Antipsychotics
   (D) Cognitive behavior therapy

References

1. American Psychiatric Association. Diagnostic and statistical manual of mental disorders. 5th ed. Arlington, VA: American Psychiatric Publishing; 2013.
2. Ruscio AM, Hallion LS, Lim CCW, et al. Cross-sectional comparison of the epidemiology of DSM-5 generalized anxiety disorder across the globe. JAMA Psychiat. 2017;74(5):465–75.
3. Weisberg RB. Overview of generalized anxiety disorder: epidemiology, presentation, and course. J Clin Psychiatry. 2009;70(Suppl. 2):4–9.
4. Bandelow B, Boerner JR, Kasper S, Linden M, Wittchen H-U, Möller H-J. The diagnosis and treatment of generalized anxiety disorder. Dtsch Arztebl Int. 2013;110(17):300–9.
5. Zhang X, Norton J, Carrière I, et al. Preliminary evidence for a role of the adrenergic nervous system in generalized anxiety disorder. Sci Rep. 2017;7:42676.
6. Mochcovitch MD, da Rocha Freire RC, Garcia RF, Nardi AE. A systematic review of fMRI studies in generalized anxiety disorder: evaluating its neural and cognitive basis. J Affect Disord. 2014;167:336–42.
7. Delvecchio G, Stanley JA, Altamura AC, Brambilla P. Metabolic alterations in generalized anxiety disorder: a review of proton magnetic resonance spectroscopic studies. Epidemiol Psychiatr Sci. 2017;26(6):587–95.
8. Fricchione G. Clinical practice. Generalized anxiety disorder. N Engl J Med. 2004;351 (7):675–82.
9. Vargas-Prada S, Coggon D, Ntani G, et al. Descriptive epidemiology of somatizing tendency: findings from the CUPID study. PLoS One. 2016;11(4):e0153748.
10. Hollifield MA. Somatoform disorders. In: Sadock BJ, Sadock VA, editors. Kaplan and Sadock's comprehensive textbook of psychiatry. 8th ed. New York: Lippincott Williams & Wilkins; 2005. p. 1808.
11. Haftgoli N, Favrat B, Verdon F, et al. Patients presenting with somatic complaints in general practice: depression, anxiety and somatoform disorders are frequent and associated with psychosocial stressors. BMC Fam Pract. 2010;11(1):67.
12. Atmaca M, Sirlier B, Yildirim H, Kayali A. Hippocampus and amygdalar volumes in patients with somatization disorder. Prog Neuro-Psychopharmacol Biol Psychiatry. 2011;35 (7):1699–703.
13. van der Kruijs SJ, Bodde NM, Vaessen MJ, et al. Functional connectivity of dissociation in patients with psychogenic non-epileptic seizures. J Neurol Neurosurg Psychiatry. 2012;83 (3):239–47.
14. Ejareh Dar M, Kanaan RA. Uncovering the etiology of conversion disorder: insights from functional neuroimaging. Neuropsychiatr Dis Treat. 2016;12:143–53.
15. Sheppes G, Suri G, Gross JJ. Emotion regulation and psychopathology. Annu Rev Clin Psychol. 2015;11(1):379–405.
16. D'Agostino A, Covanti S, Rossi Monti M, Starcevic V. Reconsidering emotion dysregulation. Psychiatry Q. 2017;88(4):807–25.
17. Kring S. Emotion regulation and psychopathology: a transdiagnostic approach to etiology and treatment. New York: Guilford Press; 2010.
18. Berking M, Wupperman P. Emotion regulation and mental health: recent findings, current challenges, and future directions. Curr Opin Psychiatry. 2012;25(2):128–34.
19. Hilt L, Hanson J, Pollac S. Emotion dysregulation. In: Brown B, Prinstein M, editors. Encyclopedia of adolescence. Wyman, MA: Elsevier; 2011. p. 160–9.
20. Davidson RJ. Anxiety and affective style: role of prefrontal cortex and amygdala. Biol Psychiatry. 2002;51(1):68–80.
21. Farb NAS, Anderson AK, Segal ZV. The mindful brain and emotion regulation in mood disorders. Can J Psychiatr. 2012;57(2):70–7.
22. Cisler JM, Olatunji BO, Feldner MT, Forsyth JP. Emotion regulation and the anxiety disorders: an integrative review. J Psychopathol Behav Assess. 2010;32(1):68–82.

23. Tull MT, Jakupcak M, Roemer L. Emotion suppression: a preliminary experimental investigation of its immediate effects and role in subsequent reactivity to novel stimuli. Cogn Behav Ther. 2010;39(2):114–25.
24. Novick-Kline P, Turk CL, Mennin DS, Hoyt EA, Gallagher CL. Level of emotional awareness as a differentiating variable between individuals with and without generalized anxiety disorder. J Anxiety Disord. 2005;19(5):557–72.
25. De Gucht V, Heiser W. Alexithymia and somatization: quantitative review of the literature. J Psychosom Res. 2003;54(5):425–34.
26. Drevets WC, Price JL, Furey ML. Brain structural and functional abnormalities in mood disorders: implications for neurocircuitry models of depression. Brain Struct Funct. 2008;213 (1-2):93–118.
27. Newberg AB, Moss AS, Monti DA, Alavi A. Positron emission tomography in psychiatric disorders. Ann N Y Acad Sci. 2011;1228(1):E13–25.
28. Baldwin DS, Hou R, Gordon R, Huneke NTM, Garner M. Pharmacotherapy in generalized anxiety disorder: novel experimental medicine models and emerging drug targets. CNS Drugs. 2017;31(4):307–17.
29. Altamura AC, Serati M, Buoli M, Dell'Osso B. Augmentative quetiapine in partial/ nonresponders with generalized anxiety disorder: a randomized, placebo-controlled study. Int Clin Psychopharmacol. 2011;26(4):201–5.
30. Katzman MA. Current considerations in the treatment of generalized anxiety disorder. CNS Drugs. 2009;23(2):103–20.
31. Demartini B, Batla A, Petrochilos P, Fisher L, Edwards MJ, Joyce E. Multidisciplinary treatment for functional neurological symptoms: a prospective study. J Neurol. 2014;261 (12):2370–7.
32. Lehn A, Gelauff J, Hoeritzauer I, et al. Functional neurological disorders: mechanisms and treatment. J Neurol. 2016;263(3):611–20.
33. Benatti B, Camuri G, Dell'Osso B, et al. Which factors influence onset and latency to treatment in generalized anxiety disorder, panic disorder, and obsessive–compulsive disorder? Int Clin Psychopharmacol. 2016;31(6):347–52.
34. Mansur RB, Cunha GR, Asevedo E, et al. Perinatal complications, lipid peroxidation, and mental health problems in a large community pediatric sample. Eur Child Adolesc Psychiatry. 2017;26(5):521–9.
35. Testa A, Giannuzzi R, Daini S, Bernardini L, Petrongolo L, Gentiloni SN. Psychiatric emergencies (part III): psychiatric symptoms resulting from organic diseases. Eur Rev Med Pharmacol Sci. 2013;17(Suppl 1):86–99.

Conversion Disorders Across Psychiatry and Neurology

13

Tiziana Carandini, Andrea Arighi, and Elio Scarpini

Abstract

The term conversion disorders refers to a variety of neurological symptoms genuinely experienced by the patient but inconsistent with an underlying identifiable neurologic cause. Conversion disorders have a high prevalence among population, leading to frequent hospitalization and causing significant distress and disability in patients. In the last years, a better understanding of the possible phenotypes and underlying psychopathologic mechanisms in conversion disorders has been provided. A complete neurologic examination represents the main part of the clinical assessment in patients with conversion disorders and the main tool to differentiate them from organic neurological conditions. The diagnosis of conversion disorders is based on the demonstration of positive functional signs, together with the exclusion of signs of disease. Management of conversion disorders is difficult, and many patients are reluctant to accept their diagnosis. Neurologists, psychiatrists, psychologists, and physiotherapists need to work together in order to help the patients to understand their symptoms and to heal. The following chapter provides a detailed explanation about clinical assessment, psychopathologic mechanisms, and current treatment of conversion disorders, with the aim of helping physician in the diagnosis and differential diagnosis with organic diseases. A clinical case is also presented to supply a concrete example of how insidious can be the management of a conversion disorder.

Keywords

Conversion disorders · Functional disorders · Hoover sign

T. Carandini · A. Arighi (✉) · E. Scarpini
Neurodegenerative Disease Unit, Fondazione Cà Granda IRCCS Ospedale Maggiore Policlinico, University of Milan, Milan, Italy
e-mail: andrea.arighi@policlinico.mi.it

A. C. Altamura, P. Brambilla (eds.), *Clinical Cases in Psychiatry: Integrating Translational Neuroscience Approaches*,
https://doi.org/10.1007/978-3-319-91557-9_13

13.1 Introduction

The term conversion disorder or functional neurological symptom disorder (FNSD) is referred to those neurological symptoms that are genuinely experienced (not intentionally produced or feigned) by the patient but that are inconsistent with an underlying identifiable neurologic or medical cause. These symptoms cause significant distress or impairment in social, occupational, or other important areas of functioning [1].

The first reports of conversion disorders date back to ancient times. Hippocrates coined the term *hysteria*—from the Greek word *hysterika* (uterus)—to describe symptoms (like anxiety, tremors, paralysis, convulsions) that were thought to be caused by uterine pathology in females. The etiology of hysteria started to move from the uterus to the brain, thus possibly affecting both sexes, from the end of the seventeenth century. However, it was only in the nineteenth century that the concept of conversion disorder came to prominence, thanks to the studies on hysteria by the French neurologist Jean-Martin Charcot and his pupil Joseph Babinski (Fig. 13.1). In 1895, Charcot's student Sigmund Freud postulated with Joseph Breuer the so-called conversion hypothesis, according to which emotional charge deriving from painful experiences would be consciously repressed to relieve pain but subconsciously converted into neurological symptoms [2]. During the last 20 years, several biological studies and randomized controlled trials have been performed in order to investigate conversion disorder [3]. Nowadays many authors prefer the more neutral and acceptable term FNSD [1], meaning that the symptoms arise from abnormal nervous system functioning in the absence of structural pathology.

Conversion disorders are one of the most common reasons for referral to medical attention and represent around 30–50% of outpatient visits in primary and secondary care [4, 5]. Estimated incidence is 4–12 per 100,000 per year [6]. They have been reported in patients of all ages, and they appear to be more common in rural settings, lower socioeconomic status, and among military personnel [7]. The comparative incidence among men and women is not known for certain [7]. However, Stone et al. suggested that females are more likely to experience psychogenic non-epileptic seizures than psychogenic weakness [8]. Patients with conversion disorders have been found to be as disabled as patients with corresponding organic disease and even more disabled. Their prognosis is considered poor [9, 10].

13.2 New Classification (DSM-V and ICD11): The Neurological Positive Diagnosis

In the last 10 years, the classification of conversion disorders has been revisited. Lately, the new *Diagnostic and Statistical Manual of Mental Disorders, 5th Edition* (DSM-V) made substantive changes in the definition of conversion disorder, adding the term FNSD and clarifying the criteria to make a correct diagnosis (Box 13.1) [1]. It moved from the DSM-IV classification based on a psychological etiology (following Breuer and Freud's conversion hypothesis) to one based on the

Fig. 13.1 André Brouillet's painting "Une leçon clinique à la Salpêtrière" (1887) illustrates Charcot demonstrating a case of "hysteria" on a young lady, supported by Joseph Babinski (Pierre Aristide André Brouillet, 1887, Paris Descartes University, Paris)

demonstration of typical presenting neurological signs [3, 11]. Moreover DSM-V removed the requirement to identify a recent psychological factor and to exclude symptoms that are intentionally produced or feigned by the patient.

Box 13.1 DSM-V Criteria for Conversion Disorder (or Functional Neurological Symptom Disorder)

1. One or more symptoms of altered voluntary motor or sensory function.
2. Clinical findings provide evidence of incompatibility between the symptom and recognized neurological or medical conditions.
3. The symptom or deficit is not better explained by another medical or mental disorder.
4. The symptom or deficit causes clinically significant distress or impairment in social, occupational, or other important areas of functioning or warrants medical evaluation.

The 10th edition of the International Classification of Diseases (ICD-10) classified conversion disorders exclusively in the psychiatry section as dissociative motor, seizure, and sensory disorders (F 44.4) [9]. However, in the 11th edition (ICD-11), which is currently being revised and due by 2018, conversion disorders should appear for the first time as a separate category in the neurology section as well as in the psychiatry section [3].

DSM-V and ICD-11 classifications testify to the emerging belief that conversion disorders are now considered a field in which both neurology and psychiatry (both mind and brain) are equally important to diagnose and treat these disorders [4].

13.3 Phenotypes

Conversion disorder can present with a variety of neurological symptoms that may be episodic or sustained, both acute and chronic. Non-epileptic seizures represent the most common and reported phenotype, followed by functional sensory (including hemisensory or other sensory disturbances or visual alterations) and functional motor symptoms (including weakness, mixed motor/sensory movement disorder, gait or speech disturbances) [5]. A list of symptoms that can be found in patients with conversion disorder is provided in Box 13.2.

Box 13.2 Possible Phenotypes of Conversion Disorder

Non-epileptic seizures

Motor functional symptoms: unilateral weakness, monoparesis, paraparesis, pseudoptosis, tongue deviation, cauda equina syndrome (leg weakness and urinary retention)

Movement disorders: tremor, dystonia, gait disorder, myoclonus, parkinsonism, hemifacial spasm

Speech alterations: stuttering speech, speech arrests, foreign accent syndrome, hypophonia, dysphonia

Swallowing symptoms: globus sensation

Sensory functional symptoms: hemisensory syndrome (patient feels "cut in half"), anesthesia, paresthesias

Visual symptoms: loss of acuity, intermittent blurring of vision, functional binocular diplopia, monocular diplopia, voluntary nystagmus, complete blindness

Others: dizziness, hearing loss, cognitive symptoms (poor concentration and memory, impaired fluency, word-finding difficulty)

13.4 Clinical Assessment

The assessment of conversion disorders requires a detailed history-taking and a neurological examination, as well as laboratory, neurophysiology, and neuroimaging tests [7]. It is important to investigate symptoms in all organ systems and evaluate any possible psychopathological comorbidities. Patients with conversion disorders have high rates of depression and anxiety but are often reluctant and worried about these questions; thus it is better to leave questions about psychiatric symptoms till the end of the consultation [7, 12].

A complete neurologic examination represents the main part of the clinical assessment in patients with conversion disorder. The famous painting by Brouillet shows Jean-Martin Charcot during a lesson at the Salpêtrière, demonstrating the "hysterical arc-de-cercle" (the "arching" position of the spinal column due to a hyperextension similar to opisthotonus) on a young lady (Marie "Blanche" Wittmann, known historically as the Queen of Hysterics), supported by Joseph Babinski (Fig. 13.1). The latter would later spend much of his career searching for objective clinical signs that cannot be mimicked by the patient, consciously or unconsciously [13].

As assessed by the DSM-V [1], the diagnosis of conversion disorder is based on the demonstration of positive functional signs—mostly related to inconsistency or incongruity—and on the exclusion of signs of disease. Examples of positive signs in conversion disorders are reported in Table 13.1.

Patients with conversion disorders, particularly with functional motor symptoms or dissociative seizures, commonly experience depersonalization (the feeling of being disconnected from the body) or derealization (the feeling of being disconnected from the surroundings) at the onset of their symptoms or attacks [4]. Depersonalization and derealization are always important to investigate, although patients are usually reluctant since they worry that these symptoms sound "crazy." They find it difficult to describe them and may just say they feel "dizzy" [4, 15].

"La belle indifférence," an apparent lack of concern about the nature or implications of symptoms or disability, has been traditionally considered an important feature in patients with conversion disorder. However, its role has been revised because it has no validity in discriminating conversion disorder from recognizable neurologic disease [15, 16].

All functional movement disorders are typically characterized by a rapid onset, variability in frequency, amplitude, or distribution and improvement with distraction [15]. The diagnosis is reinforced in presence of a Bereitschaftspotential (BP), a premovement potential seen in voluntary movement during electroencephalogram (EEG) [3]. The BP, also called the premotor potential, reflects activity in the motor and supplementary motor areas leading up to voluntary muscle movement, thus representing the cortical contribution to the planning of volitional movement.

Functional gait alteration is excessively unsteady or slow (with a marked delay in initiation and subsequent "foot sticking" but without the subsequent improvement typical of extrapyramidal disorders). Patients tend to fall toward or away from doctor and may present sudden knee buckling, preventing themselves from falling before they touch the ground [15]. A dragging gait with external or internal hip rotation is characteristic of functional weakness [12].

Video EEG recordings provided useful information about non-epileptic seizures. Objective signs that help distinguish dissociative from generalized epileptic seizures are long-duration events (longer than 2 min), fluctuation of course, side-to-side head or body movements, closed eyes during an attack, crying during or immediately after the attack, and memory recall of being in a generalized attack. Conversely, onset during sleep and stertorous breathing are more suggestive of epileptic activity [3, 4, 17].

Table 13.1 Examples of clinical signs in conversion disorder

Motor signs	
Arm stabilization test	
Drift without pronation	Arms stretched out, palms up in supinated position, fingers adducted, eyes closed for 10 seconds: a downward drift and no pronation are seen in functional paresis
Non digiti quinti sign	Same maneuver: no abduction movement of the 5th finger is seen in a functional paresis
Irregular drift	Same maneuver: the arm drifts regularly in an organic paresis but irregularly in a functional paresis
Inconsistency of direction	Same maneuver: there is an oscillating movement of the arm (downward-upward-downward) in functional paresis
Collapsing weakness	Same maneuver: the arm collapses from the normal position with a light touch (by the examiner [E]) in functional paresis
Nonconcavity of the palm of hand	Same maneuver: no movement of the outstretched straight hand in a flexed/concave position is seen in functional paresis
Weakness not distal > proximal	Same maneuver: a greater weakness is presented in distality than proximity (pyramidal distribution) = organic. More weakness proximal than distal = functional paresis
Sternocleidomastoid test	Patient [P] is asked to do a forced head rotation against E's resistance. In organic hemiparesis, sternocleidomastoid is usually spared (as bilateral innervated). Sign positive if weakness of rotation to the ipsilateral side appears
Give-way weakness	During muscle strength testing, a normal strength is developed and then suddenly collapses in functional paresis
Co-contraction	Observation during muscle strength testing (or with surface electromyography): simultaneous contraction of agonist and antagonist resulting in no/little movement (= sign positive)
Arm-drop test	E puts P's arm over the head (P lying supine) and drops the weak arm: in organic paresis the arm hits the face; in functional paresis a voluntary movement allows avoiding the face
Drift against gravity	P lying supine, arm stabilization with an angle of 45° toward the head: in organic paresis arm drifts with gravity; in functional paresis the arm is lifted against gravity and drifts in the former manner of the usual arm stabilization test
Spinal injury test	P in supine position asked to lift up his knees; if not possible E lifts them up. Sign positive if P keeps leg up after lifted by E, sign negative if leg drops in abduction
Babinski trunk-thigh test	P in supine position, arms across chest, asked to sit up. In organic paresis, the paretic limb rises above the bed, and the contralateral shoulder comes forward. In functional paresis, no asymmetry is seen
Leg stabilization	
Mingazzini: drift without extension	P in supine position, legs bent 90° at the hip and knees, eyes closed for 5 s: in organic paresis there is a hip + knee extension with the downward drift; in functional paresis only the lower part of the leg drifts (knee bent) without an extension movement of the hip
Mingazzini: irregular drift	Same maneuver: the leg drifts regularly in an organic paresis but irregularly in a functional paresis (= sign positive)
Barré: drift without contraction	Leg stabilization in prone position, legs bent 90° at the knees: in functional paresis, rapid drift without contraction of the hamstring muscles

(continued)

Table 13.1 (continued)

Motor signs	
Hoover's sign	(A) P supine, E's hand under paretic leg (under heel): P exerts max force downward (B) Same but P exerts max force upward with contralateral leg (against E's resistance) Comparison of felt pressure in E's hand under paretic side heel. Hoover positive if strength in condition B > A. Hoover negative if strength B = or < A
Sensory signs	
Nonanatomical sensory loss	Diminished sensation fitting a "nondermatomal pattern," for example, anteriorly but not posteriorly delineated truncal deficit, unilateral glove or sock distribution, sharp midline delineation in one limb (hemilimb distribution) or face/trunk
Systematic failure	Sign positive if patients always fail in a discriminative task (e.g., pin or prick/cold-hot/up- or down-going joint)
Splitting the midline	Sign positive if exact splitting of sensation in the midline
Splitting of vibration sense	Sign positive if difference in the sensation of a tuning fork placed over the left compared to the right side of the sternum, chin, or frontal bone
Yes/no test	P with eyes closed is asked to say "yes" when he feels E touching and "no" when he does not
Description of gait	
Falls always in direction of support	During walking P falls always to the side of E (or another hold, e.g., wall, furniture, etc..)
Excessive slowness	Resembling slow motion, incompatible with a neurological disease
Hesitation	P's initiation of intended movement is either delayed or impossible; the feet seem to stick on the ground; not overcome as the first step is taken (like in Parkinson)
Walking on ice	A walking pattern mimicking ice skating or "as if" on slippery ground
Robot walk	Robotic, staccato, stiff, and square-cut walking pattern
Noneconomic posture	A walking pattern that requires considerable effort as well as balance to maintain the posture (e.g., walking with knees flexed)
Sudden knee buckling	Knee buckling during stance or walk but usually with no falls
Leg dragging	The leg is dragged at the hip behind the body instead of performing a circumduction
Tremulousness	Body tremor with up and down shaking of the body (flexion/extension of the knees), not compatible with an orthostatic tremor
Flailing arms	Exaggerated large amplitude movements of the arms during walking—seemingly to keep the balance
Bizarre excursion of the trunk	Large amplitude body sway building up after a silent latency of a few seconds
Sudden side steps	P will display a big displacement in his trajectory with sudden side steps, without falling
Staggering long distances to obtain support from opposite walls	Very instable gait but no falls, as subject will find a support, even if far out of reach

(continued)

Table 13.1 (continued)

Description of gait	
Psychogenic Romberg	Constant falls toward or away from the observer, large amplitude body sway building up after a silent latency of a few seconds and improvement of balance with distraction
Visual signs	
Convergence spasm	Tendency for the convergence reflex to be transiently overactive, either unilaterally or bilaterally. Lateral gaze restriction can sometimes be present, but the presence of miosis helps to confirm the diagnosis
Fogging test	Placing lenses of progressively increasing power before the "unaffected" eye, while the vision is tested (with both eyes open) until the "unaffected" eye becomes sufficiently blurred for the E to be certain that the "affected" eye is reading
Tubular fields	Psychogenic field defects remain unchanged in width when tangent screen testing is performed at varying distances while maintaining equivalence by adjusting target size
General signs	
Expressive behavior	Suffering and strained facial expression, mannered posture of the hand, moaning, hyperventilation

Adapted from Daum et al. [14]

As regards functional visual symptoms, many tests could give an objective measure of acuity [18]. The most commonly used test in patients with suspected functional monocular visual problems is the fogging test. Lenses of progressively increasing power are placed before the "unaffected" eye, while the vision is tested (with both eyes open); eventually the "unaffected" eye will be sufficiently blurred for the examiner to be certain that the "affected" eye is in fact reading [4, 12, 18]. Patients complaining of a loss of visual field can be tested at bedside by looking for "tubular fields." Psychogenic field defects remain unchanged in width when tangent screen testing is performed at varying distances while maintaining equivalence by adjusting target size and are consequently known as "tubular fields." In the presence of true loss of field, the area of constricted field expands with increasing test distance [18].

13.5 Mechanism of Functional Disorders

Many biological, psychological, and social factors may predispose patients to conversion disorder and then precipitate or perpetuate the symptoms (for a complete review, see Stone et al., Neurol Clin, 2011) [12].

Psychodynamic models state Freud's conversion hypothesis and emphasize abnormal interpersonal relationships that develop due to problematic relationships or traumatic experiences [5]. Conversely, cognitive-behavioral theory (CBT) states that conversion symptoms are generated by factors such as processing of perception and behavior outside of awareness, selective attentional bias that amplifies minor physiological stimuli and asymmetries, dissociation, and maladaptive thoughts [19]. Recently, several functional imaging analyses have studied the neural basis of functional motor disorders [20–25]. In summary, they found an alteration in areas involved in planning, execution, and interpretation/attribution of movement, moderated by those areas involved in emotional regulation [3]. Moreover, frontal cortical and limbic activation associated with emotional stress may act via inhibitory basal ganglia-thalamocortical circuits to produce a deficit of conscious sensory or motor processing [26].

13.6 Therapy

It is now well accepted that a good explanation to the patient of the nature of his functional symptoms and the sharing of his physical signs are the main prerequisites for successful treatment of a conversion disorder [12]. It is important to use an appropriate terminology, in order not to alienate the patients, to accurately show them the inconsistency of their positive clinical signs, to emphasize that you believe their symptoms are genuine and not "made up" or "crazy," to explain that their symptoms are common and potentially reversible, and to arrange a follow-up [4, 15, 27]. A constructive time-consuming explanation can be therapeutic itself [28].

Physiotherapy and physical treatments have a favorable impact on patients with functional motor disorders [3, 27, 29]. Recent randomized controlled trials found that both CBT and hypnosis may be beneficial in conversion disorders [12, 30, 31]. Conversely, psychodynamic psychotherapy lacks empirical support, although it is historically popular in the treatment of functional neurological symptoms [12]. Other treatments that have been evaluated as potentially helpful in conversion disorders are transcutaneous electrical nerve stimulation, transcranial magnetic stimulation, biofeedback, and therapeutic sedation.

As regards pharmacotherapy, there is little evidence of its benefit in conversion disorders. Antidepressants can be used starting at a low dose when there are symptoms known to respond to these agents, such as depression, anxiety, pain, or insomnia [28]. The literature reports a case successfully treated with haloperidol [32] and one that benefited from electroconvulsive therapy (ECT) [33].

Patients with conversion disorders usually refer first to a neurologist due to the characteristics of their symptoms. Once the diagnosis is suspected, many neurologists hand care back to the family doctor or refer the patient to a psychiatrist. However, it is important that the neurologist should play an active role in the management of patients with conversion disorders. Where unimproved, patients should be referred to a psychiatrist with a careful explanation, a trial of antidepressants, and physiotherapy [12].

13.7 Case Presentation

A 46-year-old man presented to the emergency room of our clinic due to a sudden episode of nonfluent aphasia, lasting 15 min and then completely resolving. Basal CT scan was negative, and he was admitted to our neurological department with the suspected diagnosis of transitory ischemic attack.

Patient was known to have a 3-year history of progressively worsening paraparesis and urinary incontinence. Symptoms developed subacutely 2 months after the surgical exeresis of an abdominal lipoma and gradually led to the loss of walking and to a severe impairment in his social and occupational life. He previously underwent a spine MRI study, with no evidence of signs of myelopathy. In parallel, patient referred a history of epilepsy with desultory episodes of pleomorphic seizures (absence, focal, and tonic-clonic generalized), for which he was currently under antiepileptic treatment with phenobarbital. He was also known for depression and anxiety, treated with benzodiazepines, but he was not regularly followed by a psychiatrist. Moreover, he was receiving a strong analgesic therapy with opioids, acetaminophen, and nonsteroidal anti-inflammatory drugs due to chronic abdominal pain.

During hospitalization, the speech disturbance relapsed several times. Symptoms typically presented after eating and lasted about 15 min. Patient referred that he was able to feel the onset of the "attack" claiming that "during digestion I feel like I'm being disconnected from my body." During attack his speech was discontinued, poor, and slowed, with frequent anomie. Patient was not able to move his mouth properly but was once able to write a note correctly: "I'm having an aphasia attack because I'm digesting." He was suffering very much and worried about his symptoms.

At neurological examination, patient had a severe paraparesis, being unable to move his legs with the exception of movements of flexion and extension of the feet. He was confined to a wheelchair and suffered from urinary incontinence with the constant need of self-catheterization. Nevertheless, legs had a normal muscular tropism and tone, deep tendon reflexes were normal, and the plantar reflex caused a flexion of the toes of both feet. Sensibility and other neurologic domains were preserved.

Blood tests resulted normal. We repeated a brain and spine MRI and a MRI angiography, which excluded pathologic lesions in the central nervous system, both acute and chronic, and vascular alterations. Neurophysiologic exams did not evidence an involvement of the central and peripheral nervous system. Particularly, both the motor and somatosensory evoked potentials were normal, and the electromyography showed only a poor activation of the motor unit potential, possibly on a voluntary basis. We also performed an EEG (both standard and after sleep deprivation), which resulted normal. Lastly a urodynamic testing did not reveal an underlying neuroanatomical cause for the patient's urinary incontinence.

In conclusion, the patient presented with transient episodes of nonfluent aphasia occurring during digestion, associated with a history of paraparesis, urinary incontinence, and pleomorphic seizures. Due to these symptoms, patient manifested significant distress and impairment in social and occupational life. All the instrumental exams performed resulted normal, and the clinical evaluation showed incompatibility between the symptoms and recognized neurological or medical conditions. In accordance with DSM-V, a diagnosis of conversion disorder was made [1]. Notably, no specific clinical signs, as listed in Table 13.1, were used to support the diagnosis.

After diagnosis, patient received an extensive explanation of his symptoms and his condition. Physiotherapy was started and patient was referred to a psychologist. Patient underwent structured clinical interview for DSM-IV axis I and II disorders (SCID-I and SCID-II), which confirmed the diagnosis of conversion disorder and excluded a personality disorder, respectively. Despite several time-consuming talks, patient refused the diagnosis and claimed the urgent need of further medical investigations. Given his reluctance and hostility, together with the long history of his illness, we referred the patient to a psychiatrist, but he never attended the scheduled visit. After discharge, patient was lost at the neurological follow-up.

13.8 Literature Review

This case-report regards a middle-aged man presenting to the emergency department because of transitory episodes of speech impairment; he was hospitalized with the suspected diagnosis of transitory ischemic attack. However, the accurate evaluation of the "aphasic attacks" that recurred during the following days, together with the negative results of the instrumental exams performed, leads us to a diagnosis of functional speech disorder.

Functional cognitive disorders represent a common but understudied part of conversion disorders. As reported by Pennington and colleagues, one-third of patients of less than 60 years referred for cognitive symptoms had a final diagnosis of functional cognitive disorder [34]. Delis and Wetter divided patients with cognitive functional disorders in two categories: those having cogniform disorder and those with cogniform condition, according to whether or not their symptoms affected their daily performance and social life [34, 35]. The majority of patients complain of poor concentration and memory alterations, but nonneurogenic speech disorders are also reported, with impaired fluency, jumbling of words when speaking, word-finding difficulty, and variability in speed of response [36]. Few studies are available about functional speech disorders: although a young/middle-aged female predominance is suggested, no information on the incidence or prevalence and no classification system are available [37]. Mendez has recently reviewed functional speech disorders in detail. Generally functional aphasias are "nonfluent" with abnormal quantity and flow of language, agrammatism, and relatively preserved comprehension, naming, and repetition [37]. The patient described above had recurrent transitory episodes of speech disturbance characterized by impaired fluency and word-finding difficulty. These episodes could be differentiated from organic aphasia for several reasons: (1) they occurred only in certain moments of the day and particularly after eating; (2) during these episodes, the patient retained comprehension, reading, and writing; and (3) as frequently experienced in conversion disorders [4], the patient felt depersonalization (the feeling of being disconnected from the body) at the onset of his attacks.

The speech alteration was not the only manifestation of patient's conversion disorder. Indeed he was also known for a long history of paraparesis and urinary incontinence, gradually developed after a trivial surgical abdominal intervention that could not be related to his presumed spinal problem. Inconsistency of his motor symptoms was noted, i.e., patient was able to move from the bed to the wheelchair with no aid, and no sign of disuse was observed in his muscles (such as muscular atrophy or spasticity). A useful test that can be performed in patients with suspected functional motor symptoms is electromyography, searching for poor activation. This term refers to the reduced firing of motor unit potentials (MUPs) during voluntary contraction. In contrast to reduced recruitment, in which a few MUPs fire at rapid rates, with poor activation, the MUPs fire slowly with a normal rate of recruitment. Poor activation can be related to pain (not complained of by our patient), upper motor neuron lesions (not detected with MRI and motor evoked potentials), or poor cooperation possibly due to a functional disorder.

Moreover, despite that the patient did not experience any seizure during hospitalization, his history of pleomorphic epilepsy (absence, focal and tonic-clonic generalized), without a previous pathologic EEG, led us to suspect associated functional epilepsy. Notably, the patient was taking antiepileptic treatment without a precise medical indication, with all the possible side effects of the drug.

Lastly, our patient suffered from depression and anxiety, symptoms that are commonly reported in patients with conversion disorders [7].

In the assessment of our patient, no specific clinical signs, as listed in Table 13.1, were used to support the diagnosis of conversion disorder. This underlines how fundamental are the neurologist's experience and the knowledge of semeiotics in the assessment of conversion disorders. Together with the neurologist, the psychologist confirmed the diagnosis of conversion disorder and excluded a personality disorder, using the SCID-I and SCID-II scales, respectively.

Despite the best efforts of physicians, the patient was firmly reluctant to accept the diagnosis of conversion disorder and did not attend the scheduled visits of follow-up. This underlines how difficult the management of conversion disorder may be. As reported in literature, a substantial proportion of patients with conversion disorders are not able to accept or understand the diagnosis, making their treatment impossible [4, 28]. The website www.neurosymptoms.org is a useful tool, created by a neurologist to give patients a better understanding of their symptoms and to help them to comprehend the difference between real neurological and functional non-organic symptoms.

In conclusion, conversion disorders represent a common problem that causes significant distress and disability. In the last years, a better understanding of the possible phenotypes, underlying psychopathologic mechanisms, and new treatments have been provided. Given the difficult management of conversion disorders, neurologists, psychiatrists, psychologists, and physiotherapists need to work together [3] in order to help the patients to understand their symptoms and to heal.

Key Points

- The term conversion disorder or functional neurological symptom disorder is referred to those neurological symptoms that are genuinely experienced (not intentionally produced or feigned) by the patient but that are inconsistent with an underlying identifiable neurologic or medical cause.
- Conversion disorders are one of the most common reasons for referral to medical attention and represent around 30–50% of outpatient visits in primary and secondary care.
- DSM-V and ICD-11 classifications testify to the emerging belief that conversion disorders are now considered a field in which both neurology and psychiatry are equally important to diagnose and treat these disorders.
- Conversion disorder can present with a variety of neurological symptoms (non-epileptic seizures, sensory, motor, and cognitive symptoms) that may be episodic or sustained, both acute and chronic.
- The assessment of conversion disorders requires a detailed history-taking and a neurological examination, as well as laboratory, neurophysiology, and neuroimaging tests.
- Patients with conversion disorders commonly experience depersonalization or derealization at the onset of their symptoms or attacks.
- The neurologist's experience and a knowledge of semeiotics and positive functional signs in the assessment of conversion disorders are fundamental for diagnosis.
- A good explanation to the patient of the nature of his functional symptoms and the sharing of his physical signs are the main prerequisites for successful treatment of a conversion disorder.
- Given the difficult management of conversion disorders, neurologists, psychiatrists, psychologists, and physiotherapists need to work together in order to help the patients to understand their symptoms and to heal.

Self-Assessment Questionnaire

1. In DSM-V and ICD-11 classifications, conversion disorders are considered:
   (A) A field of psychiatry
   (B) A field of neurology
   (C) **A field in which both neurology and psychiatry are equally important**
   (D) A field of psychology
2. The most common phenotype of conversion disorders is:
   (A) Swallowing alteration
   (B) **Non-epileptic seizure**
   (C) Complete blindness
   (D) Word-finding difficulty
3. In Daum's proposed classification of clinical positive signs in conversion disorders, Hoover's sign is considered:

(A) **Highly reliable**
(B) Reliable
(C) Suggestive
(D) Not suggestive

4. Poor activation at electromyography can't be:
   (A) Related to pain
   (B) Related to poor cooperation possibly due to a functional disorder
   (C) Related to upper motor neuron lesion
   (D) **A normal finding**
5. The best therapeutic approach to functional symptoms is:
   (A) Antidepressants
   (B) Neuroleptics
   (C) Electroconvulsive therapy
   (D) **A good explanation to the patient of the nature of his functional symptoms and the sharing of physical signs**

References

1. American Psychiatric Association. Diagnostic and statistical manual of mental disorders. 5th ed (DSM-V). Arlington: American Psychiatric Association; 2013.
2. Breuer J, Freud S. Studies in hysteria; 1895.
3. Lehn A, Gelauff J, Hoeritzauer I, et al. Functional neurological disorders: mechanisms and treatment. J Neurol. 2016;263(3):611–20.
4. Stone J. Functional neurological disorders: the neurological assessment as treatment. Pract Neurol. 2016;16(1):7–17.
5. Stone J, Carson A, Duncan R, et al. Who is referred to neurology clinics? – the diagnoses made in 3781 new patients. Clin Neurol Neurosurg. 2010;112(9):747–51.
6. Carson AJ, Brown R, David AS, et al. Functional (conversion) neurological symptoms: research since the millennium. J Neurol Neurosurg Psychiatry. 2012;83(8):842–50.
7. Tsui P, Deptula A, Yuan DY. Conversion disorder, functional neurological symptom disorder, and chronic pain: comorbidity, assessment, and treatment. Curr Pain Headache Rep. 2017;21 (6):29.
8. Stone J, Sharpe M, Binzer M. Motor conversion symptoms and pseudoseizures: a comparison of clinical characteristics. Psychosomatics. 2004;45(6):492.
9. Stone J, Hallett M, Carson A, Bergen D, Shakir R. Functional disorders in the neurology section of ICD-11: a landmark opportunity. Neurology. 2014;83(24):2299–301.
10. Carson A, Stone J, Hibberd C, et al. Disability, distress and unemployment in neurology outpatients with symptoms "unexplained by organic disease". J Neurol Neurosurg Psychiatry. 2011;82:810–3.
11. American Psychiatric Association. Diagnostic and statistical. Manual of mental disorders. 4th ed. Text revision (DSM-IV). Washington, DC: American Psychiatric Association; 2000.
12. Stone J, Carson A. Functional neurologic symptoms: assessment and management. Neurol Clin 2011;29(1):1–18, vii.
13. Drouin E, Drouin G, Péréon Y. The Babinski sign. Lancet Neurol. 2017;16(3):180.
14. Daum C, Gheorghita F, Spatola M, et al. Interobserver agreement and validity of bedside 'positive signs' for functional weakness, sensory and gait disorders in conversion disorder: a pilot study. J Neurol Neurosurg Psychiatry. 2015;86(4):425–30.
15. Stone J, Carson A, Sharpe M. Functional symptoms and signs in neurology: assessment and diagnosis. J Neurol Neurosurg Psychiatry. 2005;76(Suppl 1):i2–12.

16. Stone J, Smyth R, Carson A, et al. La belle indifférence in conversion symptoms and hysteria: systematic review. Br J Psychiatry. 2006;188:204.
17. Avbersek A, Sisodiya S. Does the primary literature provide support for clinical signs used to distinguish psychogenic non-epileptic seizures from epileptic seizures? J Neurol Neurosurg Psychiatry. 2010;81:719–25.
18. Beatty S. Non-organic visual loss. Postgrad Med J. 1999;75(882):201–7.
19. Stone J, Carson A, Sharpe M. Functional symptoms in neurology: management. J Neurol Neurosurg Psychiatry. 2005;76(Suppl I):i13–21.
20. Boeckle M, Liegl G, Jank R, Pieh C. Neural correlates of conversion disorder: overview and meta-analysis of neuroimaging studies on motor conversion disorder. BMC Psychiatry. 2016;16(1):195.
21. Hassa T, Sebastian A, Liepert J, Weiller C, Schmidt R, Tüscher O. Symptom-specific amygdala hyperactivity modulates motor control network in conversion disorder. NeuroImage Clin. 2017;15:143–50.
22. Hallett M, Stone J, Carson A. Imaging studies of functional neurologic disorders. Funct Neurol Disord. 2016;139:73.
23. Morris LS, To B, Baek K, et al. Disrupted avoidance learning in functional neurological disorder: implications for harm avoidance theories. NeuroImage Clin. 2017;16:286–94.
24. Aybek S, Nicholson TR, Zelaya F, et al. Neural correlates of recall of life events in conversion disorder. JAMA Psychiat. 2014;71(1):52–60.
25. Scott RL, Anson JG. Neural correlates of motor conversion disorder. Mot Control. 2009;13 (2):161–84.
26. Harvey SB, Stanton BR, David AS. Conversion disorder: towards a neurobiological understanding. Neuropsychiatr Dis Treat. 2006;2(1):13–20.
27. Stone J, Edwards M. Trick or treat? Showing patients with functional (psychogenic) motor symptoms their physical signs. Neurology. 2012;79(3):282–4.
28. Stone J. Functional neurological symptoms. J R Coll Physicians Edinb. 2011;41(1):38–41.
29. Nielsen G, Stone J, Edwards MJ. Physiotherapy for functional (psychogenic) motor symptoms: a systematic review. J Psychosom Res. 2013;75:93–102.
30. Goldstein LH, Chalder T, Chigwedere C, et al. Cognitive-behavioral therapy for psychogenic nonepileptic seizures: a pilot RCT. Neurology. 2010;74(24):1986–94.
31. Moene FC, Spinhoven P, Hoogduin KA, et al. A randomized controlled clinical trial on the additional effect of hypnosis in a comprehensive treatment program for in-patients with conversion disorder of the motor type. Psychother Psychosom. 2002;71:66–76.
32. Masuda D, Hayaski Y. A case of dissociative (conversion) disorder treated effectively with haloperidol: an adolescent boy who stopped talking. Seishin Igaku (Clin Psychiatry). 2003;45:663–5.
33. Cybulska EM. Globus hystericus – a somatic symptom of depression? The role of electroconvulsive therapy and antidepressants. Psychosom Med. 1997;59(1):67–9.
34. Pennington C, Newson M, Hayre A, Coulthard E. Functional cognitive disorder: what is it and what to do about it? Pract Neurol. 2015;0:1–9.
35. Delis DC, Wetter SR. Cogniform disorder and cogniform condition: proposed diagnoses for excessive cognitive symptoms. Arch Clin Neuropsychol. 2007;22:589–604.
36. Stone J, Pal S, Blackburn D, Reuber M, Thekkumpurath P, Carson A. Functional (psychogenic) cognitive disorders: a perspective from the neurology clinic. J Alzheimer's Dis. 2015;48: S5–S17.
37. Mendez MF. Non-neurogenic language disorders: a preliminary classification. Psychosomatics. 2018;59(1):28–35.

14 Long-Acting Injection for Psychotic Disorder

M. C. Mauri and C. Di Pace

Abstract

We present the case of a patient (27 years old) with a diagnosis of psychotic disorder who started a long-acting therapy after unsuccessful antipsychotic oral treatment. After the first admission, he was treated with haloperidol, but a few months later, he discontinued therapies due to extrapyramidal side effects. A new hospitalization was necessary for reacutization of his psychotic symptoms characterized by severe incongruous laughter, agitation, hostility, and delusions of persecution. In the psychiatric unit, olanzapine was started. By the second day, his psychotic presentation cleared with exception of mild residual perplexity and social isolation. By the fourth day of olanzapine treatment, patient agreed to start long-acting injectable olanzapine with the goal to eventually discontinue the oral olanzapine to provide a safeguard for nonadherence, and he was successfully discharged home.

In the presented case, we modified the dosage and the frequency of the injections on the basis of clinical picture, adapting the long-acting therapy to patient's symptomatology with a good clinical response. Given the complex nature of symptoms presentation and medication regimens, some patients may benefit from personalized treatments. The new long-acting injectable options provide additional flexibility in terms of increasing the time interval between injections.

Keywords

Long-acting injections · Antipsychotic · Depot · Psychotic disorder · Olanzapine

M. C. Mauri (✉) · C. Di Pace
Department of Neurosciences and Mental Health, Fondazione IRCCS Ca' Granda Ospedale Maggiore, Policlinico, University of Milan, Milan, Italy
e-mail: maurimc@policlinico.mi.it

A. C. Altamura, P. Brambilla (eds.), *Clinical Cases in Psychiatry: Integrating Translational Neuroscience Approaches*,
https://doi.org/10.1007/978-3-319-91557-9_14

14.1 Introduction

The management of patients with psychotic disorders should extend beyond acute psychotic symptoms control to include prevention of relapse, amelioration of negative and cognitive symptoms, and improvement in the patient's overall functional capacity and quality of life [1]. Maintained antipsychotic therapy is a key element in relapse prevention because of covert noncompliance to treatment [2]. Poor compliance is the most predictive factor of rehospitalization. Depot preparations of the conventional neuroleptics are recognized for the important role they play in the management of patients with psychotic disorders; they reduce the daily fluctuations in plasma drug concentrations and thus can maximize both the efficacy and tolerability of treatment [3].

This is reflected in reduced rates of relapse and rehospitalization in patients receiving depot neuroleptics compared to those treated with oral formulations [3].

Usually, the use of long-acting injections (LAIs) has been reserved for patients suspected to be poorly adherent and uncooperative or those with refractory illness. Individuals in the early stages of their disorder may be ideal candidates for treatment with an antipsychotic LAI. In psychotic disorders, the initial treatment period represents a critical window which might determine the illness trajectory [4]. These individuals are sensitive to medication side effects and often have limited insight or acceptance of their illness, which in turn contributes to the poor adherence, high-treatment discontinuation rates, and relapse rates. Also, many individuals in the early stages of the disease pathway do not accept the gravity of their illness, and there can be a false sense of treatment being unnecessary or an unwanted imposition. Additionally, many clinicians may have the preconceived view that patients are unwilling to accept injections in the early stages of their disease. This presumption of rejection could be viewed as physician prejudice, in an era when fully informed patient choice is advocated [5].

Recently, many authors have demonstrated the effectiveness and the clinical advantages of LAI even in early schizophrenia [5, 6]. Usually, medication-naïve individuals are acutely sensitive to antipsychotics in terms of responsiveness as well as of side effects, including extrapyramidal symptoms and weight gain [7]. However, it is generally thought that LAIs have a more acceptable side-effect profile in comparison with their oral counterparts due to their differences in pharmacokinetics, and any concerns over debilitating side effects may be due to dosing errors; peaks and troughs in drug concentrations can be minimized via the dose averaging of LAIs, reducing the risk of some adverse effects of these medications [8].

Heres et al. [9] found that the three main factors influencing their choice not to prescribe a LAI for subjects with a psychotic episode were: (a) limited availability of different second-generation antipsychotic depot drugs, (b) the frequent rejection of the depot offer by patients, and (c) the patient's skepticism based on an inexperience of relapse, demonstrating the importance of a patient-centered approach when discussing LAI as a treatment option. The prescription of a LAI should also involve a collaborative psychosocial approach concentrating on the individual's needs and involving the multidisciplinary team in order to optimize outcomes. This approach would be consistent with first-episode or early intervention services [10].

14.2 Emerging Long-Acting Antipsychotics: What's New, What's Different, and What's Next?

Since the introduction of risperidone long-acting injection in 2003, three additional second-generation antipsychotics have become available in a long-acting injectable formulation: paliperidone, olanzapine, and aripiprazole. Although these different depot options can help with adherence and thus encourage better treatment outcomes, they differ in terms of specific indications, approved injection sites, needle gauge, injection volume, injection interval, requirements for oral supplementation, availability of prefilled syringes, storage needs, and postinjection observation period, as well as potential drug-drug interactions and commonly encountered adverse reactions [11].

There are several new and emerging medication interventions for both the acute and maintenance treatment phases of schizophrenia. Recently approved are two new dopamine-receptor partial agonists, brexpiprazole and cariprazine, as well as two new long-acting injectable antipsychotic formulations, aripiprazole lauroxil and 3-month paliperidone palmitate. The new long-acting injectable options provide additional flexibility in terms of increasing the time interval between injections.

Risperidone microspheres require a period of overlap of 3 weeks with oral risperidone. It has a 2-week dosing interval.

Olanzapine pamoate does not need to overlap with oral olanzapine. It has a small risk of postinjection syndrome (0.07% of injections): symptoms include sedation, confusion, agitation, anxiety, aggressiveness, dizziness, ataxia, and extrapyramidal symptoms. This risk limits olanzapine pamoate use. After injection, the patient must be monitored for 3 h by a health-care professional.

Paliperidone palmitate does not need overlap with oral paliperidone and requires two separate loading dose injections during the first week.

The 3-month paliperidone palmitate (PPM-3) formulation can only be used if the patient has been receiving 1-month paliperidone palmitate injections for at least 4 months. It is administered four times a year, providing the longest interval of any approved LAI.

Aripiprazole monohydrate requires a period of overlap of 2 weeks with oral aripiprazole. Available as a lyophilized powder which needs to be reconstituted.

Aripiprazole lauroxil requires a period of overlap of 3 weeks with oral aripiprazole, available as a prefilled syringe that does not require reconstitution. Aripiprazole lauroxil is not available in Italy.

The first aripiprazole LAI formulation, Abilify Maintena®, was approved in early 2013 at the recommended dose of 400 mg IM injection every 4 weeks. Aripiprazole lauroxil (Aristada®) is a newer LAI aripiprazole formulation, which was FDA approved in October 2015. Aripiprazole lauroxil is an *N*-acyloxymethyl prodrug that undergoes a two-step bioconversion in the plasma from the lauroxil to an intermediate *N*-hydroxymethyl-aripiprazole via enzyme-mediated hydrolysis. The *N*-hydroxymethyl-aripiprazole then undergoes a hydrolysis reaction to aripiprazole. Aripiprazole lauroxil is available at a dose of 441 mg (deltoid or gluteal), 662 mg, and 882 mg (only gluteal) corresponding to aripiprazole LAI 300 mg, 450 mg, and 600 mg (corresponding to oral aripiprazole 10 mg/day, 15 mg/day, and 20 mg/day), respectively. While these two formulations are similar with respect to dosing/administration intervals, they are different with regard to their formulations which influence some of their pharmacokinetic parameters including $T_{1/2}$ and T_{max} [12].

Considering the prevention of relapses, the effectiveness of newer LAIs (aripiprazole, olanzapine, paliperidone, and risperidone) and older LAIs (haloperidol, fluphenazine, flupenthixol) is similar [13, 14] (Boxes 14.1 and 14.2).

Box 14.1 First-generation antipsychotics available as long-acting injectable medications

Drug	Starting dose (mg)	Maintenance dose (mg)
Haloperidol decanoate	50	50–200 every 3–4 weeks
Fluphenazine decanoate	12.5	12.5–50 every 2–3 weeks
Flupenthixol decanoate	20	50–300 every 2–4 weeks
Zuclopenthixol decanoate	100	200–500 every 1–4 weeks

Second-generation antipsychotics available as long-acting injectable medications

Drug (Brand name)	Manufacturer	Available formulations	Injection interval	Comments
Aripiprazole monohydrate (Abilify Mantenna)	Otsuka/ Lundbeck	300, 400 mg vials, prefilled syringes	400 mg once/ month	Requires a period of 2 weeks of overlap with oral aripiprazole.
Aripiprazole lauroxil (Aristada)	Alkermes	441, 662, 882 mg prefilled syringes	441–882 mg once/month 882 mg q 6 weeks	The 882 mg dose can be administered every 6 weeks. Requires a period of 3 weeks of overlap with oral aripiprazole.
Olanzapine pamoate (Zyprexa Relprew)	Lilly	210, 300, 405 mg vials	150–300 mg q2 weeks 300–405 mg once/month	Requires monitoring post injection (3 h)
Paliperidone palmitate (Invega Sustenna, Xeplion)	Janssen	39, 78, 117, 156 or 234 mg prefilled syringes	117 mg once/ month	Oral supplementation not necessary.

(continued)

Box 14.1 (continued)

Paliperidone palmitate (Invega Trinza)	Janssen	273, 410, 546, 819 mg prefilled syringes	410 mg q3 months	Use in patients already treated with Invega Sustenna
Risperidone microspheres (Risperdal Consta)	Janssen	12.5, 25, 37.5 or 50 mg vials	25 mg q2 weeks	Requires a period of 3 weeks of overlap with oral risperidone

Box 14.2 Potential advantages and disadvantages of long-acting drugs

Potential advantages	Potential disadvantages
• Early identification of nonadherence • Providing a mechanism for monitoring adherence with injections • No need to remember to take medication every day • Regular interactions between patient and medical staff • Reduced relapse frequency and rehospitalization rates • Clear attribution of the cause of relapse or non-response, discriminating between nonadherence or lack of response • Reduce the risk of accidental or deliberated overdose • Treating patients with more stable plasma concentrations than oral medications • Avoidance of first-pass metabolism—better relationship between dose and blood level of drug • Lower and less frequent peak plasma level—reduced side effects	• Slow-dose titration • Longer time to achieve steady-state levels • Less flexibility of dose adjustment • Delayed disappearance of distressing and/or severe side effects • Pain at the injection site can occur, and leakage into the subcutaneous tissue and/or the skin may cause irritation and lesions (especially for oily long-acting injectable) • Burden of frequent travel to outpatient clinics or home visits by community nurses for their administration • Risperidone long-acting injectable needs refrigeration, which may be cumbersome in some latitudes • Perception of stigma

Box 14.3 Recommended dosage scheme between oral olanzapine and olanzapine LAI

Oral dose	Starting dose	Maintenance dose
10 mg	210 mg/2 weeks; 405 mg/4 weeks	150 mg/2 weeks; 300 mg/4 weeks
15 mg	300 mg/2 weeks	210 mg/2 weeks; 405 mg/4 weeks
20 mg	300 mg/2 weeks	300 mg/2 weeks

14.3 Case Presentation

A.S., a 27-year-old Italian male (height 185.0 cm, weight 80 kg) with diagnosis of psychotic disorder not otherwise specified, was referred to our inpatient clinic for the first time in October 2015 for a psychotic episode with aggressiveness and psychomotor agitation.

No previous psychiatric history had been reported before this episode. He had a family history for unipolar major depressive disorder (maternal grandmother). He interrupted studies before college. He works in a mechanical workshop with his father. He had a normal childhood and upbringing. He was shy and reserved, but he had friends during childhood/adolescence.

Substance abuse was reported by the patient: he had used cannabis daily since he was 15 years old and sporadically since he was 17 years old; he had drunk alcohol occasionally since he was 15 years old. His medical history includes previous varicocele, appendectomy, and bacterial pneumonia. The subpsychotic symptoms appeared gradually, and the time frame was not clear. After the first admission (October 2015), he was treated with haloperidol (titrated up to 6 mg/day); after few months he discontinued therapies due to extrapyramidal side effects.

In September 2016 a new hospitalization was necessary: the patient was brought to the emergency room by his father for psychotic reacutization, exhibiting severe incongruous laughter, agitation, hostility, and delusions of persecution.

His family said that he became irritable, sleepless, and aggressive toward his ex-girlfriend, and he presented a delusional jealousy. Furthermore, in the last months, the family referred reduced social drive, loss of motivation, lack of interest, and difficulties in work activities (he missed several days of work), with a decrease in quality of life.

During hospital stay, the patient was disorganized, paranoid, and believed that he was being judged by the public. He was socially isolative with flat affect and neglecting his personal hygiene. He denied any perceptual disturbances. There was no history of mood symptoms or medical illness. The patient's physical examination, laboratory values, revealed no abnormalities. His toxicology screen was negative.

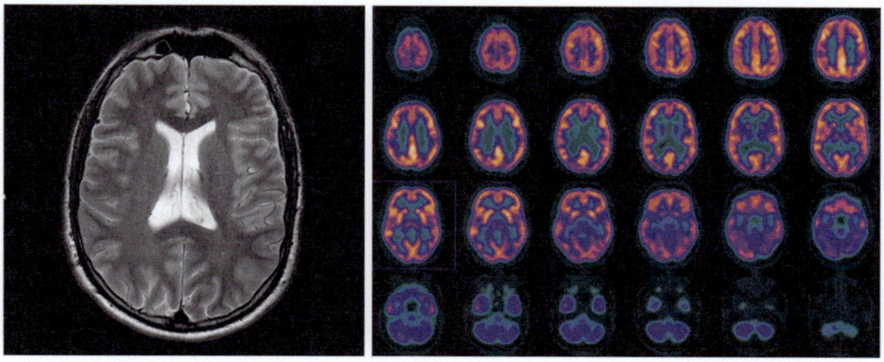

Fig. 14.1 MRI (left) and FDG-PET (right) images

Table 14.1 Neurocognitive evaluation

Test	Normal Score	Score	Comment
Language			
Verbal fluency	v.n. ≥ 31.68	40.50	Normal
Memory			
Verbal memory	v.n. ≥ 33.01	41.00	Normal
Motor proficiency			
Token task	v.n. ≥ 68.77	57.25	**Deficit**
Symbol-coding task	v.n. ≥ 40.49	35.25	**Deficit**
Frontal proficiency			
Working memory	v.n. ≥ 14.93	22.75	Normal
Tower of London	v.n. ≥ 12.37	15.00	Normal

The magnetic resonance imaging (MRI) and fluorodeoxyglucose positron emission tomography (FDG-PET) showed no significant modification in the normal and symmetric brain metabolism (Fig. 14.1). The brief assessment of cognition in schizophrenia (BACS) was assessed, showing a deficient motor proficiency (Table 14.1). Wechsler Adult Intelligence Scale (WAIS) intelligence quotient (IQ) was 88. In the psychiatric unit, treatment with olanzapine was initiated. The patient responded well. By the second day, his psychotic presentation cleared with an exception of mild residual perplexity and social isolation. By the fourth day of olanzapine treatment, A.S. agreed to start long-acting injectable olanzapine with the goal of eventually discontinuing the oral olanzapine to provide a safeguard for nonadherence, and he was successfully discharged home with a diagnosis of schizoaffective disorder.

Subsequently, olanzapine LAI, 300 mg intramuscularly (IM), was administered into the gluteal muscle every 2 weeks in day hospital. At every injection olanzapine plasma levels (PL) were determined (Fig. 14.2). Some weeks after initiating olanzapine LAI, a marked improvement in psychiatric symptoms was noted.

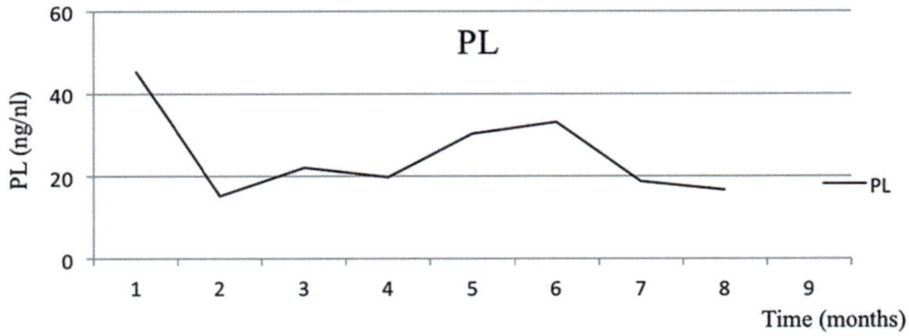

Fig. 14.2 Patient's olanzapine plasma levels during olanzapine LAI treatment

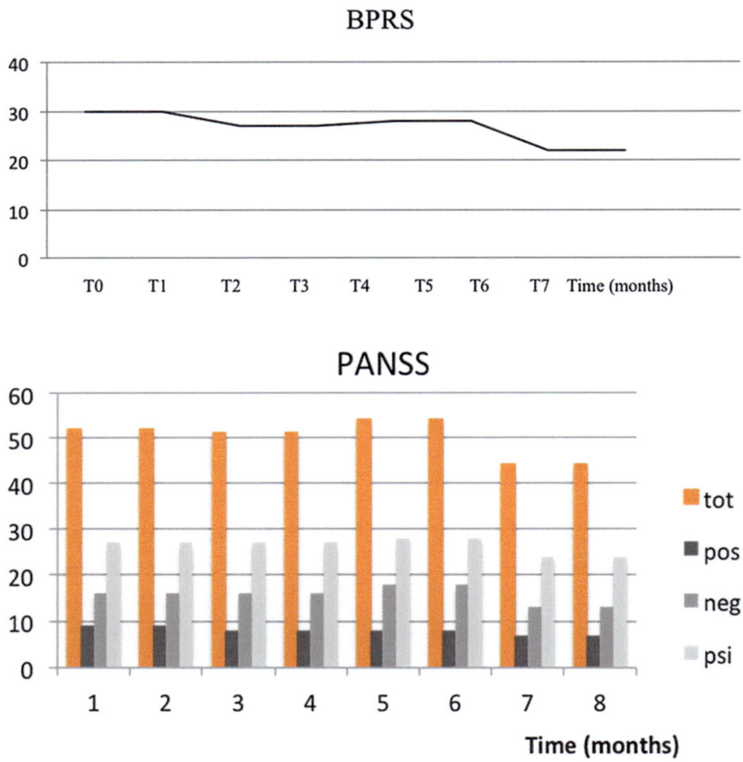

Fig. 14.3 BPRS and PANSS total score (mean values ± SD) time course

Between September and March 2017, A.S. showed a good adherence and clinical stabilization as demonstrated by clinical rating scales BPRS (the Brief Psychiatric Rating Scale) and PANSS (Positive and Negative Syndrome Scale) (Fig. 14.3). A.S. revealed perception disturbance during his stay in the inpatient unit.

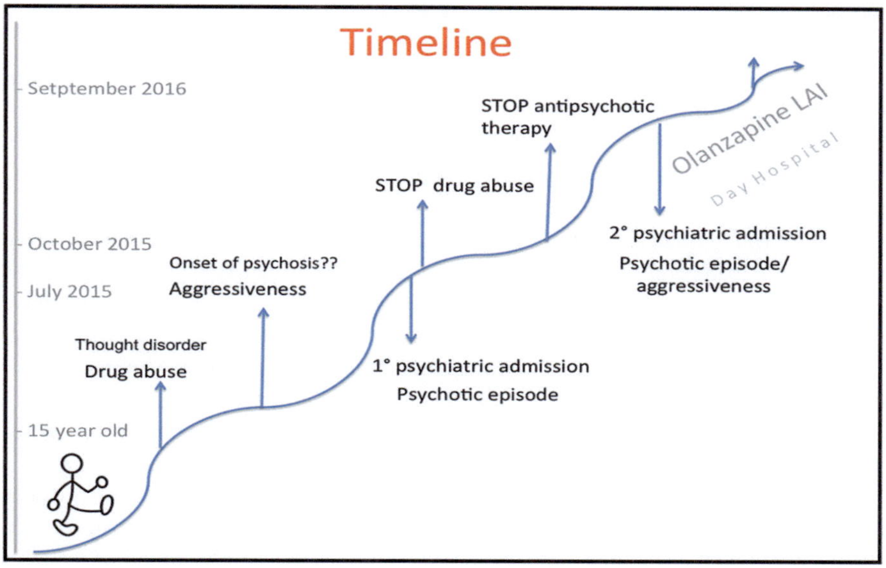

Fig. 14.4 Timeline of course of illness

In April 2017 at follow-up appointment, A.S. presented acute worsening of apathy, social isolation, loss of interest in activities, and hypersomnia. The plan was to reduce the olanzapine LAI frequency from 300 mg IM every 2 weeks to 405 mg every 4 weeks.

In June 2017, since patient complained of worsening irritability and suspiciousness, approximately 5 days prior to his next injection, his current regimen (olanzapine LAI 405 mg every 4 weeks) was modified (olanzapine LAI 405 mg every 3 weeks) for this wearing off of symptoms. These pharmacological changes resulted in a good stabilization of clinical symptomatology (Fig. 14.4). As of July 2017, he continues to tolerate the olanzapine LAI and reports not needing to use oral additional therapies.

14.4 Literature Review

We presented a clinical case of a young patient with a diagnosis of psychotic disorder who started a long-acting therapy after unsuccessful antipsychotic oral treatment.

The guidelines for depot antipsychotic treatment that were developed by a European neuropsychopharmacology consensus conference recommend that: "any patient for whom long-term antipsychotic treatment is indicated should be considered for depot drugs" [15].

Many physicians are still reluctant to use long-acting antipsychotics initially. According to literature, there are several studies in support of depot antipsychotic use as first-line treatment for patients with schizophrenia, which may improve adherence and thereby lower risk of relapse, suicide, and rehospitalization [16].

In the presented case, we modified the dosage and the frequency of the injections on the basis of the clinical picture, adapting the long-acting therapy to patient's symptomatology with a good clinical response. Given the complex nature of symptoms presentation and medication regimens, some patients may benefit from personalized treatments.

The dosing strategies for LAI antipsychotics include relatively standardized conversions from oral formulations to recommended administration intervals. The injection intervals for LAIs depend on the specific agent and are based on extensive pharmacokinetic studies of patients. Although not extensively reported in the literature, there are clinical reports of some patients requiring shorter dosing intervals or higher than approved doses of LAI antipsychotics with unclear reasons [12].

Regarding olanzapine LAI, there is a broad range in time between injections. Most patients continue to receive the same initial dose instead of switching to a maintenance dose. This may suggest that some clinicians are not reassessing the dose after the initial starting dose because the patient was stabilized on olanzapine oral before beginning olanzapine LAI [17]. Mauri et al. [18] demonstrated efficacy of olanzapine LAI in maintenance treatment of schizophrenia also at lower dosage. The recommended initial and maintenance doses equivalent to those for oral olanzapine are given in Boxes 14.3 and 14.4.

Our case report suggests further investigations of dosing and frequency strategies during a long-acting treatment.

Furthermore, there are multiple studies showing the efficacy of oral atypical antipsychotics in the treatment of bipolar I disorder (BP-I); however, literature data on the use of LAIs in BP-I are lacking. Adherence remains a significant challenge in the treatment of patients with BP-I. Available data on atypical LAI antipsychotics in BP-I are largely derived from controlled studies of risperidone LAI; thus, additional studies on the potential benefits of LAIs in bipolar disorder are needed, including comparisons with oral formulations. Recently aripiprazole LAI has demonstrated efficacy as maintenance treatment for BP-I by reducing the risk of recurrence of mood episodes [19].

Box 14.4 ***Olanzapine pamoate:*** **recommendations**

Patients should remain under the supervision of properly qualified staff at a health-care center for at least 3 h after each injection so that signs and symptoms of an olanzapine overdose may be detected.

Before olanzapine-LA therapy is initiated, patients should first be treated with oral olanzapine to determine its tolerability and their response.

There is no need to supplement with oral olanzapine.

Olanzapine-LA should not be used in patients who are elderly or have renal insufficiency unless an effective and well-tolerated dosing regimen for oral

(continued)

Box 14.4 (continued)
olanzapine has been established. For these patients, a lower initial dose (150 mg every 4 weeks) should be considered.

A dosage reduction should be considered when more than one factor is present that could trigger a slowing of the metabolism (female sex, geriatric age, no tobacco habit). Increasing the dosage, if indicated, should be done with caution in these patients.

Because the pamoate salt of olanzapine dissolves slowly to facilitate its steady slow release, which is not complete until approximately 6–8 months after the last injection, a doctor's supervision is required when switching to another antipsychotic drug considered medically appropriate—especially during the first 2 months after interrupting the olanzapine-LA therapy.

Key Points

- Depot neuroleptics were seen to reduce relapse rate and rehospitalization in comparison to oral formulations.
- Long-acting injections (LAIs) had been reserved for patients suspected to be poorly adherent and uncooperative or those with refractory illness: actually, individuals in the early stages of their disease may be ideal candidates for treatment with an antipsychotic LAI.
- The new long-acting injectable options provide additional flexibility in terms of increasing the time interval between injections.
- The effectiveness of newer LAIs, aripiprazole, olanzapine, paliperidone, and risperidone, and older LAIs haloperidol, fluphenazine, flupenthixol, is similar.

Self-Assessment Questionnaire

1. Who should receive LAIs?
   (A) **Consider LAIs for patients with recent-onset schizophrenia and those with risk factors for medication nonadherence**
   (B) Patients with poor insight
   (C) Patients with severe symptoms
   (D) Only patients with chronic schizophrenia
2. Are the newer LAIs more effective than the older LAIs in terms of prevention of relapses?
   (A) **The effectiveness of newer LAIs (aripiprazole, olanzapine, paliperidone, and risperidone) and older LAIs (haloperidol, fluphenazine, flupenthixol) is similar.**
   (B) The newer LAIs are more effective
   (C) The older LAIs are more effective
   (D) Aripiprazole LAI is less effective

3. Which of the following statements is true?
   (A) **Any patient for whom long-term antipsychotic treatment is indicated should be considered for depot drugs**
   (B) The presence of hallucinations is necessary for receiving long-acting therapy
   (C) Risperidone long acting has a small risk of postinjection sedation syndrome
   (D) Olanzapine long acting must overlap with oral supplementation
4. The factors influencing the choice not to prescribe a LAI for first-episode psychosis were:
   (A) **Limited availability of different second-generation long-acting antipsychotics**
   (B) Rejection of the depot by patient's family
   (C) The inexperience of the clinicians with long-acting treatment
   (D) The necessity to receive depot by a health-care professional in a community setting
5. Which of the following antipsychotic medications does not come in a long-acting injectable formulation?
   (A) Risperidone
   (B) Olanzapine
   (C) Haloperidol
   (D) **Quetiapine**

References

1. Buoli M, Caldiroli A, Panza G, Altamura AC. Prominent clinical dimension, duration of illness and treatment response in schizophrenia: a naturalistic study. Psychiatry Investig. 2012;9(4): 354–60. https://doi.org/10.4306/pi.2012.9.4.354.
2. Buoli M, Kahn RS, Serati M, Altamura AC, Cahn W. Haloperidol versus second-generation antipsychotics in the long-term treatment of schizophrenia. Hum Psychopharmacol. 2016;31(4): 325–31. https://doi.org/10.1002/hup.2542.
3. Altamura AC, Colacurcio F, Mauri MC, Moro AR, De Novellis F. Haloperidol decanoate in chronic schizophrenia: a study of 12 months with plasma levels. Prog Neuro-Psychopharmacol Biol Psychiatry. 1990;14(1):25–35.
4. Altamura AC, Serati M, Buoli M. Is duration of illness really influencing outcome in major psychoses? Nord J Psychiatry. 2015;69(6):403–17. https://doi.org/10.3109/08039488.2014.990919.
5. Taylor M, Ng KY. Should long-acting (depot) antipsychotics be used in early schizophrenia? A systematic review. Aust N Z J Psychiatry. 2013;47(7):624–30.
6. Correll CU, Citrome L, Haddad PM, Lauriello J, Olfson M, Calloway SM, Kane JM. The use of long acting injectable antipsychotics in schizophrenia: evaluating the evidence. J Clin Psychiatry. 2016;77(Suppl 3):1–24.
7. Strassnig M, Miewald J, Keshavan M. Weight gain in newly diagnosed first-episode psychosis patients and healthy comparisons: one-year analysis. Schizophr Res. 2007;93:90–8.
8. Taylor D. Psychopharmacology and adverse effects of antipsychotic long-acting injections: a review. Br J Psychiatry. 2009;195(Suppl 52):S13–9.
9. Heres S, Reichhart T, Hamann J. Psychiatrists' attitude to antipsychotic depot treatment in patients with first-episode schizophrenia. Eur Psychiatry. 2011;26:297–301.
10. Weiden PJ, Buckley PF, Grody M. Understanding and treating 'first-episode' schizophrenia. Psychiatr Clin North Am. 2007;30:481–510.

11. Citrome L. New second-generation long-acting injectable antipsychotics for the treatment of schizophrenia. Expert Rev Neurother. 2013;13(7):767–83. https://doi.org/10.1586/14737175.2013.811984.
12. Eum S, Schneiderhan ME, Brown JT, Lee AM, Bishop JR. Pharmacogenetic evaluation to assess breakthrough psychosis with aripiprazole long-acting injection: a case report. BMC Psychiatry. 2017;17(1):238.
13. Castillo EG, Stroup TS. Effectiveness of long-acting injectable antipsychotics: a clinical perspective. Evid Based Ment Health. 2015 May;18(2):36–9. https://doi.org/10.1136/eb-2015-102086.
14. McEvoy JP, Byerly M, Hamer RM, et al. Effectiveness of paliperidone palmitate vs. haloperidol decanoate for maintenance treatment of schizophrenia: a randomized clinical trial. JAMA. 2014;311:1978–87.
15. Kane JM, Aguglia E, Altamura AC, Ayuso Gutierrez JL, Brunello N, Fleischhacker WW, Gaebel W, Gerlach J, Guelfi JD, Kissling W, Lapierre YD, Lindström E, Mendlewicz J, Racagni G, Carulla LS, Schooler NR. Guidelines for depot antipsychotic treatment in schizophrenia. European Neuropsychopharmacology Consensus Conference in Siena, Italy. Eur Neuropsychopharmacol. 1998;8(1):55–66.
16. Khan AY, Salaria S, Ovais M, Ide GD. Depot antipsychotics: where do we stand? Ann Clin Psychiatry. 2016;28(4):289–98.
17. Case MG, Karagianis JL, Schuh KJ, Hoffmann VP. Real-world use patterns of olanzapine long-acting injection in the United States: comparison to the recommended dosing strategy. Curr Med Res Opin. 2013;29(11):1533–7.
18. Mauri MC, Maffini M, Di Pace C, Reggiori A, Paletta S, Moliterno D, Rovera C, Altamura AC. "Long-acting" olanzapine in maintenance therapy of schizophrenia: a study with plasma levels. Int J Psychiatry Clin Pract. 2015;19:99–105.
19. Calabrese JR, Sanchez R, Jin N, Amatniek J, Cox K, Johnson B, Perry P, Hertel P, Such P, Salzman PM, McQuade RD, Nyilas M, Carson WH. Efficacy and safety of Aripiprazole once-monthly in the maintenance treatment of bipolar I disorder: a double-blind, placebo-controlled, 52-week randomized withdrawal study. J Clin Psychiatry. 2017;78(3):324–31.

Cognitive Enhancement in the Early Phases of Psychosis 15

Elisabetta Caletti and Francesca Siri

Abstract

Major psychoses, including schizophrenia (SKZ), bipolar disorder (BD), and major depression (MD) with psychotic features, include the continuity of positive symptoms ranging from mild to severe. Literature suggests that cognitive deficits and brain changes, associated with long duration of illness, are markers of vulnerability of psychoses, with important consequences on day-to-day functioning and quality of life. Working memory, attention, executive functions, and theory of mind are the most altered cognitive domains in psychosis. Based on such evidence, neurocognition may be considered a main focus of treatment, in particular at the onset of psychosis.

We have defined a cognitive intervention program for young psychotic patients (aged 18–40) in which a computerized cognitive training (CR), associated with social skills training or mindfulness groups, integrates medical/physical treatment. Here we present the case of a 25-year-old single male patient with a diagnosis of SKZ and cognitive deficits. The patient was asked to complete pre- and post-group assessment to allow formal evaluation of the treatment program, as well as neurocognitive testing.

After the engagement in this intervention (CR associated with mindfulness), a reduction of psychological stress and improved social functioning were obtained. With regard to neurocognition, we observed an enhancement of executive functions (in particular working memory, planning capacity, and decision-making abilities) and increased performances on ecological tests, showing the actual impact of the integrated CR on patients' life.

E. Caletti (✉) · F. Siri
Department of Neurosciences and Mental Health, Fondazione IRCCS Ca' Granda Ospedale Maggiore Policlinico, University of Milan, Milan, Italy
e-mail: calettielisabetta@gmail.com

A. C. Altamura, P. Brambilla (eds.), *Clinical Cases in Psychiatry: Integrating Translational Neuroscience Approaches*,
https://doi.org/10.1007/978-3-319-91557-9_15

Keywords
Psychosis · Schizophrenia · Bipolar disorder · Cognitive impairments · Cognitive training

15.1 Introduction

The diagnosis of major psychoses, including schizophrenia (SKZ), bipolar disorder (BD), and major depression (MD) with psychotic features, has been progressively overtaken by a transdiagnostic phenotype encompassing non-affective and affective psychosis, in which psychosis could be defined as a continuity of positive symptoms ranging from mild to severe [1–3]. The DSM-5 [4] classifies two different categories based on the severity of psychopathology: "non-affective psychoses," referring to psychotic disorders with delirium, hallucinations, thought disorganization, abnormal motor behavior, and negative symptoms (i.e., anhedonia, anti-socialness) as key features, and "affective psychoses" such as psychotic disorders with a prevalent mood component (e.g., BD type I with psychotic characteristics) in which delusions and hallucinations may occur in any moment of the episode and may be congruent or not to mood. Frequently associated risk factors for psychosis are age, gender, and suffering during prenatal and perinatal period, environmental influences, motor and neurological development abnormalities and delays, formal thinking alterations, childhood environment instability, childhood trauma, cranial trauma, substance abuse, and environmental stressors [5–7]. Findings by Simonsen et al. [8] suggested and supported a dimensional approach in which neurocognitive dysfunctions in BD and SKZ are more determined by history of psychosis than by a diagnostic category. Neurodegeneration affecting these subjects is one of the most replicated facts in scientific researches [9, 10], and this hypothesis highlights the negative effects of illness [11]. Buoli et al. [12] highlighted that cognitive impairment and brain changes associated with long duration of major psychosis are implicated both in aging and in the progression of disorders. Overall, literature suggests that cognitive deficits are a marker of vulnerability of major psychoses, with important consequences on day-to-day functioning, quality of life, and adaptation to a real-world context [13]. Common characteristics of a "prepsychotic" personality include impairments in short-term memory, working memory, visuospatial memory, attention, difficulty of concentration, poor social competence, and hypersensitivity to outside stimuli [14–17].

Structural hippocampal alterations seem to correlate with cognitive impairment such as memory deficit, difficulties in learning, and abstract reasoning (e.g., [18, 19]). A greater hippocampal neurodegeneration can be predictive of psychosis, given that the processes of remodeling seem to take place during onset of illness [20]. Interestingly, the hippocampus is very vulnerable to neuronal loss [21, 22]: but it is also characterized by great neuroplasticity, so it is sensitive to environmental, clinical, and therapeutic phenomena. In SKZ, working memory, attention, and executive functions are the most altered cognitive domains [23–26]. In BD, there is significant cognitive interindividual variability [27], including deficient working memory performance [28], inability to withstand interferences and to program actions, associative learning [29, 30], sustained attention, psychomotor speed, inhibitory response, set-shifting ability, verbal memory [31, 32], and theory of mind (ToM) impairment [33]. Specifically, ToM dysfunction could lead to poor ability to reflect upon and ascribe, to oneself and others, mental states, including desires, beliefs, knowledge, intentions, and feelings [34]. Based on such evidence, neurocognition may be considered a main focus of treatment, in particular at the onset of psychosis [35–37]. In general, understanding the existing cognitive overlaps between SKZ and BD will help to plan effective cognitive remedial strategies for distinct patient groups. Some specialists have examined various exhaustive cognitive batteries in order to capture the advantages and disadvantages of each tool [38] and found a more appropriate brief assessment. The Italian validated version of the Brief Assessment of Cognition in Schizophrenia (BAC-S, [39, 40]) may be an ideal solution for our psychotic patients. In addition, to check the generalization of the results of cognitive remedy treatments and thus the actual impact on patients' life, a number of "ecological" tests could be useful too [34, 41, 42]. Some specific TOM tasks (e.g., the Reading the Mind in the Eyes; the Faux Pas test), which are able to grasp changes in the patient's social interaction [43, 44], could also complete testing.

15.2 Cognitive Remedies

15.2.1 Identifying Early Psychotic Disturbances

Specific interventions in addition to accurate pharmacological therapy need consequently to be applied: psychoeducation, psychosocial therapies, cognitive rehabilitation, community treatment, etc. The aim is to offer a timely and targeted treatment, a "cognitive" remedy, which allows delaying or moderating the consequences of the disease by improving the patient's quality of life. Benefits could be reduction of morbidity, maintenance of psychosocial skills, conservation of family and social supports, less need of hospitalization, and faster healing process with better prognosis. If people begin to receive treatment only when the disease is already well-established and consolidated, the effects on personality structure have already affected its development, the social and affective network, and worsened cognitive functionality. Before starting remediation therapies, it is important to check whether the patient has understood the meaning of a particular learning pathway for his or her life. If the patient is not motivated or does not understand the direction of treatment, he or she will not be able to change behavior.

15.2.2 Cognitive Intervention Strategies

Providing cognitive remediation in addition to psychiatric rehabilitation may contribute to greater improvement in both cognitive and social functioning [45, 46], but it is crucial to understand which interventions have the greatest impact on the long-term functioning of psychotic patients.

Restorative/reparative and compensatory interventions are the main models of cognitive remedial strategies. Specifically, the restorative/reparative model is based on neural plasticity and focused on the concrete repair of compromised neural processes and therefore on the ability of the brain to develop and evolve throughout life; it is both top-down and bottom-up: relearning and repetition of tasks (retraining) [47–49]. Cognitive remediation therapy (CRT) has often been used in SKZ, a restorative/reparative intervention consisting of a set of behavioral-based interventions aimed at improving cognitive processes; commonly CRT is subdivided into individual sessions for a total duration of 40 h (normally 3 days a week), primarily targeting attention, memory, executive functions, metacognition, and social cognition [50, 51].

An intensive cognitive training may induce a significant plasticity in the cortical system of individuals with SKZ: their activation patterns appeared more similar to those of healthy individuals than they did at baseline, and this activation was correlated with functional improvements [52]. FMRI studies after CRT show functional improvements in reality examination [53], attention, and working memory, with greater activation of the prefrontal cortex [54, 55] and in the parietal lobe [56] in relation to reasoning. Moreover, activations of the bilateral cortical regions associated with memory process have been observed [57].

Several authors have suggested the effectiveness of the social skills training (SST) program in psychotic patients [50, 58–62], so an integrated intervention, with activities aimed at increasing social skills such as interpersonal relationships and emotional management, seems to be crucial [63] in order to overcome reported poor performance in everyday life [64, 65] and to support patients' transition from the hospital routine to daily living activities [66, 67].

Another important complement to cognitive remediation could be the mindfulness-based intervention: it decreases the impact of, and time needed to recover from, negative emotional events [68, 69]. Reductions of stress and clinical problems have been extensively shown [70–72], and a recent pilot study provides evidence for the tolerability and feasibility of brief mindful cognitive enhancement training for psychosis [73]. Cognitive factors like concentration, attention, and non-judging acceptance toward whatever one is experiencing in the present moment may increase willingness to tolerate uncomfortable emotions and sensations [74, 75].

15.3 A Model of Intervention

The Psychiatric Equipe of the Department of Neurosciences and Mental Health, Ospedale Maggiore Policlinico in Milan, has defined a cognitive intervention program for young psychotic patients (aged 18–40, maximum 5 years of illness) in which computerized cognitive training (CR), associated with SST or mindfulness groups, may integrate medical/physical treatment. It can be applied every time there are identifiable areas subject to possible improvement in functioning (lack of positive coping strategies, over- or underactive, mild/moderate anxiety/depression). Exclusion criteria are current drug abuse and intellectual impairment. The program is co-facilitated by two psychologists, a resident psychiatrist and a psychiatric rehabilitation therapist. Each patient is followed in a stepping program that includes a succession of exercises offered in an order of increasing complexity and related both to basic neurocognitive activities and social skills enhancement or mindfulness.

15.3.1 Pre-group Treatment Evaluation/Questionnaires, Scales, and Neurocognitive Testing

All patients attending the program are asked to complete pre-group and post-group measures to permit formal evaluation of the treatment program (Table 15.1), as well as neurocognitive testing (Table 15.2). Evaluation is generally from the repeat of the questionnaires and neuropsychological tests (this is usually arranged at the end of the group by the psychologist/resident psychiatrist).

Table 15.1 Evaluation scales

Scales	Scope of investigation
Positive and Negative Syndrome Scale (PANSS) [76]	Severity of positive and negative symptoms
Young Mania Rating Scale (YMRS) [77]	Manic symptoms
Hamilton Rating Scale for Depression (HDRS) [78]	Depressive symptoms
Global Assessment of Functioning (GAF) [79]	General functioning
Manchester Short Assessment Quality of Life Scale (MANSA) [80]	Quality of life
Short Form-36 Health Survey (SF-36) [81]	Health status
Socioeconomic status (SES) [82]	Socioeconomic status
Client Sociodemographic and Service Receipt Inventory—European Version (CSSRI-EU) [83]	Sociodemographic data
Five Facets Mindfulness Questionnaire [84]	Mindfulness facets: observing, describing, acting with awareness, non-judging, and nonreactivity to inner experience

Table 15.2 Neurocognitive testing

BAC-S	The Brief Assessment of Cognition in Schizophrenia [39, 40]
Iowa gambling task (IGT)	Initially applied in neurological setting to patients with prefrontal cortical lesions [85]. IGT recreates experimental controlled situations as naturally as possible, reproducing conditions in decision-making in which a situation remains uncertain
Multiple Errand Test (MET-HV)	Patient is required to complete a set of real-world tasks (e.g., purchasing specific items, collecting and writing down specific information) within the hospital structure. Tasks are performed within the constraints of specified rules [34, 42]
Hotel task	Participants have to imagine that they are working in a hotel playing a part of each of the five jobs that are done each day in this trade (all tasks must be completed within 15 minutes, but this is of course not enough time to complete all of the tasks) [86]
Reading the Mind in the Eyes Test	ToM task [87, 88]. The participants have to choose between four different possible adjectives the one which is the most applicable and appropriate to describe the mood of the person of whom the patient sees only the eyes
Faux Pas Test	ToM advanced adult version [44, 89]. Participants are required to read 20 stories, 10 of which contain awkward "faux pas" situations that they must be able to identify

15.3.2 Cognitive Training and Groups

In our model, CR is done using the Italian version of the COGPACK software (http://www.markersoftware.com/). Advantages compared to a traditional pen and paper assessment are flexibility of tasks; possibility of individualizing a set of different exercises, depending on the cognitive function they are targeting; and the possibility of an immediate feedback (progression in performance requests is modulated on individual response). Approximately 40 sessions are planned for each patient individually. CR is carried out 2–3 times a week, 45 min each, preferably in the morning when the levels of vigilance and attention are higher. Meetings are preferably conducted in the same room, which should not contain objects that capture the attention of patients, to avoid distraction. Patients are guided by an operator who helps them by providing the appropriate instructions for the activities and any technical problems and also gives positive feedback to strengthen strategies in solving the tasks. For patients treated with antipsychotic therapy, the same pharmacological treatment is maintained throughout the duration of the training. CR is supplemented with SST or mindfulness skills training (MST) groups. SST is a psychoeducational training which allows the patient to acquire psychosocial skills. Open groups with a weekly schedule, each of 1.5 h, are run by a leader and an assistant (it is preferable that the group leader should not be one of the individual patient-therapists). From both integrated interventions, consisting generally of nine 2-h sessions held on consecutive weeks, we expect primary outcomes geared to improving cognitive and secondary functions such as social and interpersonal abilities, work, quality of life, and tasks that are useful in everyday life.

15.4 Case Presentation

A 25-year-old single male patient (high school graduate) was admitted in May 2015 to our psychiatric unit for persecutory behavior (the patient was convinced that other people were observing him at home) and social withdrawal. The patient was born at term without any complications; the first phases of somatic and psychic development were normal. His psychiatric anamnesis was positive: his mother committed suicide at the age of 50. He had poor dietary habits. He reported occasional use of cannabis, MDMA, and cocaine in the past and alcohol and smoking 10 cigarettes/day for 6 years.

The patient was at his fourth hospitalization in the psychiatric diagnostic and treatment facilities. Previous hospitalizations were in 2012 and 2013, when the patient was diagnosed with psychosis NOS and mixed abuse NOS, receiving Clopixol, gabapentin, and Rivotril as therapy. In addition the patient kept taking olanzapine long-acting injection every 28 days at the day hospital facilities. Since September 2014, his treating physicians had reported a deterioration of patient's conditions and an acute exacerbation of negative symptoms. He was then hospitalized in the psychiatric unit for 15 days and then discharged in June 2015 with a diagnosis of schizophrenia, chronic with acute exacerbation. Aripiprazole therapy was prescribed, and since discharge, he has been admitted into the Comunità Riabilitativa ad Alta Assistenza (CRA), a high-assistance rehabilitation community, to continue medical treatment. Mental state examination at discharge was the following:

Appearance and grooming sufficient.
The quality of mimicry good, although at times the patient seemed perplexed.
Voice tone regular.
Volition and planning scant and superficial.
Affectivity was flat and emotions were withdrawn.
Anti-conservative and/or other-directed ideation absent; hypnotic profile normal.
Consciousness appears clear as well as orientation in space, time, the self, and other parameters examined.

15.4.1 Individual COGPACK Training

CR offers exercises with several difficulty levels to target visual-motor functioning, comprehension, attention, memory, language use, and skills training. It also allows clinicians and users to determine the initial difficulty level; moreover, the program provides feedback that allows patients to track progress over time. It is possible to adapt the difficulty of the tasks as well as change the whole package (i.e., *adaptive, exercise variation, progressive*) (Fig. 15.1).

Fig. 15.1 Example: CR 2A package

15.4.2 MST Group

MST group aims are awareness and observation, which include observing the "uncomfortable" sensations happening in our mind, rather than avoiding or suppressing them, and looking with kindness upon whatever should arise. In our case MST has been integrated with some physiotherapeutic components of cognitive behavioral therapy (CBT), including specific topics on the processes of pain and physiology, thoughts and emotions, functions, and behavior.

Indeed a number of vicious cycles develop with chronic illness conditions, which actually lead to increased discomfort and decreased tolerance. These cycles include increasing muscle tension, guarding, avoidance, anxiety, anger, and depression which can augment the risk of increasing suffering, distress, and disability, thus decreasing quality of life (see Fig. 15.2).

Each MST session adheres to the following format:

- Short didactic presentation by therapist
- Practice of mindfulness exercises

Homework tasks discussion as described below.

The following is an example of first session:

Simple introduction to awareness, followed by a focus on breathing exercise:

- Discuss posture—alert versus not alert (straight back, rest hands in lap, feet on ground). Discomfort is inevitable, and practicing mindfulness when we feel challenged is the best way to develop these skills. So if we feel discomfort, what is asked is to simply notice it, observe it, be curious about it, make room for it, and then return to focus on breathing and on the present moment.

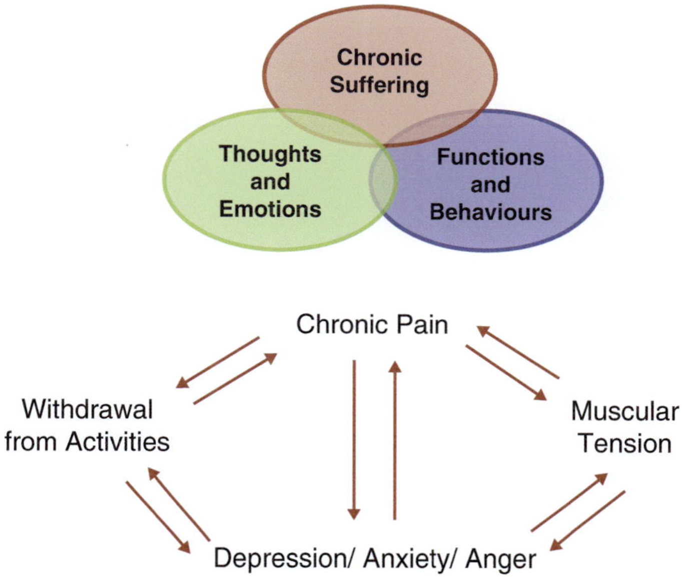

Fig. 15.2 Vicious cycles

- Ask participants to close their eyes during "awareness exercise": notice the feeling of being on the chair, feet on the floor, and clothes on the skin. After a few minutes, the exercise will be stopped. Then ask patients for feedback.

"Breathing exercise" concerning breathing sensations.

Specific body parts can be involved. After a few minutes, stop and get feedback again. Asking each participant to provide feedback can be quite useful, especially about certain topics.

- Focus on trying, not on the results.
- Distraction (e.g., feelings/sounds/thoughts, etc.) and mind wandering are normal, but it is important to bring awareness back from wherever it has gone. Struggling to keep it focused on one thing is the difficult part. Thus patients can be encouraged to stop the fight and merely bring awareness back to the sensation of breathing and to the present moment whenever they become aware that they have got lost.
- Redirect the purpose from relaxation to awareness gently. Relaxation occurs often with these exercises, but it is not the explicit purpose of the exercise. Any decrease in suffering that has occurred can be treated in the same way.

MST groups are on a 2-h weekly basis, run by an experienced psychotherapist and an assistant. MST equipment required is handouts and worksheets for each week; mindfulness exercise CDs—copy of sound files saved on the shared drive; folders; whiteboard markers; spare paper; and pens.

15.4.3 Results

Mindfulness group	Feb to March 2017
Cognitive training (COGPACK)	Feb to April 2017

Q.I. estimation	
RAVEN	98
TIB	QIT 106,368 QIV 112,509 QIP 105,586

Clinical and psychosocial evaluation	T0	T1 (3 months)
PANSS	71	72
YMRS	3	3
HDRS	3	3
GAF	40	40
SES	30	30
OQ-45 total score	13	10
• Symptoms	6	4
• Relationship	3	3
• Social role	6	5
FFMQ total score	131	155
Neurocognitive testing		
BAC-S		
– Verbal memory	AS = 33 ES = 0	AS = 29 ES = 0
– Working memory	AS = 15.75 ES = 1	AS = 22.75 ES = 3
– Motor speed (token task)	AS = 40.25 ES = 0	AS = 54.25 ES = 0
– Attention and speed information processing (symbol-coding task)	AS = 24.25 ES = 0	AS = 32.25 ES = 0
– Verbal fluency	AS = 28.25 ES = 0	AS = 30.25 ES = 0
– Executive functions (Tower of London)	AS = 15.75 ES = 2	AS = 17.75 ES = 3
MET-HV		
– Completed tasks/3	2	2
– Task failures	1	1
– Inefficiencies	0	1
– Rule breaks /6	0	0
– Interpretation failure	0	0
– Total errors	1	1

(continued)

Hotel task		
– Task correctly attempted/5	3	5
– Time deviations (s)	276	0
Faux Pas test	15/20	14/20
Reading the Mind in the Eyes Test	18/36	22/36
IOWA	−28	70

AS adjusted score, *ES* equivalent score

The impact of CR on brain plasticity has also been investigated, exploring specific structural and functional indices, both as predictors and as marker of CR effect through fMRI. We report both examinations (see Figs. 15.3 and 15.4).

Examination excludes parenchymal focal lesions of the under- and over-the-middle cerebral parenchyma. Center axis lines on the axis. Brain stroke in the boundaries. Globally occurring hypophysis gland with small T1 high intensity: the finding appears substantially unchanged with respect to a previous examination (30/05/2013).

Examination excludes subclinical and parenteral focal lesions. The appearance of the pituitary gland, already presented at a previous examination of 17 January 2017, is investigated.

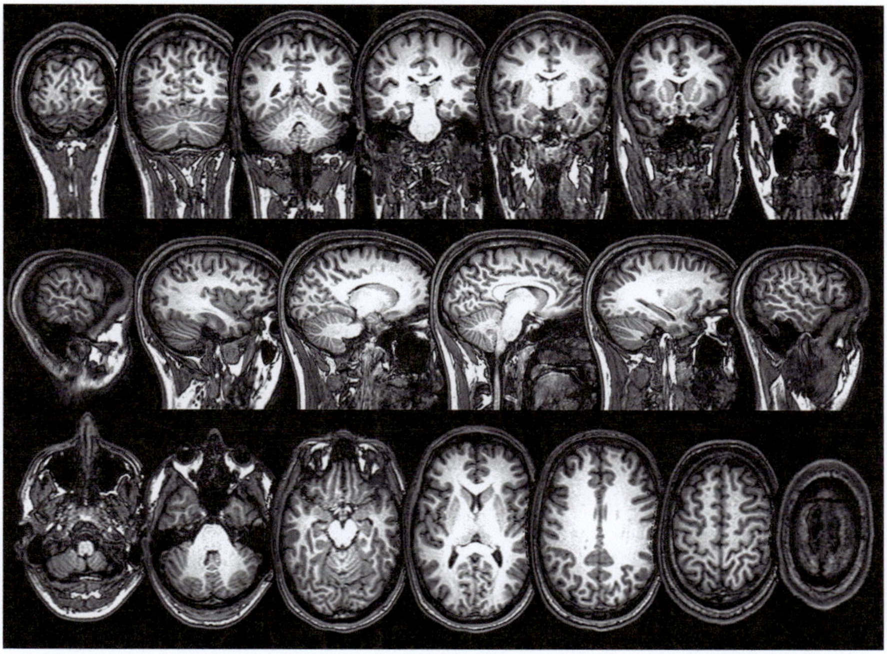

Fig. 15.3 T0

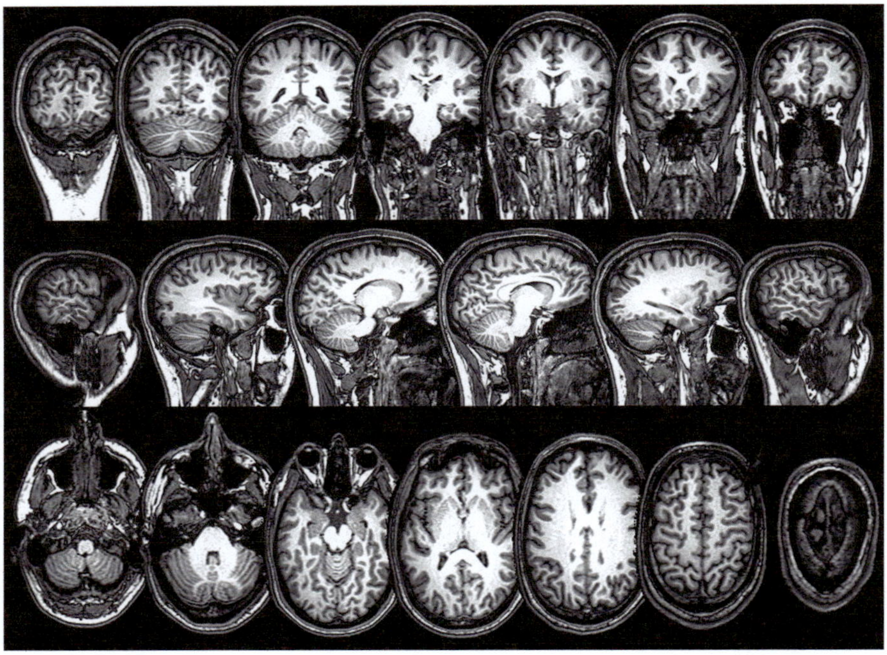

Fig. 15.4 T1

15.4.4 Conclusions

In CRA the patient seems calm and with an appropriate behavior; he cooperates with treatment. Speech is neither spontaneous nor fluent, although sufficiently informative.

His mood looks congruent, with no anxiety, and his attention, memory, and comprehension are good. The patient is still reluctant to discuss the content of his thoughts, and he seems to be listening only occasionally. Insight about his medical condition is partial.

However, after treatment, we observed a reduction of psychological stress and an improved social functioning (OQ-45 symptoms and social role subscales).

In ToM tasks (eyes test) the patient shows a slight "mentalizing" improvement (18→22), even if, qualitatively speaking, he is now more inclined to sociality, as reported by CRA psychiatrists and educators, showing an opening out in negative symptoms. Conversely, clinical scales do not show substantial changes in psychopathological dimensions.

As regards neurocognitive testing, we observed a good improvement of executive functions, in particular of planning capacity (TOL), working memory, and decision-making (IGT). An increase in ecological test performances shows the actual impact of the integrated CR on patients' life (mainly consisting in no more violations/deviations from the time allowed in the hotel task). Among the mindfulness five facets measured through the FFMQ, Mr. LF showed improvement in observation (22→34), describing (30→34), nonjudgment (27→28), and nonreacting (17→30).

Key Points

- The diagnosis of psychoses, including Schizophrenia (SKZ), Bipolar Disorder (BD) and Major Depression (MD) with psychotic features, has been progressively overcame by a transdiagnostic phenotype encompassing nonaffective and affective psychosis, in which psychosis could be defined as a continuity of positive symptoms ranging from mild to severe.
- Literature suggests that cognitive deficits and brain changes, associated with long duration of illness, are markers of vulnerability of psychoses, with important consequences on day-to-day functioning and quality of life.
- In SKZ, working memory, attention, and executive functions are the most altered cognitive domains. In BD, there is significant cognitive interindividual variability, including deficient working memory performance, inability to withstand interferences and to program actions, associative learning, sustained attention, psychomotor speed, inhibitory response, set-shifting ability, verbal memory, and theory of mind (ToM) impairment.
- Providing cognitive remediation in addition to psychiatric rehabilitation may contribute to greater improvement in both cognitive and social functioning. Cognitive remediation therapy (CRT) is a restorative/reparative intervention consisting of a set of behavioral-based interventions aimed at improving cognitive processes; commonly is subdivided into individual sessions for a total duration of 40 h (normally 3 days a week), primarily targeting attention, memory, executive functions, metacognition, and social cognition. In order to overcome poor performance in everyday life, an integrated intervention seems to be crucial.
- Our Psychiatric Equipe has defined a cognitive intervention program for young psychotic patients (aged 18–40) in which a computerized cognitive training (CR), associated with Social Skills Training or Mindfulness groups, could integrate medical/physical treatment, in order to overcome reported poor performance in everyday life.

Self-Assessment Questionnaire

1. The most altered cognitive domains in psychosis are:
   (A) Language, attention and executive functions
   (B) **Memory, attention, executive functions, and theory of mind**
   (C) Memory, executive functions, processing speed and theory of mind
   (D) Perception, executive functions, theory of mind and attention
2. Why is it crucial to apply an integrated cognitive intervention?
   (A) **In order to overcome poor performance in everyday life**
   (B) An integrated cognitive intervention could be unnecessary
   (C) In order to improve metacognition, reasoning and behaviour
   (D) In order to engage the patient in more than one activity

3. What are the main facets of mindfulness? How could they be assessed?
   (A) Observe, relax, non-judge, remember. The Four Facets Mindfulness Questionnaire
   (B) **Observe, describe, act with awareness, non-judge, and non-react to inner experience. The Five Facets Mindfulness Questionnaire**
   (C) Be quiet, observe, non-judge, improve yourself, remind. The Mindfulness Questionnaire
   (D) Observe, non-judge, relax, remember, improve yourself. The Five Facets Mindfulness Questionnaire
4. Which of the following statements about CR treatment is incorrect?
   (A) 40 sessions are planned for each patient individually
   (B) CR is carried out 2–3 times a week, 45 min each
   (C) CR target visual-motor functioning, comprehension, attention and memory
   (D) **Language stimulation is not provided**
5. What are the advantages of a computerized cognitive training?
   (A) Flexibility of task, immediate feedback, individual program
   (B) Flexibility of task, cost reduction, exercises improving reaction time
   (C) **Flexibility of tasks, a set of different exercises, and immediate feedback**
   (D) A set of different exercises, immediate feedback, cost reduction, individual program

References

1. Taylor HE, Stewart SL, Dunn G, Parker S, Bentall RP, Birchwood M, Morrison AP. Psychopathology and affect dysregulation across the continuum of psychosis: a multiple comparison group study. Early Interv Psychiatry. 2014;8(3):221–8.
2. Van Os J, Linscott RJ, Myin-Germeys I, Delespaul P, Krabbendam L. A systematic review and meta-analysis of the psychosis continuum: evidence for a psychosis proneness–persistence–impairment model of psychotic disorder. Psychol Med. 2009;39(2):179–95.
3. Van Os J, Reininghaus U. Psychosis as a transdiagnostic and extended phenotype in the general population. World Psychiatry. 2016;15:118–24.
4. American Psychiatric Association. Diagnostic and statistical manual of mental disorders. 5th ed. Arlington, VA: APA; 2013.
5. Keskinen E, Marttila A, Marttila R, Jones PB, Murray GK, Moilanen K, Koivumaa-Honkanen-H, Mäki P, Isohanni M, Jääskeläinen E, Miettunen J. Interaction between parental psychosis and early motor development and the risk of schizophrenia in a general population birth cohort. Eur Psychiatry. 2015;30(6):719–27.
6. Laurens KR, Luo L, Matheson SL, Carr VJ, Raudino A, Harris F, Green MJ. Common or distinct pathways to psychosis? A systematic review of evidence from prospective studies for developmental risk factors and antecedents of the schizophrenia spectrum disorders and affective psychoses. BMC Psychiatry. 2015;15(1):205.
7. Yung AR, Phillips LJ, Yuen HP, McGorry PD. Risk factors for psychosis in an ultra high-risk group: psychopathology and clinical features. Schizophr Res. 2004;67(2):131–42.
8. Simonsen C, Sundet K, Vaskinn A, Birkenaes AB, Engh JA, Færden A, Jónsdóttir H, Ringen PA, Opjordsmoen S, Melle I, Friis S, Andreassen OA. Neurocognitive dysfunction in bipolar and schizophrenia spectrum disorders depends on history of psychosis rather than diagnostic group. Schizophr Bull. 2009;37:73–83.
9. Archer T. Neurodegeneration in schizophrenia. Expert Rev Neurother. 2010;10:1131–41.

10. Puri BK. Progressive structural brain changes in schizophrenia. Expert Rev Neurother. 2010;10:33–42.
11. Altamura A, Buoli M, Pozzoli S. Role of immunological factors in the pathophysiology and diagnosis of bipolar disorder: comparison with schizophrenia. Psychiatry Clin Neurosci. 2014;68(1):21–36.
12. Buoli M, Serati M, Caldiroli A, Cremaschi L, Altamura AC. Neurodevelopmental versus neurodegenerative model of schizophrenia and bipolar disorder: comparison with physiological brain development and aging. Psychiatr Danub. 2017;29:24–7.
13. Bora E. Differences in cognitive impairment between schizophrenia and bipolar disorder: considering the role of heterogeneity. Psychiatry Clin Neurosci. 2016;70(10):424–33.
14. Addington J, Penn D, Woods SW, Addington D, Perkins DO. Social functioning in individuals at clinical high risk for psychosis. Schizophr Res. 2008;99(1):119–24.
15. Bartók E, Berecz R, Glaub T, Degrell I. Cognitive functions in prepsychotic patients. Prog Neuro-Psychopharmacol Biol Psychiatry. 2005;29(4):621–5.
16. Ferencz A, Bartók E, Berecz R, Glaub T, Degrell I. Cognitive functions in prepsychotic and negative-symptom schizophrenic patients. Neuropsychopharmacol Hung. 2005;7(2):61–5.
17. Howes OD, Nour MM. Dopamine and the aberrant salience hypothesis of schizophrenia. World Psychiatry. 2016;15(1):3–4.
18. Heckers S. Neuroimaging studies of the hippocampus in schizophrenia. Hippocampus. 2001;11:520–8.
19. Zipparo L, Whitford TJ, Hodge MAR, Lucas S, Farrow TF, Brennan J, Gomes L, Williams LM, Harris AW. Investigating the neuropsychological and neuroanatomical changes that occur over the first 2–3 years of illness in patients with first-episode schizophrenia. Prog Neuro-Psychopharmacol Biol Psychiatry. 2008;32:531–8.
20. Pantelis C, Velakoulis D, Wood SJ, Yücel M, Yung AR, Phillips LJ, Sun DQ, McGorry PD. Neuroimaging and emerging psychotic disorders: the Melbourne ultra-high risk studies. Int Rev Psychiatry. 2007;19:371–81.
21. Sala M, Perez J, Soloff P, Di Nemi SU, Caverzasi E, Soares JC, Brambilla P. Stress and hippocampal abnormalities in psychiatric disorders. Eur Neuropsychopharmacol. 2004;14:393–405.
22. Velakoulis D, Wood SJ, McGorry PD, Pantelis C. Evidence for progression of brain structural abnormalities in schizophrenia: beyond the neurodevelopmental model. Aust N Z J Psychiatry. 2000;34:2.
23. Addington J, Addington D. Neurocognitive and social functioning in schizophrenia: a 2.5 year follow-up study. Schizophr Res. 2000;44:47–56.
24. Keefe RS, Fenton WS. How should DSM-V criteria for schizophrenia include cognitive impairment? Schizophr Bull. 2007;33:912–20.
25. Keefe RS, Harvey PD. Cognitive impairment in schizophrenia. In: Novel antischizophrenia treatments. Berlin, Heidelberg: Springer; 2012. p. 11–37.
26. Kuperberg G, Heckers S. Schizophrenia and cognitive function. Curr Opin Neurobiol. 2000;10:205–10.
27. Martino DJ, Strejilevich SA, Scápola M, Igoa A, Marengo E, Ais ED, Perinot L. Heterogeneity in cognitive functioning among patients with bipolar disorder. J Affect Disord. 2008;109:149–56.
28. Watson S, Thompson JM, Ritchie JC, Nicol Ferrier I, Young AH. Neuropsychological impairment in bipolar disorder: the relationship with glucocorticoid receptor function. Bipolar Disord. 2006;8:85–90.
29. Brambilla P, Perlini C, Bellani M, Tomelleri L, Ferro A, Cerruti S, Marinelli V, Rambaldelli G, Christodoulou T, Jogia J, Dima D, Tansella M, Balestrieri M, Frangou S. Increased salience of gains versus decreased associative learning differentiate bipolar disorder from schizophrenia during incentive decision making. Psychol Med. 2013;43:571–80.
30. Brambilla P, Cerruti S, Bellani M, Perlini C, Ferro A, Marinelli V, Giusto D, Tomelleri L, Rambaldelli G, Tansella M, Diwadkar VA. Shared impairment in associative learning in

schizophrenia and bipolar disorder. Prog Neuro-Psychopharmacol Biol Psychiatry. 2011;35:1093–9.
31. MacQueen GM, Glenda M, Memedovich KA. Cognitive dysfunction in major depression and bipolar disorder: assessment and treatment options. Psychiatry Clin Neurosci. 2017;71 (1):18–27.
32. Robinson LJ, Thompson JM, Gallagher P, Goswami U, Young AH, Ferrier IN, Moore PB. A meta-analysis of cognitive deficits in euthymic patients with bipolar disorder. J Affect Disord. 2006;93:105–15.
33. Bora E, Yucel M, Pantelis C. Theory of mind impairment in schizophrenia: meta-analysis. Schizophr Res. 2009;109:1–9.
34. Caletti E, Paoli RA, Fiorentini A, Cigliobianco M, Zugno E, Serati M, Orsenigo G, Grillo P, Zago S, Caldiroli A, Prunas C, Giusti F, Consonni D, Altamura AC. Neuropsychology, social cognition and global functioning among bipolar, schizophrenic patients and healthy controls: preliminary data. Front Hum Neurosci. 2013;7:661.
35. Buchanan RW, Davis M, Goff D, Green MF, Keefe RS, Leon AC, Nuechterlein KH, Laughren T, Levine R, Strover E, Fenton W, Marder SR. A summary of the FDA-NIMH-MATRICS workshop on clinical trial design for neurocognitive drugs for schizophrenia. Schizophr Bull. 2005;31:5–19.
36. Gold JM. Cognitive deficits as treatment targets in schizophrenia. Schizophr Res. 2004;72:21–8.
37. Hyman SE, Fenton WS. What are the right targets for psychopharmacology? Science. 2003;299:350–1.
38. Kraus MS, Keefe RS. Cognition as an outcome measure in schizophrenia. Br J Psychiatry. 2007;191(50):46–51.
39. Anselmetti S, Poletti S, Ermoli E, Bechi M, Cappa S, Venneri A, Smeraldi E, Cavallaro R. The brief assessment of cognition in schizophrenia. Normative data for the Italian population. Neurol Sci. 2008;29:85–92.
40. Keefe RS, Goldberg TE, Harvey PD, Gold JM, Poe MP, Coughenour L. The brief assessment of cognition in schizophrenia: reliability, sensitivity, and comparison with a standard neurocognitive battery. Schizophr Res. 2004;68:283–97.
41. Altamura AC, Caletti E, Paoli RA, Cigliobianco M, Zugno E, Grillo P, Prunas C, Caldiroli A, Zago S. Correlation between neuropsychological and social cognition measures and symptom dimensions in schizophrenic patients. Psychiatry Res. 2015;230:172–80.
42. Torralva T, Roca M, Gleichgerrcht E, Bekinschtein T, Manes F. A neuropsychological battery to detect specific executive and social cognitive impairments in early frontotemporal dementia. Brain. 2009;132:1299–309.
43. Baron-Cohen S, Jolliffe T, Mortimore C, Robertson M. Another advanced test of theory of mind: evidence from very high functioning adults with autism or Asperger syndrome. J Child Psychol Psychiatry. 1997;38:813–22.
44. Stone VE, Baron-Cohen S, Knight RT. Frontal lobe contributions to theory of mind. J Cogn Neurosci. 1998;10:640–56.
45. Donohoe G, Dillon R, Hargreaves A, Mothersill O, Castorina M, Furey E, Fagan AJ, Meaney JF, Fitzmaurice B, Hallahan B, McDonald C, Wykes T, Corvin A, Robertson IH. Effectiveness of a low support, remotely accessible, cognitive remediation training programme for chronic psychosis: cognitive, functional and cortical outcomes from a single blind randomised controlled trial. Psychol Med. 2018;48(5):751–64.
46. Iwata K, Matsuda Y, Sato S, Furukawa S, Watanabe Y, Hatsuse N, Ikebuchi E. Efficacy of cognitive rehabilitation using computer software with individuals living with schizophrenia: a randomized controlled trial in Japan. Psychiatr Rehabil J. 2017;40:4–11.
47. Kani AS, Shinn AK, Lewandowski KE, Öngür D. Converging effects of diverse treatment modalities on frontal cortex in schizophrenia: a review of longitudinal functional magnetic resonance imaging studies. J Psychiatr Res. 2017;84:256–76.

48. Velligan DI, Kern RS, Gold JM. Cognitive rehabilitation for schizophrenia and the putative role of motivation and expectancies. Schizophr Bull. 2006;32:474–85.
49. Wykes T, Reeder C. Cognitive remediation therapy for schizophrenia: an introduction. Brunner-Routledge: New York; 2005.
50. Morin L, Franck N. Rehabilitation interventions to promote recovery from schizophrenia: a systematic review. Front Psych. 2017;8:100.
51. Wykes T, Reeder C, Corner J, Williams C, Everitt B. The effects of neurocognitive remediation on executive processing in patients with schizophrenia. Schizophr Bull. 1999;25:291.
52. Vinogradov S, Fisher M, de Villers-Sidani E. Cognitive training for impaired neural systems in neuropsychiatric illness. Neuropsychopharmacology. 2012;37:43.
53. Subramaniam K, Luks TL, Fisher M, Simpson GV, Nagarajan S, Vinogradov S. Computerized cognitive training restores neural activity within the reality monitoring network in schizophrenia. Neuron. 2012;73:842–53.
54. Haut KM, Lim KO, MacDonald A. Prefrontal cortical changes following cognitive training in patients with chronic schizophrenia: effects of practice, generalization, and specificity. Neuropsychopharmacology. 2010;35:1850–9.
55. Wykes T, Brammer M, Mellers J, Bray P, Reeder C, Williams C, Corner J. Effects on the brain of a psychological treatment: cognitive remediation therapy. Br J Psychiatry. 2002;181:144–52.
56. Bor J, Brunelin J, d'Amato T, Costes N, Suaud-Chagny MF, Saoud M, Poulet E. How can cognitive remediation therapy modulate brain activations in schizophrenia? An fMRI study. Psychiatry Res Neuroimaging. 2011;192:160–6.
57. Pu S, Nakagome K, Yamada T, Ikezawa S, Itakura M, Satake T, Ishida H, Nagata I, Mogami T, Kaneko K. A pilot study on the effects of cognitive remediation on hemodynamic responses in the prefrontal cortices of patients with schizophrenia: a multi-channel near-infrared spectroscopy study. Schizophr Res. 2014;153:87–95.
58. Bellack AS, Mueser KT, Gingerich S, Agresta J. Social skills training for schizophrenia. 2nd ed. A step-by-step guide. New York: The Guilford Press; 2004.
59. Bellack AS, Hersen M. Research and practice in social skills training. New York: Springer; 1979.
60. Galderisi S, Piegari G, Mucci A, Acerra A, Luciano L, Rabasca AF, Santucci F, Valente A, Volpe M, Mastantuono P, Maj M. Social skills training and computerized cognitive training in patients with schizophrenia. Eur Psychiatry. 2007;22:13–4.
61. Kopelowicz A, Liberman RP, Zarate R. Recent advances in social skills training for schizophrenia. Schizophr Bull. 2006;32:S12–23.
62. Piegari G, Galderisi S, Mucci A, Maj M. La riabilitazione cognitiva e psicosociale nei pazienti affetti da schizofrenia. Nóos. 2009;3:223–38.
63. Sacks O. L'uomo che scambiò sua moglie per un cappello: Adelphi Edizioni spa; 2013.
64. Bechi M, Riccaboni R, Ali S, Fresi F, Buonocore M, Bosia M, Cocchi F, Smeraldi E, Cavallaro R. Theory of mind and emotion processing training for patients with schizophrenia: preliminary findings. Psychiatry Res. 2012;198:371–7.
65. Lindenmayer JP, McGurk SR, Khan A, Kaushik S, Thanju A, Hoffman L, Valdez G, Wance D, Herrmann E. Improving social cognition in schizophrenia: a pilot intervention combining computerized social cognition training with cognitive remediation. Schizophr Bull. 2012;39:507–17.
66. Iwata K. Efficacy and application of cognitive rehabilitation: the importance of combining social skills training with cognitive rehabilitation. Seishin Shinkeigaku Zasshi. 2013;115:406–12.
67. Ohno H, Inoue K. The effects of cooking activities and social skills training on schizophrenics – a comparison of cognitive function and social competence. Kawasaki J Med Welf. 2014;19(2):54–63.
68. Kabat-Zinn J. Full catastrophe living: using the wisdom of your body and mind to face stress, pain, and illness. New York: Dell Publishing; 1990.

69. Kabat-Zinn J, Massion MD, Kristeller J, Person LG, Fletcher KE, Pbert L, Lenderking WR, Santorelli SF. Effectiveness of a meditation based stress reduction program in the treatment of anxiety disorders. Am J Psychiatry. 1992;149:936–43.
70. Baer RA. Mindfulness training as a clinical intervention: a conceptual and empirical review. Clin Psychol Sci Pract. 2003;10:125–43.
71. Ma SH, Teasdale J. Mindfulness-based cognitive therapy for depression: replication and exploration of differential relapse prevention effects. J Consult Clin Psychol. 2004;72:31–40.
72. Segal ZV, Williams MG, Teasdale JD. Mindfulness-based cognitive therapy for depression: a new approach to preventing relapse. New York: Guilford Press; 2002.
73. Tabak NT, Granholm E. Mindful cognitive enhancement training for psychosis: a pilot study. Schizophr Res. 2014;157:312–3.
74. Bishop SR, Lau M, Shapiro S, Carlson L, Anderson ND, Carmody J, Segal ZV, Abbey S, Speca M, Velting D, Devins G. Mindfulness: a proposed operational definition. Clin Psychol Sci Pract. 2004;11:230–41.
75. Levitt JT, Brown TA, Orsillo SM, Barlow DH. The effects of acceptance versus suppression of emotion on subjective and psychophysiological response to carbon dioxide challenge in patients with panic disorder. Behav Ther. 2004;35:747–66.
76. Kay SR, Fiszbein A, Opler LA. Positive and negative syndrome scale (PANSS). MHS; 2012.
77. Young RC, Biggs JT, Ziegler VE, Meyer DA. A rating scale for mania: reliability, validity and sensitivity. Br J Psychiatry. 1978;133:429–35.
78. Hamilton M. A rating scale for depression. J Neurol Neurosurg Psychiatry. 1960;23(1):56.
79. American Psychiatric Association. Diagnostic and statistical manual of mental disorders. 3rd ed. Revised (DSM-III-R). Washington, DC: APA; 1987.
80. Priebe S, Huxley P, Knight S, Evans S. Application and results of the Manchester short assessment of quality of life (MANSA). Int J Soc Psychiatry. 1999;45(1):7–12.
81. Ware JE Jr, Sherbourne CD. The MOS 36-item short-form health survey (SF-36): I. Conceptual framework and item selection. Med Care. 1992;30:473–83.
82. Barratt W. The Barratt simplified measure of social status (BSMSS). Indiana State University; 2006.
83. Chisholm D, Knapp MRJ, Knudsen HC, Amaddeo F, Gaite L, Van Wijngaarden B. Client socio-demographic and service receipt inventory-European version: development of an instrument for international research. Br J Psychiatry. 2000;177(39):28–33.
84. Baer RA, Smith GT, Hopkins J, Krietemeyer J, Toney L. Using self-report assessment methods to explore facets of mindfulness. Assessment. 2006;13:27–45.
85. Bechara A, Damasio AR, Damasio H, Anderson SW. Insensitivity to future consequences following damage to human prefrontal cortex. Cognition. 1994;50(1):7–15.
86. Manly T, Hawkins K, Evans J, Woldt K, Robertson IH. Rehabilitation of executive function: facilitation of effective goal management on complex tasks using periodic auditory alerts. Neuropsychologia. 2002;40:271–81.
87. Baron-Cohen S, Wheelwright S, Hill J, Raste Y, Plumb I. The "Reading the Mind in the Eyes" test revised version: a study with normal adults, and adults with Asperger syndrome or high-functioning autism. J Child Psychol Psychiatry. 2001;42(2):241–51.
88. Serafin M, Surian L. Il Test degli Occhi: uno strumento per valutare la "teoria della mente". G Ital Psicol. 2004;31(4):839–62.
89. Baron-Cohen S, O'Riordan M, Jones R, Stone V, Plaisted K. A new test of social sensitivity: detection of faux pas in normal children and children with Asperger syndrome. J Autism Dev Disord. 1999;29(5):407–18.

16 The Use of Transcranial Magnetic Stimulation (TMS) for Obsessive-Compulsive Disorder (OCD)

Bernardo Dell'Osso, Beatrice Benatti, and Chiara Arici

Abstract

We here describe a case of major depressive disorder in comorbidity with a hoarding disorder, treated with repetitive transcranial magnetic stimulation (rTMS). The patient was a 59-year-old woman with a treatment-resistant depression and a history of hoarding disorder. She had been treated with several antidepressants, belonging to different classes, with a partial or none response. The patient underwent a protocol of stimulation with high-frequency rTMS (10 Hz, 20 sessions, 1 session per day), on the left dorsolateral prefrontal cortex, 750 stimuli per session, in augmentation to the pharmacological treatment. We observed an improvement in both OCD and depressive symptoms. The rTMS had been well tolerated by the patient who did not report any side effects. In conclusion, the present clinical report shows the efficacy of rTMS in a patient with OCD in comorbidity with treatment-resistant depression.

Keywords

Repetitive transcranial magnetic stimulation · Treatment resistant depression · Obsessive-compulsive disorder · Hoarding

B. Dell'Osso (✉) · B. Benatti · C. Arici
Department of Neurosciences and Mental Health, Fondazione IRCCS Ca' Granda Ospedale Maggiore Policlinico, University of Milan, Milan, Italy
e-mail: bernardo.dellosso@unimi.it

A. C. Altamura, P. Brambilla (eds.), *Clinical Cases in Psychiatry: Integrating Translational Neuroscience Approaches*,
https://doi.org/10.1007/978-3-319-91557-9_16

16.1 Introduction

16.1.1 Obsessive-Compulsive Disorder: Epidemiology, Clinical Presentation and State of the Art

Obsessive-compulsive disorder (OCD) is an early onset and highly disabling condition, with a lifetime prevalence ranging between 1.5% and 3.5% of the general population and an equal gender distribution [1]. Prevalence rates seem to decrease with age, ranging up to 0.8% among people aged over 60 years and around 5–6% within the OCD elderly population [2–4]. Moreover, it has been reported a bimodal distribution for age at onset, with one peak at 12–14 years and another at 20–22 years [5, 6]. When compared to patients with generalized anxiety disorder and panic disorder, patients with OCD showed the earliest age at onset, suggesting a strong link between early onset, positive family history, and genetic load [7, 8].

OCD has been traditionally considered a condition with similar gender prevalence. However, some authors found that late-onset OCD was more likely to occur in females and that a significant rate of late-onset patients had a history of recent pregnancy [9]. In addition, peripartum and postpartum onsets were found to occur in 2–40% and 7–21% of OCD patients, respectively [10], the birth of a child and infant care, in fact, representing a potential source of psychological stress that may contribute to the development of OC symptoms and possibly OCD [11].

OCD symptoms are remarkably diverse, regarding both clinical presentation and severity, with patients reporting only one or, more often, many symptoms belonging to different phenotypes [12]. Studies are, however, conflicting about whether any particular phenotype of OCD is easier to treat or more likely to benefit from a particular treatment [13]. For instance, symptom presentation has received growing empirical attention, as studies have revealed that specific phenotypes exhibit different treatment response rates [14]. Checking and washing compulsions are the most common forms of ritualistic behavior in clinical samples of OCD in several different countries [15]. With respect to sociodemographic characteristics, Khanna and Mukherjee reported that patients with aggression/checking symptoms were more often young, single, and male, as well as more likely to have an early and insidious onset [16]. On the other hand, patients with contamination/washing symptoms were more frequently women and homemakers and more likely to experience OCD onset after marriage [16].

Hoarding disorder (HD) was associated with OCD, a finding supported by factor analytic studies, highlighting the prominence of hoarding behavior as a distinct symptom subtype of OCD [17]. However, in 2013, HD was classified as a distinct diagnostic entity in the last version of DSM [18], and currently it is no longer considered a symptom of OCD or obsessive-compulsive personality disorder. The prevalence of HD has been estimated at 2–5% of the general population [19].

Box 16.1 Hoarding

HD is characterized by acquiring and failing to discard a large number of objects along with difficulty in relation to keeping them organized. The resulting clutter inhibits the use of living spaces and leads to significant distress and/or impairment in day-to-day functioning [20, 21].

Before DSM-5, it was considered as an obsessive-compulsive subtype in which the accumulation was seen as a compulsive ritual. With the DSM-5, a new chapter titled "OCD and Related Disorder" has been introduced, and hoarding had been listed as a related disorder. This choice reflects the growing evidence of a link between these disorders, now separated from anxiety disorders.

DSM-5 Criteria

(A) Persistent difficulty discarding or parting with possessions, regardless of their actual value.
(B) This difficulty is due to a perceived need to save the items and to distress associated with discarding them.
(C) The difficulty discarding possessions results in the accumulation of possessions that congest and clutter active living areas and substantially compromises their intended use. If living areas are uncluttered, it is only because of the interventions of third parties (e.g., family members, cleaners, authorities).
(D) The hoarding causes clinically significant distress of impairment in social, occupational, or other important areas of functioning (including maintaining a safe environment for self and others).
(E) The hoarding is not attributable to another medical condition (e.g., brain injury, cerebrovascular disease, Prader-Willi syndrome).
(F) The hoarding is not better explained by the symptoms of another mental disorder (e.g., obsessions in obsessive-compulsive disorder, decreased energy in major depressive disorder, delusions in schizophrenia or another psychotic disorder, cognitive deficits in major neurocognitive disorder, restricted interests in autism spectrum disorder).

Specify if:

With excessive acquisition: If difficulty discarding possessions is accompanied by excessive acquisition of items that are not needed or for which there is no available space

Specify if:

With good or fair insight: The individual recognizes that hoarding-related beliefs and behaviors (pertaining to difficulty discarding items, clutter, or excessive acquisition) are problematic.

(continued)

Box 16.1 (continued)

With poor insight: The individual is mostly convinced that hoarding-related beliefs and behaviors (pertaining to difficulty discarding items, clutter, or excessive acquisition) are not problematic despite evidence to the contrary.

With absent insight/delusional beliefs: The individual is completely convinced that hoarding-related beliefs and behaviors (pertaining to difficulty discarding items, clutter, or excessive acquisition) are not problematic despite the evidence to the contrary.

16.1.2 Obsessive Compulsive and Related Disorders Treatment Guidelines

Selective serotonin reuptake inhibitors (SSRIs) are recommended first-line pharmacological interventions for OCD, while selective noradrenaline reuptake inhibitors (SNRIs), clomipramine, and other antidepressants are considered second- and third-line treatments [22]. Risperidone, olanzapine, haloperidol, and aripiprazole are considered as first-line adjunctive therapies for patients with poor response to SSRIs [22, 23].

Meta-analyses support the beneficial effects of psychological treatment for OCD, mainly cognitive behavioral therapy (CBT), generally including exposure with response prevention (ERP) [22, 24]. The combination of psychological and pharmacological treatment has been shown to be superior to medication alone, but not to CBT alone [25, 26]. These findings suggest that if pharmacotherapy is required or preferred, adding CBT to pharmacological treatment of OCD may enhance response rates and reduce relapse rates [22].

With respect to somatic therapies, a growing interest has been expressed over the last years toward brain stimulation interventions. These include noninvasive and more invasive interventions. For instance, several open trials have suggested that repetitive transcranial magnetic stimulation (rTMS) may be a promising adjunctive therapy in patients with treatment-refractory OCD [27–29]. However, results of sham-controlled trials are not univocal, and larger samples are needed to provide additional controlled evidence for the efficacy of TMS in treatment-resistant OCD [30–32]. In 2007 and 2010, two different research groups supported the efficacy of rTMS in improving comorbid depressive symptoms in patients with OCD [33, 34]. Moreover, several studies with small samples suggested that deep brain stimulation (DBS) may improve symptoms and functionality in up to two-thirds of patients with highly treatment-refractory OCD (fourth level of recommendation) [35, 36].

16.1.3 Transcranial Magnetic Stimulation: Mechanism of Action and Applications in Psychiatry

TMS is a brain stimulation technique, based on electromagnetic principles, in which magnetic fields are used to electrically stimulate targeted cortical brain areas [37].

The mechanism of action consists of an electrical flow generated into a coil with the production of a pulsating high-intensity (1.5–3 Tesla) magnetic field into targeted brain areas. The magnetic field penetrates within the different tissues at variable depths, usually no more than 2–3 cm below the stimulating coil, and reaches the brain cortex, where it is reconverted into an electrical flow [37]. Therefore, electricity interferes with the neuronal depolarization processes, enhancing or reducing cortical excitability, depending on stimulation parameters [37]. Modern devices are able to generate repetitive trains of stimuli, and this kind of stimulation is known as repetitive TMS (rTMS), commonly used for the treatment of psychiatric disorders [38].

In the last decade, TMS has been used in a wide range of neurological diseases such as migraine, tinnitus, and poststroke rehabilitation [39, 40]. Moreover, TMS has shown consistent positive results in the treatment of psychiatric disorders, such as major depression, anxiety disorders, and schizophrenia [41, 42]. However, major depressive disorder represents the only psychiatric indication approved by the International Guidelines and major International Regulatory Agencies, such as the US Food and Drug Administration [43].

Currently, two major international guidelines for the treatment of psychiatric disorders—the CANMAT [44] and the WFSBP [45]—include a section specifically dedicated to TMS. Another useful tool is represented by "TMS guidelines," published in 2009 by an international group of experts, specifically focused on the safety of TMS in psychiatric disorders [46]. Moreover, it is worth mentioning a recent publication about the first evidence-based guidelines that focuses on the clinical application of TMS on different psychiatric disorders [47]. Currently, rTMS, both at high and low frequency, applied on the dorsolateral prefrontal cortex (DLPFC) is the most commonly used protocol for treatment-resistant depression [48, 49].

With respect to the OCD treatment, a recent meta-analysis by Trevizol and colleagues including 15 RCTs (483 patients) showed heterogeneous results, in light of the different protocols of stimulation used. However, comparing active versus sham TMS, the active stimulation was found to be significantly superior for OCD symptoms [50]. These results are consistent with a previous meta-analysis by Berlim and colleagues [51] comparing 10 sham-controlled rTMS trials in OCD patients (282 patients). They concluded that low-frequency rTMS protocols targeting the orbitofrontal cortex or the supplementary motor area (SMA) seem to be the most efficacious; in addition, the efficacy of active rTMS for OCD seems to be comparable to second- or third-line pharmacological strategies for OCD without the long-term metabolic adverse effects [51].

Additionally, preliminary results showed that targeting the supplementary motor area with fMRI-guided navigation improves rTMS efficacy in patients with OCD [31]. Only 1 randomized sham-controlled study performed in 21 patients showed the potential value of this approach. In fact, after 4 weeks of treatment, the response rate in the completer sample was 67% with active and 22% with sham rTMS [52]. The clinical effects of low-frequency rTMS applied to the supplementary motor area on patients with OCD were related to an inhibitory modulation of dysfunctional motor circuits in this cortical area [53].

It should be noticed that a recent sham-controlled trial of rTMS of the DLPFC reported a significant improvement in obsessions but not in compulsions, with Y-BOCS scores reduction, as well as relief in depressive and anxiety symptoms [54].

More recently, a Canadian group published a study targeting the medial prefrontal cortex (mPFC) and applied low-frequency deep rTMS to ten patients with OCD; all patients showed a significant symptom improvement after ten sessions of rTMS that persisted 1 month following the last session of rTMS [55].

Future placebo-controlled rTMS studies in OCD patients should include larger sample sizes and be more homogeneous in terms of demographic and clinical variables, stimulation parameters, and cortical target to provide definitive evidence for the efficacy of this technique in the treatment of drug-resistant OCD.

16.2 Case Presentation

A 59-year-old woman was referred to the outpatient service of our hospital suffering from treatment-resistant major depressive disorder with a severe comorbid hoarding disorder.

There was no certainty about her psychiatric family history: she described a probable depressive episode of her father immediately after his retirement, neither diagnosed nor treated by any psychiatrist.

Patient's medical history was characterized by poliomyelitis when she was 18 months old, with a consequent delay in the acquisition of common motor skills and a residual lameness due to the shortness of the left leg.

Box 16.2 Poliomyelitis

In some pediatric patients, the poliovirus can cause a delay in the growth of a limb, while the contralateral continues growing regularly. Patients can develop a shorter leg, which forces them to limp, determining, over time, some spinal deformities, scoliosis in particular.

Moreover, when the patient was 39 years old, she developed a postpartum thyroiditis, treated with methimazole for about 18 months. This treatment led to an iatrogenic hypothyroidism, controlled by a currently ongoing therapy with levothyroxine.

Box 16.3 Postpartum Thyroiditis

The incidence of postpartum thyroiditis affects approximately 4.1% to 7% of women. This disease is characterized by autoimmune dysfunction often occurring during the first 6 months after childbirth. The disease ranges from postpartum hyperthyroidism to hyperthyroidism followed by hypothyroidism and hypothyroidism in isolation.

In this condition, lymphocytes cause destruction of the thyroid, with an initial release of thyroid hormone as the thyroid follicles are attacked, followed by hypothyroidism.

Women, in the hyperthyroid phase of postpartum thyroiditis, generally do not have severe agitation, exophthalmos, or significant vital sign changes but may have palpitations, fatigue, and **mood changes**. Postpartum thyroiditis is typically present initially as a hyperthyroid episode, followed by hypothyroidism. The thyrotoxic state (hyperthyroidism) usually occurs 2–6 months postpartum. Symptoms are mild because of the limited elevation of thyroid hormone and are usually 2–12 months in duration. Approximately 30–50% of women will remain in a hypothyroid condition and require continuous levothyroxine treatment.

Patient's psychiatric history started in childhood, with cleansing, order, and symmetry rituals, limited to her bedroom: the patient used to organize her clothes in the closet according to colors, dimension, and type of fabric. At home, she spent most of the time in her bedroom, cleaning and tidying clothes, objects, and books.

Despite the aforementioned OC symptoms, no specialist was consulted by her parents. Nevertheless, the patient kept an acceptable level of functioning and overall good quality of life. In fact, she graduated high school and then started a job as schoolteacher. When she was 30 years old, she got married and had two children when she was 36 and 39 years old. No significant worsening of the OCD symptomatology was observed in the postpartum even though, after the marriage, the compulsive rituals extended from only one bedroom to the entire home.

On the occasion of her second pregnancy, the death of her mother and the postpartum thyroiditis occurred in a short space of time. In this period, the patient developed her first major depressive episode, characterized by symptoms such as asthenia, apathy, anhedonia, guilt, and feelings of ruin. The patient started treatment with methimazole, with a simultaneous rapid improvement of her depression. However, a worsening of OC symptoms occurred, with a change in compulsive themes. In fact, the patient started accumulating an increasing number of objects, filling the entire house. She wasted a lot of money buying unnecessary things, also purchasing in multiple copies, without any logical purpose. At the same time, she could not throw away anything.

When she was 41 years old, the patient experienced an important worsening in the quality of life, given the increasing amount of accumulated objects and number of depressive recurrences. For this reason, she referred to her general practitioner who prescribed fluoxetine at an initial dosage of 20 mg/day, subsequently increased to 40 mg/day, due to lack of response.

After an initial improvement, which lasted about 3 months, the patient had several depressive recurrences, with short periods of partial remission. The depressive symptomatology progressively conditioned patient's quality of life, until she quit her job and divorced at 47 years.

When the patient was 49 years old, she had her first contact with a psychiatrist, and she was diagnosed with major depressive disorder (MDD) and obsessive-compulsive disorder (OCD). She started a treatment with venlafaxine at 75 mg/day and a cognitive behavioral therapy (CBT) specifically focused on OC symptoms. Psychotherapy, however, was soon interrupted (it lasted about 3 months) by the patient due to lack of efficacy. On the other hand, the psychopharmacological treatment was maintained for about 10 years, but the psychopathological history revealed a poor response to several first-line antidepressants, belonging to different classes—SSRIs, SNRIs, NDRIs, and tricyclics—such as venlafaxine, sertraline, escitalopram, paroxetine, and bupropion, both in monotherapy and in augmentation with olanzapine and amisulpride. Only clomipramine, up to 200 mg day, showed partial efficacy, especially on OCD symptoms.

Therefore, in light of the treatment-resistant depressive symptoms, an augmentation strategy with TMS was proposed to the patient, and, for this reason, she came to the attention of our clinic.

Box 16.4 Depression and Response to Treatment

Responder: Hamilton Depression Rating Scale (HAM-D) score reduction $\geq 50\%$ compared to baseline

Partial responder: Reduction in HAM-D score between 25% and 50% compared to baseline

Absent response: Reduction in HAM-D score $< 25\%$ compared to baseline

Remitter: HAM-D score ≤ 8

Treatment-resistant depression: Failure to ≥ 2 trials with antidepressants belonging to different pharmacological classes, given for an appropriate period of time, at standard dosages, and with compliance monitoring

During the first screening clinical interview, the patient reported several depressive symptoms, such as anhedonia, apathy, clinophilia, and terminal insomnia. Moreover, the hoarding symptoms were present and highly disabling. She was taking clomipramine 75 mg/day, zolpidem 10 mg/day, and levothyroxine 75 mcg/day. Blood tests and thyroid function were within the limit. A computed tomography (CT) of the brain was performed, in order to evaluate central nervous system (CNS) morphology. The imaging did not show any relevant abnormality (Fig. 16.1).

The screening assessment included questions about exclusion and inclusion criteria of the TMS protocol approved by the local ethical committee, and she was enrolled in a protocol of stimulation at high-frequency rTMS (10 Hz), 80% motor

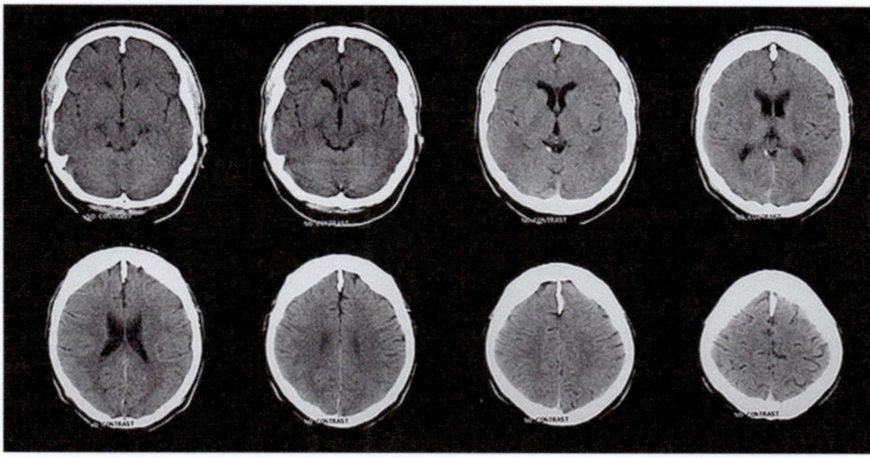

Fig. 16.1 CT showing no alteration in the posterior cranial fossa. Regular and symmetrical fourth ventricle. Normal supratentorial ventricular system. No expansive lesions in supratentorial regions. Normal cortical tropism

	HAM-D	MADRS	Y-BOCS	HAM-A
T0	24	30	23	16
T1	21	25	23	13
T2	16	20	21	10
T3	10	10	17	8
T4	7	6	15	12

Fig. 16.2 Patient's scores of psychometric scales at different time points (T0–T4)

threshold, of the left DLPFC with 5 s trains, with an interval of 25 s, 750 stimuli per session.

The pharmacological treatment remained unchanged throughout the entire duration of TMS treatment.

In Fig. 16.2, total scores of the psychometric scales at different time-points are reported.

Box 16.5 TMS Protocol (20 Applications, 4 Weeks of Treatment)

T0 (before treatment): – Clinical interview for inclusion and exclusion criteria

– Psychometric scales: Hamilton Depression Rating Scale (HAM-D), Montgomery-Asberg Depression Rating Scale (MADRS), Hamilton Anxiety Scale (HAM-A), and Yale-Brown Obsessive Compulsive Scale (Y-BOCS)

Assessment during the stimulation:

T1 (after 1 week of treatment)
T2 (after 2 weeks of treatment)
T3 (after 3 weeks of treatment)
T4 (after 4 weeks of treatment)

Phase 1: screening
Phase 2: randomization to one of the three different protocols of stimulation (high and low frequency)
Phase 3: motor threshold measuring
Phase 4: TMS application on left or right dorsolateral prefrontal cortex (depending on the type of protocol)

From T0 to T4, a global score reduction of the different psychometric scales was observed. In particular, an improvement in the depressive and obsessive scores, particularly relevant from the third week of stimulation (T2–T3), was observed. The obsessive-compulsive symptoms were initially rated as moderately severe (Y-BOCS = 23), and, at the end of the stimulation, they turned into mild (Y-BOCS = 15).

The HAM-D and MADRS scores decreased from a situation of moderate depression (T0 = 24 and 30, respectively) to remission (T4 = 7 and 6, respectively).

The treatment was well tolerated by the patient, who did not show any side effect during the stimulation and at the end of the treatment.

Figure 16.3 shows a timeline of the patient's psychiatric history.

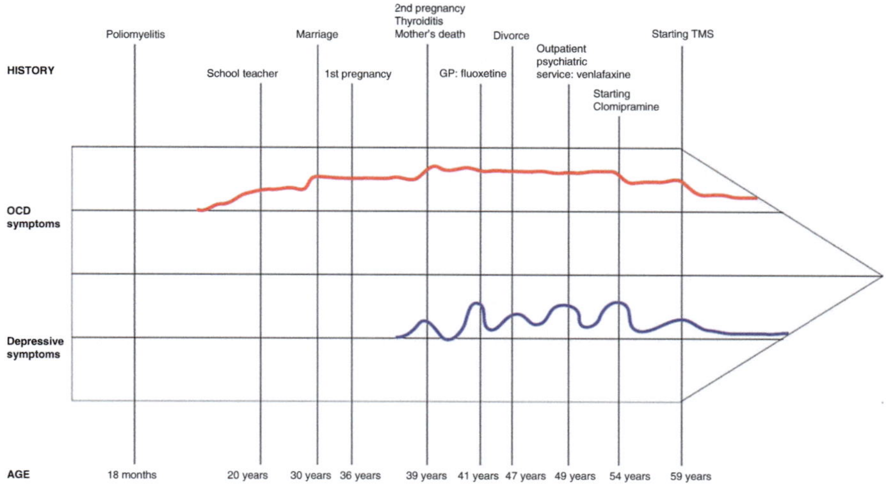

Fig. 16.3 Timeline of the psychiatric history

16.3 Discussion

The present clinical report describes the efficacy of rTMS in a complex clinical case characterized by OCD in comorbidity with treatment-resistant depression and previous hypothyroidism.

OCD symptoms began during the patient's childhood; however the full onset of the disorder, along with a significant functional impairment and reduced quality of life, manifested at the age of 30. These characteristics are consistent with the available literature about female OCD, describing a later onset compared to males although apparently not linked, in the present case, to the peripartum period [9].

OCD association with other psychiatric comorbidities has been frequently described in the literature, and major depressive disorder represents one of the most common comorbidities, both co-occurring and starting as a consequence of OCD symptoms [56].

In the clinical case set forth, patient's OCD was comorbid with a chronic depressive symptomatology. In fact, first depressive symptoms started in a context of OCD full symptomatology but were, at least initially, related to a concomitant hypothyroidism, which had been just discovered. This medical condition has been frequently linked to psychiatric disorders [57]. In the present case, even though hypothyroidism was successfully treated, the depressive disorder developed an independent course, and persisted showing a chronic relapsing course, ultimately characterized by treatment-resistance features. It should be noticed that the patient's father likely experienced at least one sub-threshold depressive episode during his life, so the family history may have played a role in the chronicization of the patient's depressive disorder [58].

In the present case, the course of OCD and depression showed an opposite trend at the beginning: while depressive symptoms were initially improved by the treatment with methimazole, OCD got worse, and symptoms increased in frequency and shifted from a cleaning/contamination subtype to a hoarding subtype. It is quite common for patients with OCD to progressively experience different or multiple obsessions/compulsions, resulting in a mixed clinical picture; in particular hoarding symptoms are frequently associated with other OCD phenotypes [59, 60].

It should be noticed, however, that after the 4-week treatment with TMS, the patient showed a progressive improvement of both depressive and OC symptomatology, with HAM-D and MADRS scores reduced from a moderate severity of illness (T0 = 24 and 30, respectively) to remission (T4 = 7 and 6, respectively) and Y-BOCS scores initially depicting a moderately severe OCD (T0 = 23) turning, at the end of the stimulation, into a mild severity. As previously mentioned, TMS has already been considered as a promising adjunctive therapy in patients with treatment-refractory OCD [27, 28], and the current clinical case seemed to support previous studies.

Comprehensively, we considered the reported case of clinical interest for different reasons. First, considering a longitudinal perspective, patient's OCD seemed to occur before the comorbid major depressive onset. Secondly, OCD symptoms were profoundly linked with hoarding manifestations which, according to the DSM-5, have been included in the chapter of obsessive-compulsive and related disorders. This is particularly relevant, as hoarding subtype of OCD or hoarding disorder "per se" has been traditionally considered a difficult-to-treat condition (either with pharmaco- or psychotherapy) [61]. In this perspective, the efficacy of TMS in such a particular OCD subtype seems to be of great clinical interest and needs further investigation. It is also important to mention that the stimulation protocol applied targeted the DLPFC which is, as already mentioned, one of the different targets implicated in the treatment with TMS of OCD patients. In fact, if this is a well-established target for the treatment of major depression, the hypothesis that it may be also a preferential target for OCD vs the SMA is still debated. Therefore, it is possible that in the clinical case, different features eventually contributed to the overall acute response to TMS, for example, the depressive comorbidity itself.

Although the whole case had many original issues that we deemed worthy of publication, we need to acknowledge that the reported TMS results were merely referred to the acute treatment and were obtained in an open setting, without any sham control. Therefore, further sham-controlled studies are warranted to specifically investigate the role of TMS in OCD treatment.

Key Points

- The association of OCD with other psychiatric comorbidities has been frequently described in the literature, and major depressive disorder represents one of the most common comorbidities.
- TMS has already been considered as a promising adjunctive therapy in patients with treatment-refractory OCD.

Self-Assessment Questionnaire

1. What is the prevalence of hoarding disorder in the general population?
   (A) 10–12%
   (B) 9%
   (C) **2–5%**
   (D) 22%
2. What kind of drugs are considered as first-line pharmacological treatment for OCD?
   (A) SNRI
   (B) Benzodiazepines
   (C) Mood stabilizers
   (D) **SSRI**
3. How could a partial responder depression be defined?
   (A) **A reduction of HAM-D scores between 25% and 50% compared to baseline**
   (B) A reduction of MADRS scores between 25% and 50% compared to baseline
   (C) HAM-D scores < 8
   (D) HAM-D score reduction $> 50\%$ compared to baseline
4. What brain area is the target for TMS in treatment-resistant depression?
   (A) Orbitofrontal cortex
   (B) **Dorsolateral prefrontal cortex**
   (C) Occipital cortex
   (D) Premotor area
5. What brain area is the target for TMS in OCD?
   (A) Dorsolateral prefrontal cortex
   (B) **Orbitofrontal cortex and supplementary motor area**
   (C) Occipital cortex
   (D) Cerebellum

References

1. Ruscio A, Stein D, Chiu W, Kessler R. The epidemiology of obsessive-compulsive disorder in the National Comorbidity Survey Replication. Mol Psychiatry. 2008;15:53–63.
2. Bassil N, Ghandour A, Grossberg GT. How anxiety presents differently in older adults. Curr Psychiatry. 2011;10:65–71.
3. Dell'Osso B, Benatti B, Rodriguez CI, Arici C, Palazzo C, Altamura AC, Hollander E, Fineberg N, Stein DJ, Nicolini H, Lanzagorta N, Marazziti D, Pallanti S, Van Ameringen M, Lochner C, Karamustafalioglu O, Hranov L, Figee M, Drummond L, Grant J, Denys D, Cath D, Menchon JM, Zohar J. Obsessive-compulsive disorder in the elderly: a report from the International College of Obsessive-Compulsive Spectrum Disorders (ICOCS). Eur Psychiatry. 2017a;45:36–40.
4. Dell'Osso B, Cremaschi L, Oldani L, Altamura AC. New directions in the use of brain stimulation interventions in patients with obsessive-compulsive disorder. Curr Med Chem. 2017b.
5. Dell'Osso B, Benatti B, Hollander E, Fineberg N, Stein DJ, Lochner C, Nicolini H, Lanzagorta N, Palazzo C, Altamura AC, Marazziti D, Pallanti S, Van Ameringen M,

Karamustafalioglu O, Drummond LM, Hranov L, Figee M, Grant JE, Zohar J, Denys D, Menchon JM. Childhood, adolescent and adult age at onset and related clinical correlates in obsessive-compulsive disorder: a report from the International College of Obsessive-Compulsive Spectrum Disorders (ICOCS). Int J Psychiatry Clin Pract. 2016;1501:1–8.
6. Rasmussen SA, Tsuang MT. Epidemiologic and clinical findings of significance to the design of neuropharmacologic studies of obsessive-compulsive disorder. Psychopharmacol Bull. 1986; 22:723–9.
7. Dell'Osso B, Camuri G, Benatti B, Buoli M, Altamura AC. Differences in latency to first pharmacological treatment (duration of untreated illness) in anxiety disorders: a study on patients with panic disorder, generalized anxiety disorder and obsessive-compulsive disorder. Early Interv Psychiatry. 2013;7:374–80.
8. Mathews CA, Badner JA, Andresen JM, Sheppard B, Himle JA, Grant JE, Williams KA, Chavira DA, Azzam A, Schwartz M, et al. Genome-wide linkage analysis of obsessive-compulsive disorder implicates chromosome 1p36. Biol Psychiatry. 2012;72:629–36.
9. Frydman I, do Brasil PE, Torres AR, Shavitt RG, Ferrao YA, Rosario MC, Miguel EC, Fontenelle LF. Late-onset obsessive-compulsive disorder: risk factors and correlates. J Psychiatr Res. 2014;49:68–74.
10. Forray A, Focseneanu M, Pittman B, McDougle CJ, Epperson CN. Onset and exacerbation of obsessive-compulsive disorder in pregnancy and the postpartum period. J Clin Psychiatry. 2010;71:1061–8.
11. Guglielmi V, Vulink NC, Denys D, Wang Y, Samuels JF, Nestadt G. Obsessive-compulsive disorder and female reproductive cycle events: results from the OCD and reproduction collaborative study. Depress Anxiety. 2014;31:979–87.
12. Rasmussen SA, Eisen JL. The epidemiology and clinical features of obsessive compulsive disorder. Psychiatr Clin North Am. 1992;15(4):743–58.
13. Mataix-Cols D, Rauch SL, Manzo PA, Jenike MA, Baer L. Use of factor-analyzed symptom dimensions to predict outcome with serotonin reuptake inhibitors and placebo in the treatment of obsessive-compulsive disorder. Am J Psychiatry. 1999;156(9):1409–16.
14. Alonso P, Menchon JM, Pifarre J, et al. Long-term follow-up and predictors of clinical outcome in obsessive-compulsive patients treated with serotonin reuptake inhibitors and behavioral therapy. J Clin Psychiatry. 2001;62:535–54.
15. Fontenelle LF, Mendlowicz MV, Soares ID, Versiani M. Patients with obsessive-compulsive disorder and hoarding symptoms: a distinctive clinical subtype? Compr Psychiatry. 2004;45(5): 375–83.
16. Khanna S, Mukherjee D. Checkers and washers: valid subtypes of obsessive compulsive disorder. Psychopathology. 1992;25(5):283–8.
17. Baer L. Factor analysis of symptom subtypes of obsessive compulsive disorder and their relation to personality and tic disorders. J Clin Psychiatry. 1994;55:18–23.
18. American Psychiatric Association, editor. Diagnostic and statistical manual of mental disorders. 5th ed. Washington, DC: American Psychiatric Association; 2013.
19. Samuels JF, Bienvenu OJ, Grados MA, Cullen B, Riddle MA, Liang K, Eaton WW, Nestadt G. Prevalence and correlates of hoarding behavior in a community-based sample. Behav Res Ther. 2008;46(7):836–44.
20. Frost RO, Hartl TL. A cognitive-behavioral model of compulsive hoarding. Behav Res Ther. 1996;34(4):341–50.
21. Frost RO, Steketee G, Tolin DF, Sinopoli N, Ruby D. Motives for acquiring and saving in hoarding disorder, OCD, and community controls. J Obsessive Compuls Relat Disord. 2015;4: 54–9.
22. Katzman MA, Bleau P, Blier P, Chokka P, Kjernisted K, Van Ameringen M, Canadian Anxiety Guidelines Initiative Group on behalf of the Anxiety Disorders Association of Canada/Association Canadienne des troubles anxieux and McGill University, Antony MM, Bouchard S, Brunet A, Flament M, Grigoriadis S, Mendlowitz S, O'Connor K, Rabheru K, Richter PM, Robichaud M, Walker JR. Canadian clinical practice guidelines for the management of anxiety,

posttraumatic stress and obsessive-compulsive disorders. BMC Psychiatry. 2014;14(Suppl 1): S1.
23. Bloch MH, Landeros-Weisenberger A, Kelmendi B, Coric V, Bracken MB, Leckman JF. A systematic review: antipsychotic augmentation with treatment refractory obsessive-compulsive disorder. Mol Psychiatry. 2006;11(7):622–32.
24. Belotto-Silva C, Diniz JB, Malavazzi DM, Valerio C, Fossaluza V, Borcato S, Seixas AA, Morelli D, Miguel EC, Shavitt RG. Group cognitive-behavioral therapy versus selective serotonin reuptake inhibitors for obsessive-compulsive disorder: a practical clinical trial. J Anxiety Disord. 2012;26:25–31.
25. O'Connor KP, Aardema F, Robillard S, Guay S, Pelissier MC, Todorov C, Borgeat F, Leblanc V, Grenier S, Doucet P. Cognitive behavior therapy and medication in the treatment of obsessive-compulsive disorder. Acta Psychiatr Scand. 2006;113:408–19.
26. Simpson HB, Liebowitz MR, Foa EB, Kozak MJ, Schmidt AB, Rowan V, Petkova E, Kjernisted K, Huppert JD, Franklin ME, et al. Post-treatment effects of exposure therapy and clomipramine in obsessive-compulsive disorder. Depress Anxiety. 2004;19:225–33.
27. Kumar N, Chadda RK. Augmentation effect of repetitive transcranial magnetic stimulation over the supplementary motor cortex in treatment-refractory patients with obsessive compulsive disorder. Indian J Psychiatry. 2011;53:340–2.
28. Mantovani A, Lisanby SH, Pieraccini F, Ulivelli M, Castrogiovanni P, Rossi S. Repetitive transcranial magnetic stimulation (rTMS) in the treatment of obsessive-compulsive disorder (OCD) and Tourette's syndrome (TS). Int J Neuropsychopharmacol. 2006;9:95–100.
29. Dell'Osso B, Altamura AC, Allen A, Hollander E. Brain stimulation techniques in the treatment of obsessive-compulsive disorder: current and future directions. CNS Spectr. 2005;10(12): 966–79.
30. Benatti B, Cremaschi L, Oldani L, De Cagna F, Dell'Osso B. Past, present and future of transcranial magnetic stimulation (TMS) in the treatment of psychiatric disorders. Evid based Psychitri Care. 2016;2:77–85.
31. Mantovani A, Simpson HB, Fallon BA, Rossi S, Lisanby SH. Randomized sham-controlled trial of repetitive transcranial magnetic stimulation in treatment-resistant obsessive-compulsive disorder. Int J Neuropsychopharmacol. 2010a;13:217–27.
32. Ruffini C, Locatelli M, Lucca A, Benedetti F, Insacco C, Smeraldi E. Augmentation effect of repetitive transcranial magnetic stimulation over the orbitofrontal cortex in drug-resistant obsessive-compulsive disorder patients: a controlled investigation. Prim Care Companion J Clin Psychiatry. 2009;11:226–30.
33. Sachdev PS, Loo CK, Mitchell PB, McFarquhar TF, Malhi GS. Repetitive transcranial magnetic stimulation for the treatment of obsessive compulsive disorder: a double-blind controlled investigation. Psychol Med. 2007;37:1645–9.
34. Sarkhel S, Sinha VK, Praharaj SK. Adjunctive high-frequency right prefrontal repetitive transcranial magnetic stimulation (rTMS) was not effective in obsessive-compulsive disorder but improved secondary depression. J Anxiety Disord. 2010;24:535–9.
35. Abelson JL, Curtis GC, Sagher O, Albucher RC, Harrigan M, Taylor SF, Martis B, Giordani B. Deep brain stimulation for refractory obsessive-compulsive disorder. Biol Psychiatry. 2005; 57:510–6.
36. Greenberg BD, Gabriels LA, Malone DA Jr, Rezai AR, Friehs GM, Okun MS, Shapira NA, Foote KD, Cosyns PR, Kubu CS, et al. Deep brain stimulation of the ventral internal capsule/ ventral striatum for obsessive-compulsive disorder: worldwide experience. Mol Psychiatry. 2010;15:64–79.
37. Fregni F, Pascual-Leone A. Technology insight: noninvasive brain stimulation in neurology-perspectives on the therapeutic potential of rTMS and tDCS. Nat Clin Pract Neurol. 2007;3(7): 383–93.
38. Fitzgerald PB, Daskalakis ZJ. A practical guide to the use of repetitive transcranial magnetic stimulation in the treatment of depression. Brain Stimul. 2012;5(3):287–96.

39. Kamble N, Netravathi M, Pal PK. Therapeutic applications of repetitive transcranial magnetic stimulation (rTMS) in movement disorders: a review. Parkinsonism Relat Disord. 2014;20(7):695–707.
40. Pinter MM, Brainin M. Role of repetitive transcranial magnetic stimulation in stroke rehabilitation. Front Neurol Neurosci. 2013;32:112–21.
41. Machado S, Paes F, Velasques B, Teixeira S, Piedade R, Ribeiro P, et al. Is rTMS an effective therapeutic strategy that can be used to treat anxiety disorders? Neuropharmacology. 2012; 62(1):125–34.
42. Montagne-Larmurier A, Etard O, Razafimandimby A, Morello R, Dollfus S. Two-days treatment of auditory hallucinations by high frequency rTMS guided by cerebral imaging: a 6 month follow-up pilot study. Schizophr Res. 2009;113(1):77–83.
43. Melkersson M. Special premarket 510(k) notification for NeuroStar TMS therapy system for major depressive disorder. Food Drug Adm. 2008.
44. Milev RV, Giacobbe P, Kennedy SH, Blumberger DM, Daskalakis ZJ, Downar J, Modirrousta M, Patry S, Vila-Rodriguez F, Lam RW, MacQueen GM, Parikh SV, Ravindran AV, CANMAT Depression Work Group. Canadian Network for Mood and Anxiety Treatments (CANMAT) 2016 Clinical guidelines for the management of adults with major depressive disorder: Section 4. Neurostimulation treatments. Can J Psychiatr. 2016;61(9):561–75.
45. Schlaepfer TE, George MS, Mayberg H, WFSBP Task Force on Brain Stimulation. WFSBP guidelines on brain stimulation treatments in psychiatry. World J Biol Psychiatry. 2010;11(1): 2–18.
46. Rossi S, Hallett M, Rossini PM, Pascual-Leone A, Safety of TMS Consensus Group. Safety, ethical considerations, and application guidelines for the use of transcranial magnetic stimulation in clinical practice and research. Clin Neurophysiol. 2009;120(12):2008–39.
47. Lefaucheur JP, André-Obadia N, Antal A, et al. Evidence-based guidelines on the therapeutic use of repetitive transcranial magnetic stimulation (rTMS). Clin Neurophysiol. 2014; 125:2150–206.
48. Rossini D, Lucca A, Magri L, Malaguti A, Smeraldi E, Colombo C, et al. A symptom-specific analysis of the effect of high-frequency left or low-frequency right transcranial magnetic stimulation over the dorsolateral prefrontal cortex in major depression. Neuropsychobiology. 2010;62(2):91–7.
49. Dell'Osso B, Oldani L, Camuri G, Dobrea C, Cremaschi L, Benatti B, Arici C, Grancini B, Altamura AC. Augmentative repetitive Transcranial Magnetic Stimulation (rTMS) in the acute treatment of poor responder depressed patients: a comparison study between high and low frequency stimulation. Eur Psychiatry. 2015;30(2):271–6.
50. Trevizol AP, Shiozawa P, Cook IA, Sato IA, Kaku CB, Guimarães FB, Sachdev P, Sarkhel S, Cordeiro Q. Transcranial magnetic stimulation for obsessive-compulsive disorder: an updated systematic review and meta-analysis. J ECT. 2016;32(4):262–6.
51. Berlim MT, Neufel NH, Van den Eynde F. Repetitive transcranial magnetic stimulation (rTMS) for obsessive-compulsive disorder (OCD): an exploratory meta-analysis of randomized and sham-controlled trials. J Psychiatr Res. 2013;47:999–1006.
52. Mantovani A, Westin G, Hirsh J, Lisanby SH. Functional magnetic resonance imaging guided transcranial magnetic stimulation in obsessive-compulsive disorder. Biol Psychiatry. 2010b;67: 39–40.
53. Mantovani A, Rossi S, Bassi BD, Simpson HB, Fallon BA, Lisanby SH. Modulation of motor cortex excitability in obsessive-compulsive disorder: an exploratory study on the relations of neurophysiology measures with clinical outcome. Psychiatry Res. 2013;210:1026–32.
54. Ma X, Huang Y, Liao L, Jin Y. A randomized double-blinded sham-controlled trial of α electroencephalogram-guided transcranial magnetic stimulation for obsessive-compulsive disorder. Chin Med J. 2014;127:601–6.
55. Modirrousta M, Shams E, Katz C, Mansouri B, Moussavi Z, Sareen J, et al. The efficacy of deep repetitive transcranial magnetic stimulation over the medial prefrontal cortex in obsessive compulsive disorder: results from an open-label study. Depress Anxiety. 2015;32(6):445–50.

56. Lochner C, Fineberg NA, Zohar J, van Ameringen M, Juven-Wetzler A, Altamura AC, et al. Comorbidity in Obsessive–compulsive Disorder (OCD): a report from the International College of Obsessive–Compulsive Spectrum Disorders (ICOCS). Compr Psychiatry. 2014;55:1513–9.
57. Placidi GP, Boldrini M, Patronelli A, Fiore E, Chiovato L, Perugi G, et al. Prevalence of psychiatric disorders in thyroid diseased patients. Neuropsychobiology. 1998;38:222–5.
58. Todd RD, Neuman R, Geller B, Fox LW, Hickok J. Genetic studies of affective disorders: should we be starting with childhood onset probands? J Am Acad Child Adolesc Psychiatry. 1993;32:1164–71.
59. Frost RO, Steketee G, Tolin DF. Comorbidity in hoarding disorder. Depress Anxiety. 2011;28: 876–84.
60. Lochner C, Stein DJ. Does work on obsessive-compulsive spectrum disorders contribute to understanding the heterogeneity of obsessive-compulsive disorder? Prog Neuro-Psychopharmacol Biol Psychiatry. 2006;30:353–61.
61. Saxena S, Maidment KM, Vapnik T, Golden G, Rishwain T, Rosen RM, Tarlow G, Bystritsky A. Obsessive-compulsive hoarding: symptom severity and response to multimodal treatment. J Clin Psychiatry. 2002;63(1):21–7.

MIX
Papier aus verantwortungsvollen Quellen
Paper from responsible sources
FSC® C105338

If you have any concerns about our products,
you can contact us on
ProductSafety@springernature.com

In case Publisher is established outside the EU,
the EU authorized representative is:
Springer Nature Customer Service Center GmbH
Europaplatz 3, 69115 Heidelberg, Germany

Printed by Libri Plureos GmbH
in Hamburg, Germany